Algorithms and Computation in Mathematics · Volume 6

Editors
E. Becker   M. Bronstein   H. Cohen
D. Eisenbud   R. Gilman

T0155789

Springer
Berlin
Heidelberg
New York
Barcelona
Hong Kong
London
Milan
Paris
Singapore
Tokyo

Mutsumi Saito   Bernd Sturmfels
Nobuki Takayama

Gröbner Deformations of Hypergeometric Differential Equations

With 14 Figures

Springer

Mutsumi Saito

Hokkaido University
Department of Mathematics
060-0810 Sapporo, Japan

e-mail:
saito@math.sci.hokudai.ac.jp

Bernd Sturmfels

University of California
Department of Mathematics
Berkeley, CA 94720, USA

e-mail:
bernd@math.berkeley.edu

Nobuki Takayama

Kobe University
Department of Mathematics
657-8501 Kobe, Japan

e-mail:
takayama@math.kobe-u.ac.jp

Library of Congress Cataloging-in-Publication Data applied for

Die Deutsche Bibliothek–CIP-Einheitsaufnahme
Saito, Mutsumi: Gröbner deformations of hypergeometric differential equations / Mutsumi Saito;
Bernd Sturmfels; Nobuki Takayama. – Berlin; Heidelberg; New York; Barcelona; Hong Kong;
London; Milan; Paris; Singapore; Tokyo: Springer, 2000
(Algorithms and computation in mathematics; Vol. 6)

Mathematics Subject Classification (1991):
13P10, 14Qxx, 16S32, 33Cxx, 34Exx, 35Axx, 68Q40

ISSN 1431-1550

ISBN 978-3-642-08534-5

© Springer-Verlag Berlin Heidelberg 2000
Softcover reprint of the hardcover 1st edition 2000

Cover design: MetaDesign plus GmbH, Berlin.

Printed on acid-free paper

Preface

In recent years, numerous new algorithms for dealing with rings of differential operators have been discovered and implemented. A main tool is the theory of Gröbner bases, which is reexamined in this book from the point of view of geometric deformations. Perturbation techniques have a long tradition in analysis; Gröbner deformations of left ideals in the Weyl algebra are the algebraic analogue to classical perturbation techniques.

The algorithmic methods introduced in this book are aimed at studying the systems of multidimensional hypergeometric partial differential equations introduced by Gel'fand, Kapranov and Zelevinsky. The Gröbner deformation of these GKZ hypergeometric systems reduces problems concerning hypergeometric functions to questions about commutative monomial ideals, and thus leads to an unexpected interplay between analysis and combinatorics.

This book contains original research results on holonomic systems and hypergeometric functions, and it raises many open problems for future research in this rapidly growing area of computational mathematics. An effort has been made to give a presentation which is both accessible to beginning graduate students and attractive to researchers in a variety of mathematical disciplines. The intended audience consists of anyone who is interested in algorithmic mathematics or in mathematical algorithms.

This book project started when the three of us met in Sapporo in August 1997. We had gotten together to work on a joint research paper on topics now contained in Chapter 4. We suddenly realized that we needed more background, almost none of which we could find in the existing literature on D-modules and linear partial differential equations. We then started to develop all the necessary basic material from scratch, and our manuscript soon turned from a draft for a research paper into a draft for a book.

We are grateful to two institutes whose support has been crucial: the Research Institute for Mathematical Sciences (RIMS) at Kyoto University hosted Bernd Sturmfels during the academic year 1997/98, and the Mathematical Sciences Research Institute (MSRI) at Berkeley hosted Nobuki Takayama during the academic year 1998/99. Mutsumi Saito visited his coauthors several times for short periods at both institutes. Bernd Sturmfels also acknowledges partial support from the U.S. National Science Foundation.

We wish to thank our friends for providing comments on earlier drafts of this book. An especially big "thank you" goes to Alicia Dickenstein, Diane Maclagan, Laura Matusevich, Greg Smith, Harrison Tsai, and Uli Walther.

This book is dedicated to our respective families, whose encouragement and support for this enterprise has been invaluable.

June 1999   *Mutsumi Saito, Bernd Sturmfels, Nobuki Takayama*

Contents

1. Basic Notions

This book provides symbolic algorithms for constructing holomorphic solutions to systems of linear partial differential equations with polynomial coefficients. Such a system is represented by a left ideal I in the Weyl algebra

$$D = \mathbf{C}\langle x_1, \ldots, x_n, \partial_1, \ldots, \partial_n \rangle.$$

By a *Gröbner deformation* of the left ideal I we mean an initial ideal $\mathrm{in}_{(-w,w)}(I) \subset D$ with respect to some generic weight vector $w = (w_1, \ldots, w_n)$ with real coordinates w_i. Here the variable x_i has the weight $-w_i$, and the operator ∂_i has the weight w_i, so as to respect the *product rule of calculus*:

$$\partial_i \cdot x_i = x_i \cdot \partial_i + 1.$$

Using techniques from computational commutative algebra, one can determine an explicit solution basis for the Gröbner deformation $\mathrm{in}_{(-w,w)}(I)$. The issue is to extend it to a solution basis of I. This problem is solved in Chapter 2 under the natural hypothesis that the given D-ideal I is *regular holonomic*. This hypothesis is valid for the D-ideals representing *hypergeometric integrals*, whose asymptotic expansions are constructed algorithmically in Chapter 5.

Our main interest lies in the systems of hypergeometric differential equations introduced by Gel'fand, Kapranov and Zelevinsky in the 1980's. Here is a simple, but important, example of a hypergeometric system for $n = 3$:

$$I = D \cdot \{ \partial_1 \partial_3 - \partial_2^2, \ x_1 \partial_1 + x_2 \partial_2 + x_3 \partial_3, \ x_2 \partial_2 + 2x_3 \partial_3 - 1 \}.$$

If $w = (1, 0, 0)$ then the Gröbner deformation of these equations equals

$$\mathrm{in}_{(-w,w)}(I) = D \cdot \{ \partial_1 \partial_3, \ x_1 \partial_1 + x_2 \partial_2 + x_3 \partial_3, \ x_2 \partial_2 + 2x_3 \partial_3 - 1 \}.$$

It is quite easy to see that the space of solutions to $\mathrm{in}_{(-w,w)}(I)$ is spanned by x_2/x_1 and x_3/x_2. Starting from these two Laurent monomials as w-lowest terms, our algorithm to be presented in Section 2.6 constructs two linearly independent Laurent series solutions to the original system I, namely,

$$-\frac{x_2}{2x_1} \pm \left(\frac{x_2}{2x_1} - \sum_{m=0}^{\infty} \frac{1}{m+1} \binom{2m}{m} \frac{x_1^m x_3^{m+1}}{x_2^{2m+1}} \right) = \frac{-x_2 \pm \sqrt{x_2^2 - 4x_1 x_3}}{2x_1}.$$

This is the familiar *quadratic formula* for expressing the two zeros of a quadratic polynomial $p(z) = x_1 z^2 + x_2 z + x_3$ in terms of its three coefficients. It is an amusing challenge to write down the analogous hypergeometric differential equations which annihilate the five roots of the general quintic

$$q(z) = x_1 z^5 + x_2 z^4 + x_3 z^3 + x_4 z^2 + x_5 z + x_6.$$

In this chapter we introduce the topics covered in this book. After treating Gröbner basics in the Weyl algebra, we review the classical Gauss hypergeometric function and how it is expressed in the Gel'fand-Kapranov-Zelevinsky (GKZ) scheme. Section 1.4 gives an introduction to holonomic systems of differential equations from the Gröbner basis point of view, and in Section 1.5 we study a special family of GKZ hypergeometric functions, namely, those which arise by integrating products of linear forms with generic coefficients.

1.1 Gröbner Bases in the Weyl Algebra

Let $\mathbf{k}$ be a field of characteristic zero, typically a subfield of the complex numbers $\mathbf{C}$. The *Weyl algebra* of dimension n is the free associative $\mathbf{k}$-algebra

$$D_n = \mathbf{k}\langle x_1, \ldots, x_n, \partial_1, \ldots, \partial_n \rangle$$

modulo the commutation rules

$$x_i x_j = x_j x_i, \quad \partial_i \partial_j = \partial_j \partial_i, \quad \partial_i x_j = x_j \partial_i \text{ for } i \neq j, \text{ and } \partial_i x_i = x_i \partial_i + 1.$$

If no confusion arises we simply drop the dimension index and write D for D_n. The Weyl algebra is isomorphic to the ring of differential operators on affine n-space $\mathbf{k}^n$. This is proved, for instance, in Coutinho's excellent text book on the Weyl algebra [26, Theorem 2.3, p.23]. The natural action of the Weyl algebra D on polynomials $f \in \mathbf{k}[x_1, \ldots, x_n]$ is as follows:

$$\partial_i \bullet f = \frac{\partial f}{\partial x_i}, \quad x_i \bullet f = x_i f. \tag{1.1}$$

Since $\mathbf{k}[x_1, \ldots, x_n]$ is also a subring of Weyl algebra D, the symbol $\bullet$ helps distinguish the action (1.1) from the product $\cdot : D \times D \to D$. For instance,

$$\partial_1^2 \bullet x_1^4 = 12x_1^2 \quad \text{but} \quad \partial_1^2 \cdot x_1^4 = x_1^4 \partial_1^2 + 8x_1^3 \partial_1 + 12x_1^2.$$

The Weyl algebra D acts by the same rule (1.1) on many $\mathbf{k}[x_1, \ldots, x_n]$-modules F, including *formal power series* $F = \mathbf{k}[[x_1, \ldots, x_n]]$, or, if $\mathbf{k} \subseteq \mathbf{C}$, *holomorphic functions* $F = \mathcal{O}^{an}(U)$ on an open subset U of $\mathbf{C}^n$.

A system of linear differential equations with polynomial coefficients can be identified with a left ideal in D. Suppose that we are given a system of linear differential equations for an unknown function $u = u(x_1, \ldots, x_n)$,

$$L_1 \bullet u = 0, \ \ldots, L_m \bullet u = 0, \ L_i \in D.$$

Then, the unknown function u also satisfies the differential equation

$$\sum_{i=1}^{m} (c_i L_i) \bullet u = 0$$

for any elements c_i in D. This implies that the system of differential equations may be expressed as

$$L \bullet u = 0, \quad L \in I$$

where I is the left ideal in D generated by $L_1, \ldots, L_m$. This point of view enables us to study differential equations through Gröbner bases for left ideals in the Weyl algebra.

Any element p of D has a unique *normally ordered expression*

$$p = \sum_{(\alpha,\beta) \in E} c_{\alpha\beta} \cdot x^\alpha \partial^\beta, \qquad (1.2)$$

where $x^\alpha = x_1^{\alpha_1} \cdots x_n^{\alpha_n}$, $\partial^\beta = \partial_1^{\beta_1} \cdots \partial_n^{\beta_n}$, $c_{\alpha\beta} \in \mathbf{k}^* = \mathbf{k} \setminus \{0\}$, and E is a finite subset of $\mathbf{N}^{2n}$. Here, $\mathbf{N} = \{0, 1, 2, \ldots\}$. In other words, we have the following natural $\mathbf{k}$-vector space isomorphism between the commutative polynomial ring in $2n$ variables and the Weyl algebra:

$$\Psi : \mathbf{k}[x, \xi] = \mathbf{k}[x_1, \ldots, x_n, \xi_1, \ldots, \xi_n] \ \to \ D, \ x^\alpha \xi^\beta \mapsto x^\alpha \partial^\beta. \qquad (1.3)$$

When doing calculations in the Weyl algebra – by hand or by computer – the isomorphism Ψ provides a useful representation of the elements. Efficient multiplication in D can be accomplished by the following *Leibnitz formula*:

Theorem 1.1.1. *For any two polynomials f and g in $\mathbf{k}[x, \xi]$ we have*

$$\Psi(f) \cdot \Psi(g) \quad = \quad \sum_{k_1, \ldots, k_n \geq 0} \frac{1}{k_1! \cdots k_n!} \cdot \Psi\left(\frac{\partial^k f}{\partial \xi^k} \cdot \frac{\partial^k g}{\partial x^k} \right).$$

Proof. Both the left hand side and the right hand side are $\mathbf{k}$-bilinear, so we may assume that f and g are monomials, say, $f = x^\alpha \xi^\beta$ and $g = x^\gamma \xi^\delta$. Clearly, we can factor out x^α and ξ^δ on both sides, so we may assume $f = \xi^\beta$ and $g = x^\gamma$. Both sides of the desired equation can be written as a product,

$$\prod_{i=1}^{n} \left(\Psi(\xi_i^{\beta_i}) \Psi(x_i^{\gamma_i}) \right) \quad = \quad \prod_{i=1}^{n} \sum_{k_i \geq 0} \frac{1}{k_i!} \cdot \Psi\left(\frac{\partial^{k_i} \xi_i^{\beta_i}}{\partial \xi_i^{k_i}} \cdot \frac{\partial^{k_i} x_i^{\gamma_i}}{\partial x_i^{k_i}} \right).$$

Hence it suffices to prove the case $n = 1$, which amounts to the formula

$$\partial^i x^j = \sum_{k=0}^{\min\{i,j\}} \frac{i(i-1) \cdots (i-k+1) \, j(j-1) \cdots (j-k+1)}{k!} x^{j-k} \partial^{i-k}. \qquad (1.4)$$

This formula can be derived from $\partial x = x\partial + 1$ by induction on i and j. $\quad \Box$

One remark on formula (1.4): throughout this book we freely use the convention that x abbreviates x_1 and ∂ abbreviates ∂_1 in the case $n = 1$.

A real vector $(u, v) = (u_1, \ldots, u_n, v_1, \ldots, v_n) \in \mathbf{R}^{2n}$ is called a *weight vector* (for the Weyl algebra) if

$$u_i + v_i \geq 0 \qquad \text{for} \quad i = 1, 2, \ldots, n.$$

Here u_i is the weight of the generator x_i, and v_i is the weight of the generator ∂_i. This condition will always be assumed in this book. The *associated graded ring* $\mathrm{gr}_{(u,v)}(D)$ of the Weyl algebra D with respect to a weight vector (u, v) is the **k**-algebra generated by

$$\{x_1, \ldots, x_n\} \cup \{\partial_i : u_i + v_i = 0\} \cup \{\xi_i : u_i + v_i > 0\}$$

with all variables commuting with each other except for $\partial_i x_i = x_i \partial_i + 1$. In fact, when u_i, v_i are integers, $\mathrm{gr}_{(u,v)}(D)$ is the associated graded ring of D with respect to the filtration $\cdots \subset F_0 \subset F_1 \subset \cdots$ defined by

$$F_m \quad = \quad \left\{ \sum_{u\alpha + v\beta \leq m} c_{\alpha\beta} x^\alpha \partial^\beta \right\}.$$

The two extreme cases of this definition are

$$\mathrm{gr}_{(u,v)}(D) = \mathbf{k}[x, \xi] \quad \text{if each coordinate of } u + v \text{ is positive;}$$
$$\mathrm{gr}_{(u,v)}(D) = D \qquad \text{if } u + v \text{ is the zero vector.}$$

For a non-zero element p in the Weyl algebra D we define the *initial form* $\mathrm{in}_{(u,v)}(p)$ of p with respect to (u, v) as follows. Let $m = \max_{(\alpha,\beta)\in E}(\alpha \cdot u + \beta \cdot v)$, select the terms of maximum weight m in the normally ordered expression (1.2), and then replace ∂_i by ξ_i for all i with $u_i + v_i > 0$. In symbols,

$$\mathrm{in}_{(u,v)}(p) \quad = \quad \sum_{\substack{(\alpha,\beta)\in E \\ \alpha\cdot u + \beta\cdot v = m}} c_{\alpha\beta} \prod_{i:u_i+v_i>0} x_i^{\alpha_i} \xi_i^{\beta_i} \prod_{i:u_i+v_i=0} x_i^{\alpha_i} \partial_i^{\beta_i} \quad \in \quad \mathrm{gr}_{(u,v)}(D).$$

For $p = 0$, we define $\mathrm{in}_{(u,v)}(p) = 0$.

A left ideal I in the Weyl algebra D will be called a *D-ideal*. The following result is an important consequence of the Leibnitz formula (Theorem 1.1.1).

Corollary 1.1.2. *Let I be a D-ideal and (u, v) any weight vector. Then the **k**-vector space*

$$\mathrm{in}_{(u,v)}(I) \quad := \quad \mathbf{k} \cdot \{\, \mathrm{in}_{(u,v)}(\ell) \mid \ell \in I \,\}$$

is a left ideal in the associated graded ring $\mathrm{gr}_{(u,v)}(D)$.

Definition 1.1.3. The ideal $\mathrm{in}_{(u,v)}(I)$ in $\mathrm{gr}_{(u,v)}(D)$ is called the *initial ideal* of a D-ideal I with respect to the weight vector (u, v). A finite subset G of D is a *Gröbner basis* of I with respect to (u, v) if I is generated by G and $\mathrm{in}_{(u,v)}(I)$ is generated by initial forms $\mathrm{in}_{(u,v)}(g)$ where g runs over G, i.e.,

$$I = D \cdot G \quad \text{and} \quad \mathrm{in}_{(u,v)}(I) = \mathrm{gr}_{(u,v)}(D) \cdot \mathrm{in}_{(u,v)}(G), \qquad (1.5)$$

$\mathrm{in}_{(u,v)}(G) := \{\, \mathrm{in}_{(u,v)}(g) \mid g \in G \,\}$.

Note that if $u + v > 0$ then $\mathrm{in}_{(u,v)}(I)$ is an ideal in the commutative polynomial ring $\mathbf{k}[x, \xi]$, while the Gröbner basis G is still a subset of the Weyl algebra D. The following examples will clarify the above definitions.

Example 1.1.4. Let $n = 1$, $(u, v) = (-1, 2)$, $F_1 = \{x^3\partial^2, x\partial^4\}$, and $I = DF_1$. The singleton $G = \{\partial^2\}$ is a Gröbner basis for I, and the initial ideal equals $\mathrm{in}_{(u,v)}(I) = \langle \xi^2 \rangle$. (In this book we use "$\langle \ldots \rangle$" for commutative polynomial ideals.) The singleton $F_2 = \{\partial^2 - x\partial^2\}$ satisfies the second condition of (1.5) but fails the first, while F_1 satisfies the first condition but fails the second. Indeed, a collection of normally ordered monomials is rarely a Gröbner basis.

It is our next goal to describe the Buchberger algorithm for computing Gröbner bases in the Weyl algebra. To this end we need to specify a total order $\prec$ on the set of normally ordered monomials $x^\alpha\partial^\beta$ in D. Such an order is called a *multiplicative monomial order* if the following two conditions hold:

1. $1 \prec x_i\partial_i$ for $i = 1, 2, \ldots, n$;
2. $x^\alpha\partial^\beta \prec x^a\partial^b$ implies $x^{\alpha+s}\partial^{\beta+t} \prec x^{a+s}\partial^{b+t}$ for all $(s, t) \in \mathbf{N}^{2n}$.

A multiplicative monomial order $\prec$ is called a *term order* (for the Weyl algebra) if $1 = x^0\partial^0$ is the smallest element of $\prec$. A multiplicative monomial order which is not a term order (sometimes called a *non-term order*) has infinite strictly decreasing chains but a term order does not (see, e.g., [27, p.70, Cor.6]). For information on frequently used term orders (*lexicographic order*, *reverse lexicographic order*, *elimination order*, *graded reverse lexicographic order*) see any book on Gröbner bases, e.g., [1], [12], [27].

The first condition $1 \prec x_i\partial_i$ in the above definition is a consequence of the relation $\partial_i x_i = x_i\partial_i + 1$. Without this assumption the order will not be compatible with multiplication; i.e. we do not have $\mathrm{in}_\prec(fg) = \mathrm{in}_\prec(f) \cdot \mathrm{in}_\prec(g)$.

Example 1.1.5. Let $n = 1$. Let $\prec$ be the total order defined by $x^\alpha\partial^\beta \prec x^a\partial^b \Leftrightarrow \beta - \alpha < b - a$ or $(\beta - \alpha = b - a$ and $\alpha > a)$. This is not a multiplicative monomial order and it is not compatible with the multiplication. For instance, $x\partial \prec 1$, and the initial term of $\partial \cdot x\partial = x\partial^2 + \partial$ with respect to the order $\prec$ is equal to $\partial \cdot 1$. (In this book, the underline __ will be used to mark the initial term with respect to a given order or the initial form for a given weight.)

Fix a multiplicative monomial order $\prec$. The *initial monomial* $\text{in}_\prec(p)$ of an element $p \in D$ is the commutative monomial $x^\alpha \xi^\beta$ in $\mathbf{k}[x, \xi]$ such that $x^\alpha \partial^\beta$ is the $\prec$-largest normally ordered monomial in the expansion (1.2) of p. For a finite set F in D, we define $\text{in}_\prec(F) = \{ \text{in}_\prec(f) \mid f \in F \}$. For a D-ideal I we define the *initial ideal* $\text{in}_\prec(I)$ to be the monomial ideal in $\mathbf{k}[x, \xi]$ generated by $\{\text{in}_\prec(p) \mid p \in I\}$. A finite subset G of D is said to be a *Gröbner basis* of I with respect to $\prec$ if I is generated by G and $\text{in}_\prec(I)$ is generated by the (commutative) monomials $\text{in}_\prec(g)$ where g runs over G.

The definitions in the previous paragraph extend naturally to elements and ideals in the associated graded rings $\text{gr}_{(u,v)}(D)$. In particular, we shall make frequent use of ordinary commutative Gröbner bases in $\mathbf{k}[x, \xi]$ with respect to term orders $\prec$. This will be relevant for part 2 in the next theorem.

As it now stands, there are two notions of Gröbner bases in D, one for weight vectors and one for multiplicative monomial orders. Theorem 1.1.6 will relate these two. Let $(u, v) \in \mathbf{R}^{2n}$ be a weight vector, and let $\prec$ be any term order. Then we define a multiplicative monomial order $\prec_{(u,v)}$ as follows:

$$x^\alpha \partial^\beta \prec_{(u,v)} x^a \partial^b \Leftrightarrow \alpha u + \beta v < au + bv \text{ or}$$

$$(\alpha u + \beta v = au + bv \text{ and } x^\alpha \partial^\beta \prec x^a \partial^b).$$

Note that $\prec_{(u,v)}$ is a term order if and only if (u, v) is a non-negative vector.

Theorem 1.1.6. *Let I be a D-ideal, $(u, v) \in \mathbf{R}^{2n}$ any weight vector, $\prec$ any term order, and G a Gröbner basis for I with respect to $\prec_{(u,v)}$. Then*

(1) *the set G is a Gröbner basis for I with respect to (u, v), and*
(2) *the set $\text{in}_{(u,v)}(G)$ is a Gröbner basis for $\text{in}_{(u,v)}(I)$ with respect to $\prec$.*

Proof. Suppose that G is not a Gröbner basis for I with respect to (u, v). Then there exists an element $f \in I$ whose initial form $\text{in}_{(u,v)}(f)$ is not in the left ideal generated by $\text{in}_{(u,v)}(G)$. Since $\prec$ is a term order, and thus has no infinite descending chains, we may further assume that the initial monomial

$$\text{in}_\prec \left(\text{in}_{(u,v)}(f) \right) = \text{in}_{\prec_{(u,v)}}(f) \in \mathbf{k}[x, \xi] \tag{1.6}$$

is minimal with respect to $\prec$ among all elements f with this property. By our assumption there exists $g \in G$ such that $\text{in}_{\prec_{(u,v)}}(g)$ divides (1.6). We can choose $c \in \mathbf{k}^*$ and $\alpha, \beta \in \mathbf{N}^n$ such that $f' := f - c x^\alpha \partial^\beta \cdot g$ has $\prec_{(u,v)}$-leading monomial $\prec$-smaller than (1.6). This implies that $\text{in}_{(u,v)}(f') = \text{in}_{(u,v)}(f) - c \cdot \text{in}_{(u,v)}(x^\alpha \partial^\beta \cdot g)$ is not in the left ideal generated by $\text{in}_{(u,v)}(G)$. This is a contradiction to the $\prec$-minimality of (1.6). Part 1 is proved.

For the proof of part 2 we consider an arbitrary (u, v)-homogeneous element $h \in \text{in}_{(u,v)}(I)$. By Corollary 1.1.2 there exists $f \in I$ such that $h = \text{in}_{(u,v)}(f)$. Formula (1.6) shows that $\text{in}_\prec(h)$ lies in the monomial ideal generated by $\text{in}_{\prec_{(u,v)}}(G)$. Moreover, since $\prec$ is a term order, we may conclude (for instance, by Theorem 1.1.7 below) that $\text{in}_{(u,v)}(G)$ actually generates $\text{in}_{(u,v)}(I)$ as a left ideal in $\text{gr}_{(u,v)}(D)$. $\qquad \square$

Theorem 1.1.6 reduces the problem of computing Gröbner bases with respect to weight vectors (u, v) to the problem of computing Gröbner bases with respect to multiplicative monomial orders. We will divide our discussion of that problem in two steps. First we study the case where $\prec$ is a term order, and next we introduce the homogenized Weyl algebra to solve the case where $\prec$ is a non-term order. The former will be carried out in this section and the latter in the next section. The latter case includes the most interesting orders $\prec_{(-w,w)}$ which arise geometrically from the action of the algebraic torus $(\mathbf{k}^*)^n$ on the Weyl algebra (see Section 2.3).

Theorem 1.1.7. *Let $\prec$ be a term order and G a Gröbner basis for its D-ideal $I = D \cdot G$ with respect to $\prec$. Any element f in I admits a standard representation in terms of G: there exist $c_1, \ldots, c_m \in D$ that satisfy*

$$f = \sum_{j=1}^{m} c_j g_j, \quad \text{where} \quad g_j \in G \quad \text{and} \quad \text{in}_\prec(c_j g_j) \preceq \text{in}_\prec(f) \text{ for all } j.$$

This implies that the first condition in (1.5) can be weakened to $G \subset I$ if we assume $(u, v) > 0$.

The proof of Theorem 1.1.7 is analogous to the familiar commutative case (see, e.g., [1], [12], [27], [32]). What we must do here, however, is to carefully define S-pairs and present the normal form algorithm in the Weyl algebra D. We fix a multiplicative monomial order $\prec$. For two normally ordered elements

$$f = f_{\alpha\beta} x^\alpha \partial^\beta + \text{lower order terms with respect to } \prec,$$

$$g = g_{ab} x^a \partial^b + \text{lower order terms with respect to } \prec$$

in D, we define the *S-pair* of f and g by

$$\text{sp}(f, g) = x^{\alpha'} \partial^{\beta'} f - (f_{\alpha\beta}/g_{ab}) x^{a'} \partial^{b'} g$$

where $\alpha'_i = \max(\alpha_i, a_i) - \alpha_i$, $\beta'_i = \max(\beta_i, b_i) - \beta_i$, $a'_i = \max(\alpha_i, a_i) - a_i$, $b'_i = \max(\beta_i, b_i) - b_i$. The multipliers for f and g are chosen to cancel the initial monomials of f and g. Note that we have used the condition $x_i \partial_i \succ 1$ to cancel the initial terms. We say that $x^\alpha \partial^\beta$ is *divisible* by $x^a \partial^b$ if $\alpha_i \geq a_i$ and $\beta_i \geq b_i$ for all i.

Let us introduce a normal form algorithm in D:

$$\text{normalForm}_\prec(f, \{g_1, \ldots, g_m\}) :=$$

$$r := f$$

$$\text{while } (\text{in}_\prec(r) \text{ is divisible by an in}_\prec(g_i)) \{$$

$$r := \text{sp}(r, g_i) \qquad \qquad (1.7)$$

$$\}$$

$$r := \text{in}_\prec(r)|_{\xi \to \partial}$$

$$+ \text{normalForm}_\prec(r - \text{in}_\prec(r)|_{\xi \to \partial}, \{g_1, \ldots, g_m\}) \qquad (1.8)$$

$$\text{return}(r)$$

Here $\mathrm{in}_\prec(r)|_{\xi\to\partial}$ means $\Psi(\mathrm{in}_\prec(r))$, with Ψ as defined in (1.3). Note that we use a recursive call at (1.8). It is used for presenting the algorithm compactly and has no deep meaning. The output f' of this algorithm is a *normal form* of f by $G = \{g_1, \dots, g_m\}$. The normal form algorithm is also called a *division algorithm*. When $\prec$ is a term order, this normal form algorithm terminates. The normal form is not always unique for general G, but when G is a Gröbner basis with respect to a term order $\prec$, the normal form is unique. In particular, the normal form of $f \in I$ is 0 by any Gröbner basis G of I.

Example 1.1.8. We present two examples of the computation of normal forms to clarify our definitions. Put $G = \{g_1, g_2\}$, $g_1 = (x\partial - 2)(x\partial - 4) = x^2\partial^2 - 5x\partial + 8$, $g_2 = (x\partial - 1)\partial^3 = x\partial^4 - \partial^3$ and let $\prec$ be the lexicographic order so that $x \prec \partial$. Then, for example, we have

$$\mathrm{normalForm}_\prec(x^2\partial^5, \{g_1, g_2\}) = 0$$

and

$$\mathrm{normalForm}_\prec(\partial^3 + x^2\partial^2, \{g_1, g_2\}) = \partial^3 + 5x\partial - 8.$$

In fact, we have

$$x^2\partial^5$$
$$\longrightarrow \boxed{x^2\partial^5} - \partial^3 g_1 = -x\partial^4 + \partial^3$$
$$\longrightarrow (\boxed{-x\partial^4} + \partial^3) + g_2 = 0$$

and

$$\partial^3 + x^2\partial^2$$
$$\longrightarrow (\partial^3 + \boxed{x^2\partial^2}) - g_1 = \partial^3 + 5x\partial - 8.$$

Here, we marked by the boxes $\boxed{\cdots}$ terms that will be reduced. Note that our normal form algorithm reduces lower order terms that are reducible.

Proof (of Theorem 1.1.7). We apply the normal form algorithm to f and G. Suppose $f \neq 0$. Since $f_0 := f \in I$ and G is a Gröbner basis, there exists g_{i_1} in G such that $\mathrm{in}_\prec(f)$ is divisible by $\mathrm{in}_\prec(g_{i_1})$. Put $f_1 := \mathrm{sp}(f, g_{i_1}) = f_0 - m_1 g_{i_1}$. Then, we have $f_1 \in I$, $\mathrm{in}_\prec(f_1) \prec \mathrm{in}_\prec(f_0)$, and $\mathrm{in}_\prec(m_1 g_{i_1}) \preceq \mathrm{in}_\prec(f_0)$. We can repeat this procedure and obtain a sequence $f_j := f_{j-1} - m_j g_{i_j}$. Since $\prec$ is a term order, this procedure terminates; there exists J such that $f_J = 0$. The sum of these $m_j g_{i_j}$ gives a standard representation. $\square$

Under our definitions of term orders, S-pairs, and the normal form algorithm, the Gröbner basis can be obtained in an analogous way to the commutative case. The Buchberger algorithm to obtain a *Gröbner basis* can be described as follows when $\prec$ is a term order.

Algorithm 1.1.9 (Buchberger's Algorithm in the Weyl algebra)

> Input: $F = \{f_1, \ldots, f_m\}$: a subset of D, $\prec$: a term order.
> Output: G : A Gröbner basis for $D \cdot F$ with respect to $\prec$.
> pair $:= \{(f_i, f_j) \mid 1 \le i < j \le m\}$
> $G := F$
> while (pair $\neq \emptyset$) {
> $\quad$ Take any element (f, f') from the set pair.$\hspace{2cm}$(1.9)
> $\quad$ pair $:=$ pair $\setminus \{(f, f')\}$
> $\quad$ $h := \mathrm{sp}(f, f')$
> $\quad$ $r := \mathrm{normalForm}_{\prec}(h, G)$$\hspace{3cm}$(1.10)
> $\quad$ if $(r \neq 0)$ {
> $\quad\quad$ pair $:=$ pair $\cup \{(g, r) \mid g \in G\}$
> $\quad\quad$ $G := G \cup \{r\}$
> $\quad$ }
> }
> return(G)

Theorem 1.1.10. *Let $F = \{f_1, \ldots, f_m\}$ be a finite subset of D. Assume that $\prec$ is a term order.*

(1) *(S-pair criterion) The set F is a Gröbner basis of $I = D \cdot F$ with respect to $\prec$ if and only if for all pairs $i \neq j$, the normal form of the S-pair $\mathrm{sp}(f_i, f_j)$ by F is zero.*
(2) *The Buchberger algorithm terminates and outputs a Gröbner basis of I with respect to $\prec$.*

The proof is analogous to the commutative case. See, e.g., [1, p.40, Theorem 1.7.4], [12, p.211, Theorem 5.48], [27, p.82, Theorem 6].

Example 1.1.11. For $n = 1$, consider $I = D \cdot \{\partial^2, x\partial - 1\}$. Let $\prec$ be any term order. Then, $G = \{\partial^2, x\partial - 1\}$ is a Gröbner basis, because

$$\mathrm{sp}(\underline{\partial^2}, \underline{x\partial} - 1) = x\partial^2 - \partial(x\partial - 1) = x\partial^2 - (x\partial + 1)\partial + \partial = 0.$$

Figure 1.1 pictures the monomials $x^a \xi^b$ which are divisible by an element of $\mathrm{in}_{\prec}(G)$. The monomials $x^c \xi^d$ which do not lie in $\mathrm{in}_{\prec}(I)$, or their Ψ-preimages $x^c \partial^d$, are called the *standard monomials* of $\mathrm{in}_{\prec}(I)$. In other words, a monomial $x^c \xi^d$ which is not divisible by any element of $\mathrm{in}_{\prec}(G)$ is a standard monomial. The standard monomials are the lattice points in the first orthant which are not dotted.

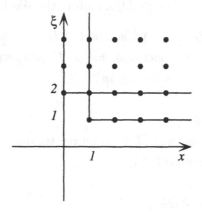

Fig. 1.1. The monomials divisible by $\mathrm{in}_{\prec}(G) = \{x\xi, \xi^2\}$.

We note that the Buchberger algorithm in this form is not efficient. The study of efficiency is an active research area in the commutative case. While many of the familiar results on Gröbner bases in $\mathbf{k}[x]$ also hold in D, there is one notable exception: *Buchberger's Criterion 1* which states that a set of polynomials whose initial monomials have disjoint support is automatically a Gröbner basis [27, Proposition 2.9.4]. (For a monomial $x^\alpha = x_1^{\alpha_1} \cdots x_n^{\alpha_n}$, we call the set $\{i \,|\, \alpha_i \neq 0\}$ the *support of the monomial.*) This does not hold in D. For example, consider $f = \underline{\partial_2} + x_1$, $g = \underline{\partial_1}$ and a term order which selects the underlined initial terms. Here Buchberger's Criterion 1 would imply that $\{f, g\}$ is a Gröbner basis, but this is clearly false since $gf - fg = 1$.

Definition 1.1.12. The *reduced Gröbner basis* G of a D-ideal with respect to a multiplicative monomial order $\prec$ is a Gröbner basis such that for any two distinct elements $g, g' \in G$, no term of $\Psi(g')$ is divisible by $\mathrm{in}_{\prec}(g)$.

Although a Gröbner basis may contain redundant elements, the reduced Gröbner basis does not contain any. The reduced Gröbner basis is uniquely determined by a D-ideal I and a term order $\prec$.

An immediate application of the computation of Gröbner bases is the elimination of variables, which will be frequently used in this book. The method for eliminating variables known from the commutative case also works in D (see, e.g. [1, §2.3], [12, p.258], [27, Theorem 2, p.113]). We present an example to illustrate the meaning of elimination in the Weyl algebra D.

Example 1.1.13. Consider the D_3-ideal I generated by $2\theta_1 + \theta_2 + 1$, $\theta_2 + 2\theta_3 - 1$, $\partial_1\partial_3 - \partial_2^2$. Here and throughout this book, θ_i means $x_i\partial_i$. The ideal I annihilates the function $f = \frac{-x_2 \pm (x_2^2 - 4x_1x_3)^{1/2}}{2x_1}$, which is the root formula for the quadratic equation. We encountered this function on the first page of Chapter 1. In order to obtain generators of $I' := I \cap \mathbf{k}\langle x_1, x_2, x_3, \partial_1 \rangle$, we compute the Gröbner basis for $\prec_{(u,v)}$, $(u,v) = (0,0,0,0,1,1)$ (eliminate ∂_2 and ∂_3). Here, $\prec$ may be any term order. Let $\prec$ be the reverse lexicographic order such that $x_1 \succ \cdots \succ \partial_3$. A Gröbner basis G for $\prec_{(u,v)}$ is

$$x_2\partial_2 + 2x_3\partial_3 - 1, \quad x_3\partial_3 - x_1\partial_1 - 1, \quad \partial_2^2 - \partial_1\partial_3,$$
$$2x_1\partial_1\partial_2 + x_2\partial_1\partial_3 + 2\partial_2,$$
$$(x_2^2 - 4x_1x_3)\partial_1\partial_3 + 2x_2\partial_2 + 2x_1\partial_1,$$
$$x_1(x_2^2 - 4x_1x_3)\partial_1^2 + 2(x_2^2 - 5x_1x_3)\partial_1 - 2x_3.$$

The ideal I' is generated by $\mathbf{k}\langle x_1, x_2, x_3, \partial_1 \rangle \cap G = \{ x_1(x_2^2 - 4x_1x_3)\partial_1^2 + 2(x_2^2 - 5x_1x_3)\partial_1 - 2x_3 \}$, which is an ordinary differential operator for the function f with respect to the variable x_1 and the parameters x_2 and x_3.

1.2 Weight Vectors and Non-term Orders

In this section, we discuss the computation of Gröbner bases for non-term orders. To this end we introduce the *homogenized Weyl algebra* $D^{(h)}$. This is the free associative $\mathbf{k}$-algebra generated by $h, x_1, \ldots, x_n, \partial_1, \ldots, \partial_n$ modulo the relations $x_ix_j = x_jx_i$, $\partial_i\partial_j = \partial_j\partial_i$, $\partial_ix_j = x_j\partial_i$ $(i \neq j)$, $\partial_ix_i = x_i\partial_i + h^2$, $hx_i = x_ih$, $h\partial_i = \partial_ih$. We call h the *homogenization variable*. The substitution $h = 1$ defines a $\mathbf{k}$-algebra homomorphism $D^{(h)} \to D$, $p \mapsto p_{|h=1}$, which will be called the *dehomogenization*.

An element $p \in D^{(h)}$ is uniquely expressed as a finite $\mathbf{k}$-linear combination

$$p = \sum_{(\lambda,\alpha,\beta)\in E} c_{\lambda\alpha\beta} \cdot h^\lambda x^\alpha \partial^\beta \quad \text{where} \quad c_{\lambda\alpha\beta} \in \mathbf{k}.$$

The *total degree* of p is $\deg(p) := \max\{\lambda + |\alpha| + |\beta| : c_{\lambda\alpha\beta} \neq 0\}$. The *homogenization* of an element $p \in D^{(h)}$ as above is the new element

$$H(p) := \sum_{(\lambda,\alpha,\beta)\in E} c_{\lambda\alpha\beta} \cdot h^{\deg(p)-|\alpha|-|\beta|} x^\alpha \partial^\beta.$$

The same formula defines the *homogenization map* $H : D \to D^{(h)}$.

A total order on the monomials $h^\lambda x^\alpha \partial^\beta$ in $D^{(h)}$ is called a *multiplicative monomial order* when the following conditions are satisfied:

1. $h^2 \prec x_i\partial_i$ for $i = 1, 2, \ldots, n$;
2. if $h^\lambda x^\alpha \partial^\beta \prec h^\ell x^a \partial^b$ then $h^{\lambda+r} x^{\alpha+s} \partial^{\beta+t} \prec h^{\ell+r} x^{a+s} \partial^{b+t}$.

A multiplicative monomial order is called a *term order* (for the homogenized Weyl algebra) if 1 is the minimal element in $D^{(h)}$. A real vector $(t, u, v) \in \mathbf{R}^{2n+1}$ is called a *weight vector* (for the homogenized Weyl algebra) when

$$2t \leq u_i + v_i \quad \text{for } i = 1, 2, \ldots, n.$$

For a multiplicative monomial order $\prec$, we can define a multiplicative monomial order $\prec_{(t,u,v)}$ by refinement of the partial order by (t, u, v) as we did in case of D. Our definition of a multiplicative monomial order implies that S-pairs and the normal form algorithm are defined as before in D. Without the hypothesis $h^2 \prec x_i \partial_i$ the normal form algorithm need not terminate:

Example 1.2.1. Let $n = 1$ and consider the lexicographic order with $h \succ \partial \succ x$. This is not a multiplicative monomial order for $D^{(h)}$. Put

$$f_1 := \underline{x} \quad \text{and} \quad f_2 := \underline{h^2 \partial} + \frac{1}{2} x \partial^2.$$

Let us apply the normal form algorithm for $x \partial^2$ and $\{f_1, f_2\}$. We obtain

$$x \partial^2 - \partial^2 f_1 = -2h^2 \partial$$
$$-2h^2 \partial + 2 f_2 = x \partial^2.$$

Therefore, the normal form algorithm may fall into an infinite loop.

Proposition 1.2.2.

(1) *The algorithm* normalForm$_\prec(g, F)$ *terminates if F consists of homogeneous elements in $D^{(h)}$ and $\prec$ is a multiplicative monomial order.*
(2) *The Buchberger algorithm in $D^{(h)}$ terminates for homogeneous inputs.*

Proof. We denote by $r_1, r_2, \ldots$ polynomials appearing in (1.7) of the normal form algorithm. Since $h^2 \prec x_i \partial_i$, we have $\text{in}_\prec(fg) = \text{in}_\prec(f)\text{in}_\prec(g)$ for any $f, g \in D^{(h)}$. Hence, by the definition of the S-pair, we have a strictly decreasing sequence, $r_1 \succ r_2 \succ \cdots$. On the other hand, there are finitely many monomials of the form $h^\lambda x^\alpha \partial^\beta$ such that $\lambda + |\alpha| + |\beta| = \deg(g)$. Hence, the normal form algorithm stops in finitely many steps. The second statement follows from the first one. □

Let I be an ideal in $D^{(h)}$ and (t, u, v) a weight vector. A finite subset G of $D^{(h)}$ is a *Gröbner basis* of I with respect to (t, u, v) if I is generated by G and $\text{in}_{(t,u,v)}(I)$ is generated by initial forms $\text{in}_{(t,u,v)}(g)$ where g runs over G. Let $\prec$ be a multiplicative monomial order in $D^{(h)}$. A finite subset G of $D^{(h)}$ is a *Gröbner basis* of I with respect to $\prec$ if I is generated by G and $\text{in}_\prec(I)$ is generated by the initial monomials $\text{in}_\prec(g)$ where g runs over G.

Proposition 1.2.3. *Let $\prec$ be a multiplicative monomial order. Suppose that F is a finite set consisting of homogeneous elements in $D^{(h)}$.*

(1) *For the input set F, the Buchberger algorithm outputs a Gröbner basis G of $I = D^{(h)} \cdot F$ with respect to $\prec$.*

(2) *Any element $f \in I$ has a standard representation*

$$f = \sum_{j=1}^{m} c_j g_j, \quad \text{where } g_j \in G \text{ and } \text{in}_{\prec}(c_j g_j) \preceq \text{in}_{\prec}(f) \text{ for all } j.$$

(3) *The reduced Gröbner basis exists for I.*

The proof follows from Proposition 1.2.2 and the same reasoning as for term orders in the Weyl algebra D. Every multiplicative monomial order $\prec$ on D induces a term order $\prec'$ on $D^{(h)}$ which respects the total degree:

$$h^\lambda x^\alpha \partial^\beta \prec' h^r x^a \partial^b \quad :\Longleftrightarrow \quad \lambda + |\alpha| + |\beta| < r + |a| + |b| \text{ or}$$
$$(\lambda + |\alpha| + |\beta| = r + |a| + |b| \text{ and } x^\alpha \partial^\beta \prec x^a \partial^b \text{ in } D).$$

Let $F = \{f_1, \ldots, f_m\}$ be any finite subset of the Weyl algebra and $I = D \cdot F$ its D-ideal. Consider its homogenization $F^h = \{H(f_1), \ldots, H(f_m)\} \subset D^{(h)}$. Clearly Proposition 1.2.2 applies to the input set F^h and the term order $\prec'$.

Theorem 1.2.4. *Let $F \subset D$ finite and $\prec$ a multiplicative monomial order in D. Let G^h be the output of the Buchberger algorithm applied to the homogeneous input F^h and the induced order $\prec'$ in $D^{(h)}$. Put $G = G^h_{|h=1}$.*

(1) *Then G is a Gröbner basis of $I = D \cdot F$ with respect to $\prec$.*
(2) *Any element of I admits a standard representation in terms of G and the order $\prec$.*

Proof. We only prove the second statement. It implies the first. Put $G^h = \{g_1^h, \ldots, g_m^h\}$. Each g_i^h is homogeneous. Let $f \neq 0$ be an element of I. Then, there exist $d_i \in D$ such that $f = \sum_{i=1}^{m} d_i g_i$ where $g_i = g_i^h_{|h=1}$. Consider the sum $\tilde{f} := \sum_{i=1}^{m} h^{e_i} H(d_i) g_i^h$ where each e_i is chosen such that $\deg(h^{e_i} H(d_i) g_i^h)$ is the same for all i. Since f is homogeneous and G^h is a Gröbner basis, f has a standard representation $\tilde{f} = \sum_{i=1}^{m} c_i g_i^h$, $\text{in}_{\prec'}(\tilde{f}) \succeq' \text{in}_{\prec'}(c_i g_i^h)$, and $\deg(\tilde{f}) = \deg(c_i g_i^h)$. The definition of the induced order $\prec'$ implies $\text{in}_{\prec}(f) \succeq \text{in}_{\prec}((c_{i|h=1}) g_i)$. $\qquad\square$

This theorem together with Theorem 1.1.6 immediately implies the following theorem.

Theorem 1.2.5. *Let $(u, v) \in \mathbf{R}^{2n}$ be a weight vector in the Weyl algebra. Take the dehomogenized set G of the previous Theorem 1.2.4 for the order $\prec_{(u,v)}$. Then, G is a Gröbner basis of $I = D \cdot G$ with respect to (u, v). Moreover, any element of I has a standard representation in terms of G and the weight vector (u, v). In other words, we have*

$$I \cap F_k = \sum_{i=1}^{m} F_{k-\text{ord}_{(u,v)}(g_i)} g_i \tag{1.11}$$

where

$$\mathrm{ord}_{(u,v)}(g_i) = u \cdot \alpha + v \cdot \beta$$

for $g_i = cx^\alpha \partial^\beta +$ (lower or equal order terms with respect to (u, v)) and $F_k = \{f \in D \mid \mathrm{ord}_{(u,v)}(f) \le k\}$, the associated filtration on D.

We call $\mathrm{ord}_{(u,v)}(g_i)$ (u, v)-*weight* (or (u, v)-*degree*) of g_i.

In summary, we present an algorithm for computing Gröbner bases with respect to arbitrary weight vectors (u, v) in the Weyl algebra D. It uses the homogenization in $D^{(h)}$ and its correctness follows from Theorem 1.2.5.

Algorithm 1.2.6
Input: $F = \{f_1, \ldots, f_m\} \subset D$ and a weight vector $(u, v) \in \mathbf{R}^{2n}$ $(u_i + v_i \ge 0)$.
Output: a Gröbner basis G of the D-ideal $D \cdot \{F\}$ with respect to (u, v).

(1) Choose any term order $\prec$ on D as a "tie breaker".
(2) Apply the Buchberger algorithm 1.1.9 to the homogenization $H(F) = \{H(f_1), \ldots, H(f_m)\}$ and the induced order $\prec'_{(u,v)}$. Let G^h be the output.
(3) Dehomogenize: $G := G^h_{|h=1}$.

Example 1.2.7. Let us apply Algorithm 1.2.6 to the input $\{\partial^3,\ x\partial - 1\}$ and the weight vector $(-1, 1)$. We take the reverse lexicographic order with $x \succ \partial$ as a tie-breaker. The homogenized input is $\{\partial^3,\ x\partial - h^2\}$ and Buchberger's algorithm outputs $\{\partial^3,\ x\partial - h^2,\ \partial^2 h^2\}$. The dehomogenization is

$$\{\partial^3,\ x\partial - 1,\ \partial^2\}.$$

Note that we have an apparent redundant element ∂^3. The Gröbner basis G obtained by the homogenization and dehomogenization process may contain redundant elements. As the following example shows, the reduced Gröbner basis does not exist in general for non-term orders.

Example 1.2.8. Let $n = 2$, $(u, v) = (-1, -1, 1, 1)$, and $\prec$ any term order. Then $I = D \cdot \{1 + x_1 + x_2, \partial_1 - \partial_2\}$ has no reduced Gröbner basis for $\prec_{(u,v)}$.

Example 1.2.9. Let us preview how Algorithm 1.2.6 is used in this book. Put $n = 4$, $(u, v) = (-1, -2, -1, 0, 1, 2, 1, 0)$. Let $\prec$ be the graded reverse lexicographic order such that $x_1 \succ \cdots \succ x_4 \succ \partial_1 \succ \cdots \succ \partial_4$ and $\mathbf{k} = \mathbf{Q}$. We fix generic rational numbers a, b, c and let

$$f_1 = \underline{\partial_2\partial_3} - \partial_1\partial_4,\ f_2 = \underline{\theta_1} - \theta_4 + 1 - c,\ f_3 = \underline{\theta_2} + \theta_4 + a,\ f_4 = \underline{\theta_3} + \theta_4 + b.$$

As we will see in Section 1.3, this is the GKZ system (1.22) representing the Gauss hypergeometric function. The output of Algorithm 1.2.6 applied to $F = \{f_1, f_2, f_3, f_4\}$, (u, v) and the tie-breaker $\prec$ consists of f_1, f_2, f_3, f_4 and $f_5 = \underline{x_4\partial_3\partial_4} + a\partial_3 + x_2\partial_1\partial_4$, $f_6 = \underline{x_4\partial_2\partial_4} + b\partial_2 + x_3\partial_1\partial_4$, $f_7 = \underline{x_4^2\partial_4^2} + (a + b + 1)x_4\partial_4 + ab - x_2x_3\partial_1\partial_4$.

From this we can verify that the initial ideal $\mathrm{in}_{(u,v)}(D \cdot F)$ is generated by $\partial_2\partial_3$ and f_2, f_3, f_4. A general discussion on this fact will be done in Theorem 3.1.3. We will observe the initial ideal determines the dominant terms of solutions for $(x_1, x_2, x_3, x_4) = (t^1, t^2, t^1, t^0)$, $t \to 0$. See Sections 1.3 and 1.5.

The existence of a standard representation is one of the most useful properties of Gröbner bases. For term orders, it is an easy application of the normal form algorithm to prove the existence of a standard representation for a Gröbner basis. The existence holds for non-term orders, too.

Theorem 1.2.10 ([80, Theorem 10.6]). *Let (u, v) be a weight vector and $G = \{g_1 \ldots, g_m\}$ be a Gröbner basis for its D-ideal $I = D \cdot G$ with respect to (u, v). Any element f in I admits a standard representation in terms of G: there exist $c_1, \ldots, c_m \in D$ that satisfy*

$$f = \sum_{j=1}^m c_j g_j, \quad \text{where} \quad g_j \in G \quad \text{and} \quad \mathrm{in}_{(u,v)}(c_j g_j) \preceq \mathrm{in}_{(u,v)}(f) \text{ for all } j.$$

In other words, the equality (1.11) holds for I, G and (u, v).

As to a proof of this theorem, we refer to [80]. In some articles, Gröbner basis G of the ideal I is defined by the conditions (1) $G \subseteq I$ and (2) $\mathrm{gr}_{(u,v)}(D) \cdot \mathrm{in}_{(u,v)}(G) = \mathrm{in}_{(u,v)}(I)$. This definition is not equivalent with ours. The proof of Theorem 1.2.10 uses the condition $D \cdot G = I$ (cf. (1.5)). Therefore, the theorem does not hold for this different definition of Gröbner basis in general.

Let $I = D \cdot \{f_1, \ldots, f_m\}$ be a D-ideal. We call the left ideal in $D^{(h)}$ generated by all the homogenized elements $H(f)$, $f \in I$, the *homogenization of I* and denote it by $H(I)$ or by $I^{(h)}$. It is not true in general the homogenization $H(I)$ is generated by $H(f_1), \ldots, H(f_m)$.

Proposition 1.2.11.

$$H(I) = \left\{ g \in D^{(h)} \,\middle|\, h^k g \in D^{(h)} \cdot \{H(f_1), \ldots, H(f_m)\} \text{ for some } k \in \mathbf{N} \right\}.$$

The proof of this proposition is easy. Let us present a method to compute $H(I)$. Let $D^{(h)}[z]$ be the ring $\mathbf{k}\langle h, x_1, \ldots, x_n, \partial_1, \ldots, \partial_n, z \rangle$ where h, x_i, ∂_j satisfy the same commutation relations with the homogenized Weyl algebra and z commutes with all the other elements.

Proposition 1.2.12. *The homogenization $H(I)$ is generated by $D^{(h)}[z] \cdot \{H(f_1), \ldots, H(f_m), zh - 1\} \cap D^{(h)}$.*

The homogenization $H(I)$ can be obtained by eliminating the variable z from the ideal in $D^{(h)}[z]$ generated by $H(f_i)$ and $zh - 1$. The proof of this proposition is analogous to that of a method of computing saturations in the commutative ring, see, e.g., [1, p.238, Proposition 4.4.1], [27, Exercise §4.4–8].

We close this section with bibliographic notes. Gröbner bases in the Weyl algebra were introduced by Castro and Galligo ([23], [35]). The homogenized Weyl algebra with infinitely many generators is called the *oscillator (Heisenberg)* algebra in mathematical physics. The homogenized Weyl algebra appeared in Takayama's system kan/sm1 (Version 2) for the purpose of efficient computation [100]. It also appears in the work of Assi, Castro, and Granger on computing irregularities of D-modules ([6], [7]).

Another homogenization technique is the V-*homogenization* due to Oaku [75], [77]. If the input set lies in $\mathbf{k}[x]$, then the V-homogenization and our homogenization in the Weyl algebra both agree with Lazard's homogenization in the polynomial ring [63]. However, for a general input, these homogenizations do not agree, that is, we have two variants of Lazard's homogenization in D. For example, let us V-homogenize the input of Example 1.2.7. The variable v is used to mark $\mathrm{ord}_{(-w,w)}(f)$ to each monomial f. The homogenized input for $(-w, w) = (-1, 1)$ is $\{ v^3 \partial^3,\ x\partial - 1 \}$. Apply the Buchberger algorithm with the lexicographic order $v \succ \partial \succ x$. Then, the output is $\{ v^2 \partial^2,\ x\partial - 1 \}$. The dehomogenization of the output is different from that of Example 1.2.7, but the dehomogenization gives a Gröbner basis with respect to $(-1, 1)$.

An analog to Mora's tangent cone algorithm ([70], [28, p.160, (3.10)]) has not yet been developed for the Weyl algebra D. This would be useful in the case $u+v = 0$. Oaku's paper [77] is primarily concerned with the computation of b-functions and annihilating ideals of powers of polynomials. His algorithm is based on results by Kashiwara and Malgrange on $\mathcal{D}$-modules. It uses the weight vector $(u, v) = (-1, 0, \ldots, 0, 1, 0, \ldots, 0)$. We will discuss his algorithm in Section 5.3. Oaku calls Gröbner bases with respect to (u, v) *involutive bases* following the standard terminology in the theory of $\mathcal{D}$-modules.

1.3 The Gauss Hypergeometric Equation

The *Gauss hypergeometric equation* is the following second order linear ordinary differential equation for an indeterminate univariate function $f(x)$:

$$\left[x(1-x)\frac{d^2}{dx^2} + (c - x(a+b+1))\frac{d}{dx} - ab \right] \bullet f = 0 \qquad (1.12)$$

Here a, b, c are complex parameters. This equation arises in various areas of mathematics such as algebraic geometry, number theory, combinatorics and mathematical physics. A substantial body of literature on Gauss' equation appeared already in the 19th century. The intrinsic reason for its importance is explained in [14, §7.4]: any equation with three regular singular points can be written in the form (1.12). A nice recent introduction is the book by Yoshida [103]. In this section we discuss series solutions and integral representations, and we translate (1.12) into a GKZ-hypergeometric system.

The differential operator in (1.12) is regarded as an element P in the Weyl algebra $D = \mathbf{k}\langle x, \partial \rangle$ over the rational function field $\mathbf{k} = \mathbf{Q}(a, b, c)$, namely,

$$P = x(1-x)\partial^2 - (a+b+1)x\partial + c\partial - ab. \qquad (1.13)$$

Using the *Euler operator* $\theta = x\partial$, we can rewrite (1.13) as follows:

$$x \cdot P = \theta(\theta + c - 1) - x(\theta + a)(\theta + b). \qquad (1.14)$$

The translation from (1.13) to (1.14) is done by the following lemma:

Lemma 1.3.1. *The following identities hold in the Weyl algebra:*

(1) $x^m \partial^m = \theta(\theta - 1) \cdots (\theta - m + 1)$,
(2) $\partial^m x^m = (\theta + 1)(\theta + 2) \cdots (\theta + m)$.

The characteristic property of the Euler operator θ is that its set of eigenvectors is precisely the set of monomials. For $a \in \mathbf{C}$, the *monomial* x^a denotes the function $\exp(a \cdot \log(x))$ for a suitable branch of the logarithm.

Lemma 1.3.2. *For any polynomial p, we have $p(\theta) \bullet x^a = p(a) \cdot x^a$. In the Weyl algebra D, we have the commutation rule $p(\theta)x^i = x^i p(\theta + i)$.*

We shall construct series solutions to the Gauss equation (1.12). Recall the existence theorem for linear ordinary differential equations [55, §12.22]:

Theorem 1.3.3. *Consider an ordinary differential equation of order m,*

$$\left[a_m(x)\frac{d^m}{dx^m} + \cdots + a_0(x) \right] \bullet f = 0, \qquad where\ a_0, \ldots, a_m \in \mathbf{C}[x].$$

Let U be a simply connected domain contained in $\{p \in \mathbf{C} \mid a_m(p) \neq 0\}$. Then, the dimension of the space of holomorphic solutions on U is equal to m.

Applying this theorem to Gauss' hypergeometric equation, we see that, for any $p \in \mathbf{C} \setminus \{0, 1\}$, the space of holomorphic solutions on a small disk centered at p is two-dimensional. When $p = 0$ or $p = 1$, the dimension of holomorphic solutions at p can be 1 or 2. However, if we consider the larger space of *logarithmic solutions* which consists of elements of the form

$$(x - p)^\lambda \cdot g_0(x) + (x - p)^\lambda \cdot g_1(x) \cdot \log(x - p) \qquad (1.15)$$

where $g_i(x)$ is holomorphic at $x = p$, then the Gauss equation (1.12) always has a two-dimensional solution space, even when $p = 0, 1$ and ∞. This fact follows from the classical existence theorem of solutions at a *regular singular point*; see [55, §16.3]. Specifically for Gauss' equation see [55, §7.232].

We begin by constructing series solutions to Gauss' hypergeometric equation at $x = 0$ under the assumption $c \notin \mathbf{Z}$. Consider a formal series solution

$$f = x^s \cdot \sum_{k=0}^{\infty} c_k x^k. \qquad (1.16)$$

We suppose $c_0 = 1$. Then, by virtue of (1.14) and Lemma 1.3.2 we have

$$xP \bullet f = \sum_{k=0}^{\infty} c_k(s+k)(s+k+c-1)x^{s+k} - \sum_{k=0}^{\infty} c_k(s+k+a)(s+k+b)x^{s+k+1} = 0.$$

Hence, we obtain the following recurrence relations for the coefficients c_k:

$$s(s+c-1)c_0 = 0,$$
$$(s+k+1)(s+k+c)c_{k+1} - (s+k+a)(s+k+b)c_k = 0, \quad k = 0, 1, 2, \ldots$$

The first of these equations is called the *indicial equation* [55, §7.21]. It determines the *exponent* s. Namely, since $c_0 = 1$, we have $s = 0$ or $s = 1 - c$.

To write the resulting two series, we introduce the *Pochhammer symbol*

$$(a)_k := \Gamma(a+k)/\Gamma(a) = a(a+1)(a+2)\cdots(a+k-1).$$

Here Γ denotes the familiar Γ-function. The recurrence relations for the coefficients c_k of the two hypergeometric series (1.16) have the following solutions

$$c_k = \frac{(a)_k(b)_k}{(1)_k(c)_k} \qquad \text{for} \quad s = 0;$$

$$c_k = \frac{(a+1-c)_k(b+1-c)_k}{(1)_k(2-c)_k} \qquad \text{for} \quad s = 1-c.$$

From these formulas one infers that the series (1.16) converge for $|x| < 1$ and hence define holomorphic functions in the open unit disk. These two series are **C**-linearly independent because their starting monomials 1 and x^{1-c} are distinct (see Section 2.5 for the general definition of starting monomials). In summary, we have the following theorem (cf. [55, §7.23]).

Theorem 1.3.4. *If $c \notin \mathbf{Z}$, then*

$$F(a, b, c; x) := \sum_{k=0}^{\infty} \frac{(a)_k(b)_k}{(1)_k(c)_k} x^k$$

and

$$x^{1-c} \cdot F(a+1-c, b+1-c, 2-c; x)$$

form a basis of solutions to the Gauss hypergeometric equation at $x = 0$.

The series $F(\alpha, \beta, \gamma; x)$ is called the *Gauss hypergeometric series*. One can also find bases of series solutions at other points expressed by $F(\alpha, \beta, \gamma; x)$. A good exercise is to construct this basis around the point at infinity $x = \infty$.

We note that the method used above works for any ordinary linear differential equation. Let us consider a general ordinary differential equation for a while. By multiplying a suitable x^m, $m \in \mathbf{Z}$, to the ordinary differential equation from the left, the equation can be written as $L \bullet f = 0$ where

$$L = \sum_{k=0}^{r} x^k p_k(\theta), \quad p_0(\theta) \neq 0.$$

The polynomial $p_0(\theta)$ is called the *indicial polynomial* of the ordinary differential equation at $x = 0$. The coefficients of the series expansion of the form (1.16) satisfy the following recurrence:

$$p_0(s)c_0 = 0, \tag{1.17}$$

$$\sum_{k=0}^{\min\{m,r\}} c_{m-k}p_k(s+m-k) = 0 \quad \text{for all } m.$$

The exponents are determined by solving the indicial equation $p_0(s) = 0$ and the coefficients c_k can be determined by the recurrence assuming genericity of the exponents.

Hypergeometric functions admit integral representations. Consider a holomorphic function of x which is given by the following type of integral:

$$I(a,b,c;p,q;x) \quad = \quad \int_p^q t^{b-1}(1-t)^{c-b-1}(1-xt)^{-a}dt.$$

Here $p, q \in \{0, 1, 1/x, \infty\}$ and we assume $\mathrm{Re}(b-1), \mathrm{Re}(c-b-1), \mathrm{Re}(-a), -\mathrm{Re}\,(c-a) \gg 0$. These positivity conditions ensure the convergence of the above integral and they imply a vanishing condition needed for the proof.

Theorem 1.3.5. *The integral $I(a,b,c;p,q;x)$ satisfies Gauss' equation (1.12).*

Proof. We shall prove this theorem in a purely "automatic" manner, by "integrating annihilating operators" via elimination in the two-dimensional Weyl algebra $D = \mathbf{k}\langle x,t,\partial,\partial_t\rangle$. An introduction to this technique can be found in [26, §20.3] and we shall develop it in full generality in Chapter 5; see in particular (5.35). Here the integrand to be considered equals

$$g(t,x) \quad = \quad t^{b-1}(1-t)^{c-b-1}(1-xt)^{-a}.$$

It satisfies the following two identities:

$$\frac{\partial g}{\partial t} = \left(\frac{b-1}{t} - \frac{c-b-1}{1-t} + \frac{ax}{1-xt}\right)g,$$

$$\frac{\partial g}{\partial x} = \left(\frac{ta}{1-xt}\right)g.$$

Hence the following two operators in D annihilate the function $g(t,x)$:

$$\ell_1 = t(1-t)(1-xt)\partial_t - (b-1)(1-t)(1-xt)$$
$$+ (c-b-1)t(1-xt) - axt(1-t),$$
$$\ell_2 = (1-xt)\partial - ta$$

These two operators generate a left ideal in D which is holonomic of rank 1. Our goal is to find an ideal member of the form

$$d_1 \ell_1 + d_2 \ell_2 \quad = \quad L + \partial_t e \tag{1.18}$$

where L does not contain t and ∂_t. We do this by computing the Gröbner basis of $\{\ell_1, \ell_2\}$ with respect to the order $\prec_w$ given by the weight vector

$$w = \begin{pmatrix} t & x & \partial_t & \partial \\ 1 & 0 & -1 & 0 \end{pmatrix}$$

and $\prec$ is an elimination order to eliminate t.

The Gröbner basis contains the following element

$$\begin{aligned}
L' = &-a^2 t \partial_t + at\partial_t + ax^3 \partial^2 + a^2 x^2 \partial + abx^2 \partial - x^3 \partial^2 - ax^2 \partial^2 \\
&+ a^2 bx - b\partial x^2 - acx\partial + x^2 \partial^2 - abx - ax^2 \partial + cx\partial - a^2 \\
&+ a - ax\partial\partial_t + x\partial\partial_t + a\partial\partial_t - \partial\partial_t.
\end{aligned}$$

By collecting terms with a left factor ∂_t, we see that the operator L' has the form desired on the right hand side of (1.18), with L a multiple of (1.13):

$$L' = (1-a)xP + \partial_t e, \quad e := -ta^2 + ta - xa\partial + x\partial + a\partial - \partial.$$

By tracing the Gröbner basis computation, we even obtain the multipliers:

$$\begin{aligned}
d_1 = &-x^3 \partial^2 - 2ax^2 \partial - a^2 x + ax, \\
d_2 = &-x^3 t^2 \partial\partial_t - ax^2 t^2 \partial_t - x^2 t^2 \partial_t - ax^3 t\partial + cx^3 t\partial - a^2 x^2 t + acx^2 t \\
&- x^3 t\partial - a2x^2 t + cx^2 t - x^2 t + x^3 t\partial\partial_t + ax^2 t\partial_t + x^2 t\partial_t + axt\partial_t \\
&- xt\partial_t + ax^3 \partial - bx^3 \partial + a^2 x^2 - abx^2 - ax^2 \partial + ax^2 - bx^2 - acx \\
&+ x^2 \partial + ax + cx - x - xa\partial_t + x\partial_t + a\partial_t - \partial_t.
\end{aligned}$$

We now apply (1.18) to the integrand g and integrate the resulting expression:

$$0 = \int_p^q \left((d_1 \ell_1 + d_2 \ell_2) \bullet g \right) dt = L \bullet \int_p^q g \, dt + [e \bullet g]_p^q.$$

The function $e \bullet g$ is zero at the points p and q, by our assumptions $\mathrm{Re}\,(b-1), \mathrm{Re}\,(c-b-1), \mathrm{Re}\,(-a), -\mathrm{Re}\,(c-a) \gg 0$. We conclude

$$L \bullet I(a,b,c;p,q;x) = L \bullet \int_p^q g \, dt = 0 \quad \text{for} \quad L = (1-a)xP.$$

Hence $I(a,b,c;p,q;x)$ is annihilated by the Gauss operator P.   □

Hypergeometric integrals whose parameters a, b, c are rational numbers appear frequently in algebraic geometry. A basic example is the integral

$$I\left(\frac{1}{2}, \frac{1}{2}, 1, p, q; x\right) = \int_p^q dt / \sqrt{t(1-t)(1-xt)}. \tag{1.19}$$

This function of x is annihilated by the particular hypergeometric operator

$$\theta^2 - x(\theta + 1/2)^2. \tag{1.20}$$

To understand this operator geometrically, pick $x \in \mathbf{C}\backslash\{0,1\}$ and let C_x be the elliptic curve in the s-t-plane defined by the following cubic equation:

$$C_x \quad : \quad s^2 = t(1-t)(1-xt).$$

We can regard C_x as a Riemann surface of genus 1. The complex vector space of holomorphic 1-forms $H^0(C_x, \Omega^1)$ is 1-dimensional and spanned by

$$\frac{dt}{s} \quad = \quad t^{-1/2}(1-t)^{-1/2}(1-xt)^{-1/2}dt$$

We now consider the family of elliptic curves $\{C_x\}$ as x varies. The *period map* of this family maps $\mathbf{C}\backslash\{0,1\}$ into the Riemann sphere $\mathbf{P}^1_{\mathbf{C}}$ as follows:

$$p \; : \; \mathbf{P}^1_{\mathbf{C}}\backslash\{0,1,\infty\} \to \mathbf{P}^1_{\mathbf{C}}, \quad x \longmapsto (p_1(x) : p_2(x)) = \left(\int_{\alpha_x} \frac{dt}{s} : \int_{\beta_x} \frac{dt}{s} \right),$$

where α_x and β_x are suitable cycles which generate the homology group $H_1(C_x, \mathbf{Z})$ (see Figure 1.2; we regard C_x as a double cover of $\mathbf{P}^1$ with branch points $0, 1, 1/x, \infty$). *Torelli's Theorem* states that two curves C_x and $C_{x'}$ are

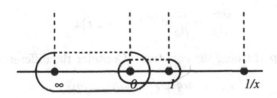

Fig. 1.2. Two cycles α_x and β_x

isomorphic if and only if they have the same periods, i.e., $p(x) = p(x')$. It is hence a natural problem to describe the image of the period map and to construct a multi-valued inverse on its image. See, e.g., [103] for details.

A key ingredient in the study of the period map is its analysis in a neighborhood of the origin $x = 0$. We sketch the basic idea. The two coordinates $p_1(x)$ and $p_2(x)$ are linearly independent solutions of the hypergeometric equation (1.20). For $c = 1$ the two series in Theorem 1.3.4 still make sense but they are identical. To construct the period map locally, we need a second linearly independent solution by considering logarithmic series as in (1.15).

Proposition 1.3.6. *A basis of solutions to (1.20) around $x = 0$ consists of*

$$\phi_1(x) = F(1/2, 1/2, 1; x)$$

and $$\phi_2(x) = \log(x) \cdot F(1/2, 1/2, 1; x) + O(x)$$

Proof. It suffices to construct a solution of the form $\phi_2(x)$. The logarithmic factor ensures that $\phi_2(x)$ will be linearly independent from $\phi_1(x)$. We present three different methods for solving (1.20) in logarithmic series at $x = 0$.

The method of perturbing the series (Frobenius method, [55, §16.1]). Let

$$f(\rho; x) = \sum_{k=0}^{\infty} c_k(\rho) x^{k+\rho}, \quad \text{where} \quad c_k(\rho) = \left(\frac{(\rho + 1/2)_k}{(\rho + 1)_k} \right)^2.$$

The coefficients $c_k(\rho)$ are chosen so that the following relation holds:

$$\left[\theta^2 - x(\theta + 1/2)^2 \right] \bullet f(\rho; x) = \rho^2 x^\rho.$$

Differentiating the right hand side with respect to ρ and taking the limit $\rho \to 0$ we obtain zero. This process commutes with differentiating with respect to x. Hence the following function

$$\phi_2(x) = \lim_{\rho \to 0} \frac{\partial}{\partial \rho} f(\rho; x)$$

is annihilated by the operator (1.20) and has the desired form. Here, we have used the identity

$$\frac{\partial x^\rho}{\partial \rho} = \frac{\partial}{\partial \rho} e^{\rho \log x} = (\log x) x^\rho.$$

The method of perturbing the equation: Consider the differential operator

$$L_\varepsilon = \theta(\theta + \varepsilon) - x(\theta + 1/2)^2,$$

where ε is a parameter. For small $\varepsilon \neq 0$, it has linearly independent solutions

$$f_1 = x^{-\varepsilon} F(1/2 - \varepsilon, 1/2 - \varepsilon, 1 - \varepsilon; x), \quad f_2 = F(1/2, 1/2, \varepsilon + 1; x).$$

Since $x^{-\varepsilon} = 1 - \varepsilon \log x + O(\varepsilon^2)$, we have the expansions

$$f_1 = F(1/2, 1/2, 1; x) - \varepsilon[\log x \cdot F(1/2, 1/2, 1; x) + O(x)] + O(\varepsilon^2)$$
$$f_2 = F(1/2, 1/2, 1; x) + \varepsilon O(x) + O(\varepsilon^2).$$

Considering the lowest term of $f_1 - f_2$ with respect to ε, we obtain ϕ_2.

The method of the linear algebra: Since $\{1, \log x\}$ is the basis of solutions of the indicial polynomial θ_x^2, as we will see in Theorem 2.5.12, there exist two linearly independent canonical series solutions f_1, f_2 such that

$$f_1 = \underline{1} + \sum_{k=1}^{\infty}(c_k + d_k \log x)x^k,$$

$$f_2 = \underline{\log x} + \sum_{k=1}^{\infty}(c'_k + d'_k \log x)x^k.$$

The underlined terms will be called starting monomials;

$$\text{Start}_{\prec_w}(f_1) = 1, \quad \text{Start}_{\prec_w}(f_2) = \log x, \quad w = 1.$$

Precise discussions of starting monomials and canonical series solutions are the main topics of Section 2.5. Let us determine the coefficients of f_1 by solving linear equations. We may assume that

$$f_1 = \sum_{k=0}^{\infty}(c_k + d_k \log x)x^k, \quad c_0 = 1, \ d_0 = 0.$$

We assume the condition $d_0 = 0$ since f_1 is a canonical solution. Since $Lf_1 = 0$ and

$$(\theta_x + a)^2 \bullet (x^k \log x) = (k + a)^2 x^k \log x + 2(k + a)x^k,$$

we have the following recurrence relation

$$\begin{pmatrix} (k+1)^2 & 2(k+1) \\ 0 & (k+1)^2 \end{pmatrix}\begin{pmatrix} c_{k+1} \\ d_{k+1} \end{pmatrix} = \begin{pmatrix} \left(k+\frac{1}{2}\right)^2 c_k + 2\left(k+\frac{1}{2}\right)d_k \\ \left(k+\frac{1}{2}\right)^2 d_k \end{pmatrix}. \quad (1.21)$$

Solving this recurrence, we have

$$d_k = 0,$$

$$c_0 = 1, \ c_1 = \left(\frac{1}{2}\right)^2, \ c_2 = \frac{\left(\frac{1}{2}\right)^2\left(\frac{3}{2}\right)^2}{2^2}, \ c_3 = \frac{\left(\frac{1}{2}\right)^2\left(\frac{3}{2}\right)^2\left(\frac{5}{2}\right)^2}{3^2 \cdot 2^2}, \dots .$$

and

$$f_1 = 1 + \left(\frac{1}{2}\right)^2 x + \frac{\left(\frac{1}{2}\right)^2\left(\frac{3}{2}\right)^2}{2^2}x^2 + \cdots.$$

Next, we determine the coefficients of f_2. We may assume

$$f_2 = \sum_{k=0}^{\infty}(c_k + d_k \log x)x^k, \quad c_0 = 0, d_0 = 1.$$

Hence, we have

$$d_1 = \left(\frac{1}{2}\right)^2, \ c_1 = -2 \cdot \left(\frac{1}{2}\right)^2 + 2 \cdot \frac{1}{2}$$

$$d_2 = \frac{\left(\frac{1}{2}\right)^2\left(\frac{3}{2}\right)^2}{2^2}, \ c_2 = \dots,$$

and

$$f_2 = \left(1 + \left(\frac{1}{2}\right)^2 x + \cdots\right) \log x + \left(-2 \cdot \left(\frac{1}{2}\right)^2 + 2 \cdot \frac{1}{2}\right) x + \cdots.$$

The linear equation (1.21) admits a unique solution at each step, because we assumed that $\log x$ does not appear in f_1 and 1 does not appear in f_2.   □

To illustrate the geometric meaning of Proposition 1.3.6, we consider the image under the period map $x \mapsto p(x) = p_2(x)/p_1(x)$ of the open domain

$U = $ (the upper half plane) $\cap$ (a small disk with the center $x = 0$).

The image $p(U)$ is a part of a hyperbolic triangle which has a vertex of angle zero. We shall *informally* describe the reason. We first consider the quotient

$$\phi(x) = \phi_2(x)/\phi_1(x) = \log x + O(x).$$

Note that $F(1/2, 1/2; 1; x)$ is real valued for $-1 < x < 1$. This local expression shows that the image of U under the map $\phi : U \to \mathbf{P}_\mathbf{C}^1$ is a horizontal strip as in Figure 1.3. Note that the angle of $\phi(U)$ at ∞ is zero.

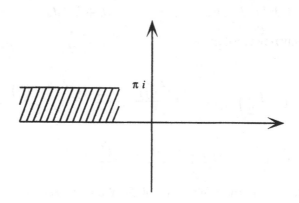

Fig. 1.3. Horizontal strip parallel to the x-axis.

Since $\{p_1, p_2\}$ and $\{\phi_1, \phi_2\}$ are bases of solutions to (1.20), the period map $p(x)$ is equal to the map $\phi(x)$ followed by a linear-fractional transformation:

$$p(x) = \frac{a\phi(x) + b}{c\phi(x) + d}, \quad \text{for some } \begin{pmatrix} a & b \\ c & d \end{pmatrix} \in GL(2, \mathbf{C}).$$

Hence $p(U)$ is the image of $\phi(U)$ by a linear fractional transformation. Such a transformation maps lines and circles to lines or circles. We conclude that $p(U)$ is a part of a hyperbolic triangle which has a vertex of angle zero. The image of the upper half plane under the period map is drawn in Figure 1.4.

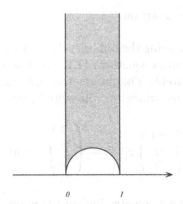

Fig. 1.4. Image of the period map

For a rigorous discussion of the conformal maps arising from bases of hypergeometric functions we refer to [14, §8.13], [103, Chapter 3, Chapter 9 (Section 5)]. A consequence is Schwarz' famous list [14, §8.14], [103, Chapter 3] of algebraic solutions to (1.12).

In this book we study hypergeometric differential equations in the formulation proposed by Gel'fand, Kapranov and Zelevinsky. On first sight their equations look rather different from (1.12) or (1.14). Consider the following system of linear partial differential equations for a function in four variables.

$$\partial_2 \partial_3 - \partial_1 \partial_4 , \quad \theta_1 - \theta_4 + 1 - c, \ \theta_2 + \theta_4 + a, \ \theta_3 + \theta_4 + b. \qquad (1.22)$$

Here $\theta_i = x_i \partial_i$ as before. What does it mean for a function $g(x_1, x_2, x_3, x_4)$ to be annihilated by (1.22) ? The last three operators express homogeneity relations which state that there exists a univariate function $f(x)$ such that

$$g(x_1, x_2, x_3, x_4) \ = \ x_1^{c-1} x_2^{-a} x_3^{-b} \cdot f\left(\frac{x_1 x_4}{x_2 x_3}\right). \qquad (1.23)$$

The first operator $\partial_1 \partial_4 - \partial_2 \partial_3$ is an encoding of Gauss' classical equation:

Proposition 1.3.7. *Suppose that (1.23) holds. Then $f(x)$ satisfies the Gauss equation (1.12) if and only if $g(x_1, x_2, x_3, x_4)$ is annihilated by $\partial_2 \partial_3 - \partial_1 \partial_4$.*

Proof. We apply the operator $\partial_2 \partial_3 - \partial_1 \partial_4$ to the expression (1.23). Using the chain rule and the product rule, and abbreviating $y = \frac{x_1 x_4}{x_2 x_3}$, we find

$$(\partial_2\partial_3 - \partial_1\partial_4) \bullet g(x_1, x_2, x_3, x_4)$$

$$= x_1^{c-1}x_2^{-a-1}x_3^{-b-1}\left((y - y^2)f''(y) - (a + b + 1)yf'(y) + cf'(y) - abf(y) \right)$$

$$= x_1^{c-1}x_2^{-a-1}x_3^{-b-1} \cdot (P \bullet f)\left(\frac{x_1 x_4}{x_2 x_3} \right).$$

This identity proves the assertion.   □

One advantage of replacing the ordinary differential equation (1.12) by the system of partial differential equations (1.22) is that it makes combinatorial structures more transparent. The parameters a, b, c are entirely encoded in the homogeneity relations, which we will write in matrix form $A \cdot \theta = \beta$ for

$$A = \begin{pmatrix} 1 & 0 & 0 & -1 \\ 0 & 1 & 0 & 1 \\ 0 & 0 & 1 & 1 \end{pmatrix}, \quad \theta = \begin{pmatrix} \theta_1 \\ \theta_2 \\ \theta_3 \\ \theta_4 \end{pmatrix} \quad \text{and} \quad \beta = \begin{pmatrix} c - 1 \\ -a \\ -b \end{pmatrix}.$$

For instance, consider the symmetry between hypergeometric series at $x = 0$ and $x = \infty$. Their *starting monomials* are annihilated by the leading forms of P with respect to the weights $(-1, 1)$ and $(1, -1)$ respectively. These *indicial polynomials* $x \cdot \text{in}_{(-1,1)}(P)$ and $\text{in}_{(1,-1)}(P)$ can be expressed in terms of the Euler operator θ:

$x = 0$: indicial polynomial $\theta(\theta + c - 1)$, series solutions $1 + \cdots,\ x^{1-c} + \cdots$

$x = \infty$: indicial polynomial $(\theta+a)(\theta+b)$, series solutions $x^{-a}+\cdots,\ x^{-b}+\cdots$

In the GKZ formulation (1.22) this corresponds to replacing the operator $\partial_2\partial_3 - \partial_1\partial_4$ by one of its two monomials, while retaining the homogeneity:

initial ideal $D\cdot\{\partial_1\partial_4,\ A\cdot\theta-\beta\}$,   solutions: $x_1^{c-1}x_2^{-a}x_3^{-b},\ x_2^{c-a-1}x_3^{c-b-1}x_4^{1-c}$.

initial ideal $D\cdot\{\partial_2\partial_3,\ A\cdot\theta-\beta\}$,   solutions $x_1^{c-a-1}x_3^{a-b}x_4^{-a},\ x_1^{c-b-1}x_2^{b-a}x_4^{-b}$.

These two ideals (systems) are the two possible *Gröbner deformations* of the hypergeometric differential equations (1.22), which is what this book is all about. Note that the supports of the starting monomials are precisely the triangles in the two possible triangulations of a unit square, as in Figure 1.5. For instance, the triangle $\{1, 3, 4\}$ in the right triangulation corresponds to the starting monomial $x_1^{c-a-1}x_3^{a-b}x_4^{-a}$. Where does the square come from ? Its vertices are the column vectors of the matrix A ! The operator $\partial_2\partial_3-\partial_1\partial_4$ encodes the property that the midpoint of the segment $\overline{23}$ equals the midpoint of the segment $\overline{14}$. In other words, it encodes the kernel $\mathbf{Z}\{(1, -1, -1, 1)\}$ of the matrix A. Later in this book, we examine the *GKZ hypergeometric system* arising from any integer matrix A, and we shall see that the relation between triangulations, leading monomials and indicial ideals holds in general.

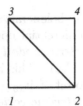   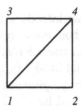

Fig. 1.5. The two Gröbner deformations of Gauss' equation

Example 1.3.8. (Some classical orthogonal polynomials)
The transformation (1.23) takes polynomial solutions f of Gauss' equation
(1.12) to Laurent polynomials g which are annihilated by $\partial_2\partial_3 - \partial_1\partial_4$. For
instance, if P_n denotes the classical *Legendre polynomial* of degree n then

$$\frac{x_2^n}{x_3^{n+1}} \cdot P_n\left(1 - 2 \cdot \frac{x_1 x_4}{x_2 x_3}\right)$$

is a Laurent polynomial solution to (1.22) for $a = -n, b = n+1, c = 1$.

Let us close this section with some additional remarks on the equation
$\partial_2\partial_3 - \partial_1\partial_4$ and bibliographic notes. For an algebraic geometer this is simply
the defining polynomial of the quadric surface in $\mathbf{P}^3$, also known as the *Segre
embedding* of $\mathbf{P}^1 \times \mathbf{P}^1$. In analysis the equation $\partial_2\partial_3 - \partial_1\partial_4$ is known as the
ultrahyperbolic differential equation, after the work of Fritz John [56]. Thus
Proposition 1.3.7 states in other words that Gauss' hypergeometric equation
is equivalent to John's ultrahyperbolic equation plus homogeneity relations.

Our sources on the Gauss hypergeometric function are mainly [14], [103].
Our proof of Theorem 1.3.5 is done in an "automatic manner". See the book
of Petkovšek, Wilf, and Zeilberger [86] for "automatic" proofs for special
function identities in general. Our proof is done in the same spirit, but is
different from theirs and is using an algorithm to get the 0-th integral of a
given $\mathcal{D}$-module (see Section 5.5). A traditional proof of Theorem 1.3.5 can be
found in, e.g., [103, 4.2.1]. The book of Yoshida [103] is an exposition of the
study by Matsumoto, Sasaki, and Yoshida on the image of the period map of
a four parameter family of $K3$ surfaces as the hypergeometric system $E(3,6)$.
Period maps described by GKZ systems such as (1.22) play a prominent role
in recent work on mirror symmetry (see, e.g., [49], [50], [95]).

1.4 Holonomic Systems

In this section we introduce the characteristic ideal, holonomic systems, the singular locus, holonomic rank, and the ring of differential operators with rational function coefficients. We define the *characteristic ideal* of a D-ideal I to be the initial ideal $\mathrm{in}_{(0,e)}(I)$ where $e = (1, 1, \ldots, 1)$. This is an ideal in the commutative polynomial ring $\mathbf{k}[x, \xi]$. The zero set of $\mathrm{in}_{(0,e)}(I)$ in affine $2n$-space is denoted by $\mathrm{ch}(I)$ and called the *characteristic variety* of I.

For experts in D-modules we note that this definition is consistent with the sheaf-theoretic approach traditionally used in algebraic analysis. Let $\mathbf{k} = \mathbf{C}$. We write $\mathcal{D}$ for the *sheaf of algebraic differential operators* on complex affine n-space $\mathbf{C}^n$. General references for the theory of $\mathcal{D}$-modules are [15], [17], [54], [91]. In these references, the characteristic variety of a $\mathcal{D}$-module is defined in terms of good filtrations for sheaves of modules.

Theorem 1.4.1. *The characteristic variety* $\mathrm{ch}(I)$ *equals the characteristic variety of the left $\mathcal{D}$-module $\mathcal{D}/\mathcal{D}I$ in the sense of the theory of $\mathcal{D}$-modules.*

This theorem is due to Oaku [76]; we refer to his paper for the proof. An essential ingredient is the flatness of $\mathcal{D}$ over the Weyl algebra D. It follows from Theorem 1.1.6 that the characteristic variety can be obtained by computing a Gröbner basis for the given ideal I with respect to the weight vector $(0, e)$. This is done using Algorithm 1.1.9. We present three examples.

Example 1.4.2. Let $n = 1$ and consider the D-ideal $I = D \cdot \{\partial^2, x\partial - 1\}$ which consists of all operators which annihilate the identity function $f(x) = x$. This is the standard example of a D-module which is projective but not free; see [93, page 301]. Note that the two given generators are a Gröbner basis. Hence the characteristic ideal of I equals $\mathrm{in}_{(0,e)}(I) = \langle \xi^2, x\xi \rangle = \langle \xi \rangle \cap \langle \xi^2, x \rangle$.

Example 1.4.3. Let $n = 2$ and $I = D \cdot \{x_1\partial_2, x_2\partial_1\}$. To get a Gröbner basis for I one must add the commutator of the two given monomial generators. The characteristic ideal of I equals $\mathrm{in}_{(0,e)}(I) = \langle x_1\xi_2, x_2\xi_1, x_1\xi_1 - x_2\xi_2 \rangle$.

Example 1.4.4. Consider the GKZ system (1.22) associated to the Gauss hypergeometric function. It follows from more general results to be presented in Theorem 4.3.8 or from a direct computation of the characteristic ideal that the four given generators are a Gröbner basis with respect to $(0, e) = (0, 0, 0, 0, 1, 1, 1, 1)$. Hence the characteristic ideal equals

$$J := \mathrm{in}_{(0,e)}(I) = \langle \xi_2\xi_3 - \xi_1\xi_4, x_1\xi_1 - x_4\xi_4, x_2\xi_2 + x_4\xi_4, x_3\xi_3 + x_4\xi_4 \rangle.$$

This ideal is unmixed of dimension 4. Its primary decomposition equals

$$\begin{aligned}
J = \ & \langle \xi_1, \xi_2, \xi_3, x_4 \rangle \ \cap \ \langle \xi_1, \xi_2, x_3, \xi_4 \rangle \ \cap \ \langle \xi_1, x_2, \xi_3, \xi_4 \rangle \ \cap \\
& \langle x_1, \xi_2, \xi_3, \xi_4 \rangle \ \cap \ \langle \xi_1, \xi_2, x_3, x_4 \rangle \ \cap \ \langle x_1, \xi_2, x_3, \xi_4 \rangle \ \cap \\
& \langle x_1, x_2, \xi_3, \xi_4 \rangle \ \cap \ \langle \xi_1, x_2, \xi_3, x_4 \rangle \ \cap \ \left(J + \langle \xi_1, \xi_2, \xi_3, \xi_4 \rangle^2 \right) \ \cap \\
& \left(J + \langle x_1 x_4 - x_2 x_3, x_1\xi_2 + x_3\xi_4, x_1\xi_3 + x_2\xi_4, x_3\xi_1 + x_4\xi_2, x_2\xi_1 + x_4\xi_3 \rangle \right).
\end{aligned}$$

This corresponds to the decomposition of the characteristic variety ch(I) into ten irreducible components each of dimension 4. The last ideal is a prime ideal isomorphic to the toric ideal of degree 6 associated with the regular 3-cube. The second-to-last ideal is not prime but $\langle \xi_1, \xi_2, \xi_3, \xi_4 \rangle$-primary.

Let us briefly explain some terms from commutative algebra which have been used in the examples and will be used in this book. Our explanation is mostly to fix notations and terminologies. The relevant notions are *Hilbert polynomial*, *dimension (Krull dimension)*, *degree*, *radical* of an ideal, *primary ideal decomposition* of an ideal, and *associated prime ideal*.

Consider an ideal $I \subseteq \mathbf{k}[x] := \mathbf{k}[x_1, \ldots, x_n]$ and the monomial ideal $\text{in}_{\prec_e}(I)$ where $e = (1, \ldots, 1)$ and $\prec$ is a term order on $\mathbf{k}[x]$. Let $p(k)$ be the number of the standard monomials which have total degree less than or equal to k. It is known that $p(k)$ is a polynomial in k for sufficiently large k. The polynomial $p(k)$ is called the *Hilbert polynomial* of I (with respect to the weight vector e). It is independent of the choice of the refining term order $\prec$. It is known that the Hilbert polynomial has the form $p(k) = \frac{m}{d!}k^d + \cdots$. The degree d of the Hilbert polynomial is the Krull dimension of I. See, e.g., [12, 9.3, p.441], [27, Chapter 9], [32, Chapter 8]. The Krull dimension is, intuitively speaking, the dimension of the zero set $V(I)$ of the ideal. We remark that the Krull dimension of I is equal to that of $\text{in}_{\prec}(I)$ for any term order $\prec$. Geometrically speaking, the invariant m is equal to the number of intersections of $V(I)$ and a sufficiently generic affine space of dimension $n - d$. When I is generated by homogeneous elements, the number m is called the *degree* of I and is denoted by degree(I) (see, e.g., Chapter 4 or [27, Exercise §9.5–12, page 465]).

An ideal of dimension 0 is called an *Artinian* ideal or a *zero-dimensional* ideal. The corresponding zero set (the set of the roots) is a finite set of points. The number of the roots counted with the multiplicity of each root in the algebraic closure of $\mathbf{k}$ is equal to $m = \dim_{\mathbf{k}} \mathbf{k}[x]/I$, which can be computed by counting the number of the standard monomials of $\text{in}_{\prec}(I)$.

The radical of an ideal I is defined by

$$\{f \in \mathbf{k}[x] \mid f^m \in I \text{ for some } m.\}$$

and is denoted by rad(I) or $\sqrt{I}$. See, e.g., [12, Chapter 8], [27, §4.2, p.174]. An ideal Q is called *primary ideal* if $fg \in Q$ and $f \notin Q$, then there exists m so that $g^m \in Q$. Any proper ideal $I \subset \mathbf{k}[x]$ has a primary ideal decomposition

$$I = Q_1 \cap \cdots \cap Q_m \qquad (1.24)$$

satisfying

1. Q_i is primary ideal.
2. No Q_i is contained in the intersection of the other Q_j's.
3. The associated prime ideals rad(Q_j) are distinct.

The primary ideal decomposition gives a geometric decomposition of $V(I)$ into irreducible pieces:

$$V(I) = V(\operatorname{rad}(Q_1)) \cup \cdots \cup V(\operatorname{rad}(Q_m)).$$

See, e.g., [12, Chapter 8], [32, Chapter 3] for algorithms and details. For a prime ideal J', a primary ideal J is called J'-*primary* or primary to J' when $\operatorname{rad}(J) = J'$. For an ideal I, a prime ideal P is called an *associated prime* of I if there exists $m \in \mathbf{k}[x]$ such that

$$P = (I : m) := \{f \in \mathbf{k}[x] \mid mf \in I\}.$$

The number of associated primes is finite for a given ideal I. The set of the associated primes is denoted by $\operatorname{Ass}(I)$. Assume that I admits a primary ideal decomposition (1.24). Then, the set $\operatorname{Ass}(I)$ agrees with the set $\{\operatorname{rad}(Q_j) \mid j = 1, \ldots, m\}$. The set of minimal primes in $\operatorname{Ass}(I)$ with respect to the inclusion relation is called the set of the *minimal primes*. An associated prime $P_i \in \operatorname{Ass}(I)$ which contains an other associated prime $P_j \in \operatorname{Ass}(I)$ is called an *embedded associated prime*. For example, consider the following primary ideal decomposition $I := \langle \xi^2, x\xi \rangle = \langle \xi \rangle \cap \langle \xi^2, x \rangle$. In this case, $\operatorname{Ass}(I) = \{\langle \xi \rangle, \langle \xi, x \rangle\}$, but the set of the minimal primes is $\{\langle \xi \rangle\}$. A primary ideal is characterized as an ideal which has only one associated prime.

The characteristic variety is the object of the *Fundamental Theorem of Algebraic Analysis (= FTAA)*. There are a weak version and a strong version.

Theorem 1.4.5. (Weak FTAA)
The characteristic ideal $\operatorname{in}_{(0,e)}(I)$ *of a proper D-ideal I has dimension* $\geq n$.

The weak FTAA first appeared in [13]. The proof of this theorem is elementary and can be found in [26, Chapter 9]. In the later section on the Gröbner fan we shall prove that the Weak FTAA holds not just for $(u, v) = (0, e)$ but it holds for all weight vectors $(u, v) \in \mathbf{R}_+^{2n}$ with $u + v > 0$. Here and throughout this book $\mathbf{R}_+$ denotes the non-negative reals. The strong version of the FTAA states that each irreducible component of the characteristic variety $\operatorname{ch}(I)$ has dimension at least n.

Theorem 1.4.6. (Strong FTAA) *Let I be a proper D-ideal. Then every minimal prime of the characteristic ideal* $\operatorname{in}_{(0,e)}(I)$ *has dimension* $\geq n$.

This theorem is due to Sato, Kashiwara and Kawai [91]. A purely algebraic proof was given by Gabber [34]. We shall state the main result of Gabber's paper in precise terms and explain why it implies Theorem 1.4.6. The *Poisson bracket* of two polynomials f and g in $\mathbf{k}[x, \xi]$ is defined as follows:

$$\{f, g\} := \sum_{i=1}^{n} \left(\frac{\partial f}{\partial \xi_i} \cdot \frac{\partial g}{\partial x_i} - \frac{\partial g}{\partial \xi_i} \cdot \frac{\partial f}{\partial x_i} \right).$$

The Poisson bracket is useful for computing the $(0, e)$-initial form of the commutator of two elements p and q in the Weyl algebra D directly from their initial forms. More precisely, we have

$$\{\text{in}_{(0,e)}(p), \text{in}_{(0,e)}(q)\} \;=\; \text{in}_{(0,e)}(pq - qp) \text{ or } 0. \qquad (1.25)$$

The identity (1.25) is derived from the Leibnitz formula (Theorem 1.1.1). An ideal J in the polynomial ring $\mathbf{k}[x, \xi]$ is said to be *integrable* if it is closed under the Poisson bracket, i.e., $f, g \in J$ implies $\{f, g\} \in J$. It follows immediately from (1.25) that the characteristic ideal $\text{in}_{(0,e)}(I)$ is integrable.

In Gabber's paper it is proved – by purely algebraic means – that the radical of the characteristic ideal is integrable as well. In other words, "the characteristic variety $\text{ch}(I)$ is integrable". This statement and its relations to symplectic geometry are discussed in [26, §11.2].

Theorem 1.4.6 is derived from the integrability of $\text{ch}(I)$ as follows. Consider any smooth point p of $\text{ch}(I)$. Then the tangent space of $\text{ch}(I)$ at p is integrable as well. However, for a linear space L to be integrable at most one of the coordinate functions x_i and ξ_i can vanish on L. Hence an integrable linear space has dimension at least n. Hence the component of $\text{ch}(I)$ containing p has dimension at least n, and the strong FTAA follows.

Remark 1.4.7.
The characteristic ideal can have embedded associated primes of dimension less than n. For instance, in both Examples 1.4.2 and 1.4.3, the irrelevant maximal ideal (which is zero-dimensional) is an associated prime of $\text{in}_{(0,e)}(I)$.

Definition 1.4.8. Let I be a non-zero D-ideal. We call I *holonomic* if its characteristic ideal $\text{in}_{(0,e)}(I)$ has dimension n. The *holonomic rank* of I is the following vector space dimension over the field $\mathbf{k}(x) = \mathbf{k}(x_1, \dots, x_n)$:

$$\text{rank}(I) \;:=\; \dim_{\mathbf{k}(x)}\big(\mathbf{k}(x)[\xi] \,/\, \mathbf{k}(x)[\xi] \cdot \text{in}_{(0,e)}(I)\big). \qquad (1.26)$$

The D-ideals in Examples 1.4.2, 1.4.3, and 1.4.4 are holonomic. Their (holonomic) ranks are 1, 1 and 2 respectively. The paradigm of a holonomic D-ideal is the principal D-ideal defined by a linear ordinary (i.e. $n = 1$) differential equation of order m:

$$I = D \cdot \{a_m(x)\partial^m + a_{m-1}(x)\partial^{m-1} + \cdots + a_0(x)\},$$

where $a_m \neq 0$. Here $\text{in}_{(0,e)}(I) = \langle a_m(x)\xi^m \rangle$, hence I is holonomic of rank m.

Proposition 1.4.9. *If I is a holonomic D-ideal, then $\text{rank}(I)$ is finite. The converse is not true: there exist non-holonomic D-ideals I with $\text{rank}(I) < \infty$.*

Proof. The affine variety in $2n$-space defined by $\text{in}_{(0,e)}(I)$ has dimension n. Consider its projection onto the $(n + 1)$-dimensional subspace with coordinates $x_1, \dots, x_n, \xi_i$. The Zariski closure of this projection has dimension $\leq n$, and hence its defining ideal $\sqrt{\mathrm{in}_{(0,e)}(I)} \cap \mathbf{k}[x_1, \ldots, x_n, \xi_i]$ is non-zero. Note that this elimination ideal is homogeneous with respect to $(0, e)$. Hence, for each $i \in \{1, 2, \ldots, n\}$ there exists a non-zero $(0, e)$-homogeneous polynomial $a_i(x_1, \ldots, x_n) \cdot \xi_i^{d_i}$ in $\mathrm{in}_{(0,e)}(I)$. We conclude that $\mathrm{rank}(I) \le \prod_{i=1}^{n} d_i < \infty$. The second assertion is proved by the following example.   □

Example 1.4.10. Let $n = 2$, let $f = x_1^3 - x_2^2$, and consider the D-ideal

$$I = D \cdot \left\{ f \partial_1 + \frac{\partial f}{\partial x_1}, \quad f \partial_2 + \frac{\partial f}{\partial x_2} \right\}.$$

The two given generators are a Gröbner basis. The characteristic ideal equals

$$\mathrm{in}_{(0,e)}(I) = \langle f(x_1, x_2) \cdot \xi_1, \ f(x_1, x_2) \cdot \xi_2 \rangle.$$

The characteristic variety $\mathrm{ch}(I)$ has codimension 1 in affine 4-space. Hence I is not holonomic. On the other hand, the holonomic rank of I is finite:

$$\mathrm{rank}(I) = 1 < \infty.$$

Note that I is a subideal of the annihilator of the rational function $1/f$.

The following algorithm can be used to decide whether a given D-ideal I is holonomic and to compute its holonomic rank $\mathrm{rank}(I)$. Let $\prec$ be any term order. Compute a Gröbner basis $\mathcal{G}$ of I with respect to the new term order $\prec_{(0,e)}$. The initial monomials of the elements in $\mathcal{G}$ generate

$$\mathrm{in}_{\prec_{(0,e)}}(I) = \mathrm{in}_{\prec}\big(\mathrm{in}_{(0,e)}(I)\big) \subset \mathbf{k}[x, \xi]. \tag{1.27}$$

The (Krull) dimension of (1.27) equals the dimension of the characteristic ideal $\mathrm{in}_{(0,e)}(I)$. It can be determined combinatorially as in [27, §9]. The following lemma is an exercise in commutative Gröbner basis theory.

Lemma 1.4.11. *Let M be the monomial ideal in $\mathbf{k}[\xi_1, \ldots, \xi_n]$ obtained by replacing each variable $x_1, \ldots, x_n$ by 1 in the minimal generators of the ideal $\mathrm{in}_{\prec_{(0,e)}}(I)$. Then*

$$\mathrm{rank}(I) = \dim_{\mathbf{k}} \mathbf{k}[\xi_1, \ldots, \xi_n]/M = \#\{\ \xi\text{-monomials not in } M\}.$$

We can decide if a given D-ideal is holonomic by computing Gröbner bases for a class of weight vectors (u, v) not just $(0, e)$. Choosing a good weight vector can make a huge difference in the running time of practical computations.

Theorem 1.4.12. *Let $(u, v) \in \mathbf{R}_+^{2n}$ be any nonnegative weight vector with $u + v > 0$. A proper D-ideal I is holonomic if and only if $\dim \mathrm{in}_{(u,v)}(I) = n$.*

We shall prove this in the end of Section 2.2 as an application of the Gröbner fan. Another way to prove this theorem is to apply the techniques of homological algebra explained in [15, p.125], [17, p.184], or [54, p.62].

If one only wishes to compute the holonomic rank of a D-ideal I, without deciding whether or not I is holonomic, then it can be more efficient in practice to replace the Weyl algebra D by the *ring of differential operators*

$$R = \mathbf{k}(x)\langle\partial\rangle := \mathbf{k}(x_1,\ldots,x_n)\langle\partial_1,\ldots,\partial_n\rangle.$$

This is the $\mathbf{k}$-algebra generated by (commutative) polynomials in $\partial_1,\ldots,\partial_n$ and rational functions in $x_1,\ldots,x_n$, subject to the commutation relations

$$\partial_i \cdot c(x) = c(x) \cdot \partial_i + \frac{\partial c(x)}{\partial x_i}, \quad c(x) \in \mathbf{k}(x). \tag{1.28}$$

It is straightforward to extend Gröbner basis theory to R with respect to the n variables $\partial_1,\ldots,\partial_n$. Any element $p \in R$ can be uniquely expressed as

$$p = \sum_{\beta \in E} c_\beta(x) \cdot \partial^\beta, \quad c_\beta(x) \in \mathbf{k}(x)^*, \ E \subset \mathbf{N}^n \text{ finite}. \tag{1.29}$$

Let $\prec$ be any term order on $\mathbf{N}^n$. (The following definitions are analogous for positive weight vectors $w \in \mathbf{R}_+^n$.) The *associated graded ring* of R equals

$$\mathrm{gr}_\prec(R) = \mathbf{k}(x)[\xi] = \mathbf{k}(x_1,\ldots,x_n)[\xi_1,\ldots,\xi_n].$$

The *initial term* of p in (1.29) with respect to $\prec$ is $\mathrm{in}_\prec(p) := c_\beta(x)\xi^\beta$ where β is the $\prec$-largest element of E. For a left ideal J in R we define

$$\mathrm{in}_\prec(J) := \langle \mathrm{in}_\prec(p) : p \in J \rangle \subset \mathbf{k}(x)[\xi].$$

A finite generating set $\mathcal{G}$ of J is a *Gröbner basis* if the initial ideal $\mathrm{in}_\prec(J)$ is generated by $\{\mathrm{in}_\prec(g) : g \in \mathcal{G}\}$. A Gröbner basis can be computed from any generating set of J by the *Buchberger algorithm*, where the notion of S-pair is modified according to the commutation rules (1.28) similarly to Section 1.1. In the case of $n = 1$, the ring R is a principal ideal domain and the Buchberger algorithm for $n = 1$ is nothing but the Euclidean algorithm that computes the greatest common divisor.

A left ideal J in R is *zero-dimensional* if R/J is a finite-dimensional $\mathbf{k}(x)$-vector space. Standard arguments show that, for any term order $\prec$ on R,

$$\dim_{\mathbf{k}(x)}(R/J) = \dim_{\mathbf{k}(x)}(\mathbf{k}(x)[\xi]/\mathrm{in}_\prec(J)) = \#\{\xi\text{-monomials not in } \mathrm{in}_\prec(J)\}.$$

Gröbner bases in the ring of differential operators R and in the Weyl algebra D are related as follows. A term order $\prec$ on D is an *elimination term order* if $\partial^\beta \prec \partial^\gamma$ implies $x^\alpha\partial^\beta \prec \partial^\gamma$ for all $\alpha \in \mathbf{N}^n$. If $\prec$ is any elimination term order on D then we write $\prec'$ for the term order on $\mathbf{N}^n$ gotten by restriction to monomials in $\partial_1,\ldots,\partial_n$. Note that every term order on D of the form $\prec_{(0,e)}$ is an elimination term order, and its restriction $\prec'_{(0,e)}$ is a term order on R which refines the total degree in $\partial_1,\ldots,\partial_n$.

Proposition 1.4.13. *Let G be a Gröbner basis of a D-ideal I with respect to an elimination term order $\prec$. Then G is a Gröbner basis of the left ideal RI in R with respect to $\prec'$. Hence a generating set of $\mathrm{in}_{\prec'}(RI)$ is gotten from the minimal generators of $\mathrm{in}_\prec(I)$ by replacing the variables $x_1, \ldots, x_n$ by 1.*

The proof of this is similar to [12, p.390, Lemma 8.93]; see also [75, p.93]. Proposition 1.4.13 implies that the generators of the monomial ideal M in Lemma 1.4.11 coincide with the generators of $\mathrm{in}_{\prec'_{(0,e)}}(RI)$. We conclude:

Corollary 1.4.14. *Let I be any D-ideal. Then*

$$\mathrm{rank}(I) \quad = \quad \dim_{\mathbf{k}(x)}(R/RI).$$

If I is holonomic then RI is a zero-dimensional ideal in R, but not conversely.

Example 1.4.10 gives a non-holonomic D-ideal I such that RI is zero-dimensional. However, the larger D-ideal $RI \cap D$ is holonomic in this case:

Theorem 1.4.15. *If $J \subset R$ is a zero-dimensional ideal then $J \cap D$ is holonomic.*

This result follows from work of Kashiwara [59] who proved that if a D-module is holonomic in the complement of an affine hypersurface then its localization along the hypersurface is holonomic on the whole of affine space. An algebraic algorithm for computing the localization of a *holonomic* ideal is given in [80] and an algorithm for computing the localization of a zero-dimensional R-ideal is given in [82]. The problem of computing the *Weyl closure* $J \cap D$ of an arbitrary R-ideal J has been solved recently by Tsai.

Example 1.4.16. Put $f_1 = (x_1^3 - x_2^2)\partial_1 + 3x_1^2$, $f_2 = (x_1^3 - x_2^2)\partial_2 - 2x_2$. The operators f_1 and f_2 annihilate the function $1/(x_1^3 - x_2^2)$. As we have seen in Example 1.4.10, the D-ideal generated by f_1 and f_2 is not holonomic, but

$$\frac{2x_1}{x_1^3 - x_2^2} f_1 + \frac{3x_2}{x_1^3 - x_2^2} f_2 = 2x_1\partial_1 + 3x_2\partial_2 + 6,$$

$$\frac{2x_2}{x_1^3 - x_2^2} f_1 + \frac{3x_1^2}{x_1^3 - x_2^2} f_2 = 2x_2\partial_1 + 3x_1^2\partial_2$$

generate a holonomic D-ideal I. Using Oaku's algorithm 5.3.15, we can check that I equals the annihilator of the rational function $1/(x_1^3 - x_2^2)$. In other words, the D-ideal I is the Weyl closure of the R-ideal $R \cdot \{f_1, f_2\}$.

In summary, we can evaluate the holonomic rank as follows.

Algorithm 1.4.17
Input: Generators of a D-ideal I.
Output: The holonomic rank of I.

1. Compute a Gröbner basis G of I with respect to any term order in

$$R = \mathbf{k}(x_1, \ldots, x_n)\langle \partial_1, \ldots, \partial_n \rangle.$$

2. Determine the number r of standard monomials $\{\partial^\alpha\}$ with respect to G. Then, the number r is the holonomic rank.

In the rest of this section we shall concentrate on the analytic meaning of holonomic systems and holonomic rank. We assume $\mathbf{k} = \mathbf{C}$ in the following discussion. First consider the easiest case of an ordinary differential equation with polynomial coefficients. Let $n = 1$ and I the D-ideal generated by

$$p \quad := \quad a_m(x) \cdot \partial^m + \cdots + a_1(x) \cdot \partial + a_0(x), \quad \text{where} \quad a_m \neq 0.$$

The set of complex roots of the polynomial $a_m(x)$ is denoted by $\mathrm{Sing}(I)$ and called the singular locus of I. Let U be a simply connected domain in the complex plane $\mathbf{C}$ which is disjoint from $\mathrm{Sing}(I)$. A holomorphic function f on U is a *solution to I* if $p \bullet f = 0$. The classical existence theorem (Theorem 1.3.3) states that the $\mathbf{C}$-vector space of holomorphic solutions to I on U has dimension $m = \mathrm{rank}(I)$. Thus the holonomic rank counts the number of linearly independent holomorphic solutions in a neighborhood of any non-singular point $x \in \mathbf{C} \backslash \mathrm{Sing}(I)$.

The singularities $x \in \mathrm{Sing}(I)$ of an ordinary differential equation come in two flavors: regular and irregular. We wish to express this important distinction in Gröbner basis notation. Let $I = D \cdot \{p\}$ as above. The initial ideal $\mathrm{in}_{(-1,1)}(I)$ is again a D-ideal. Its holonomic rank satisfies the inequality

$$\mathrm{rank}(\mathrm{in}_{(-1,1)}(I)) \quad \leq \quad \mathrm{rank}(I). \qquad (1.30)$$

Equality holds in (1.30) if and only if

$$\mathrm{lowdeg}(a_m(x)) - \mathrm{lowdeg}(a_i(x)) \ \leq \ m - i \quad \text{for} \quad i = 0, 1, \ldots, m-1, \ (1.31)$$

where $\mathrm{lowdeg}(a_i)$ denotes the largest integer ℓ such that x^ℓ divides $a_i(x)$. This is seen by inspecting all terms $x^j \partial^i$ of p. But the conditions (1.31) are precisely the classical *Fuchs conditions* for regular singularities. We conclude:

Theorem 1.4.18. (Fuchs' Theorem, [55, §15.3]) *The following conditions are equivalent for a sufficiently small open neighborhood U of the origin:*

(1) $\mathrm{rank}(\mathrm{in}_{(-1,1)}(I)) = \mathrm{rank}(I)$.
(2) (Polynomial growth condition) *There exists a number N such that every holomorphic solution f on the universal covering of $U \backslash \{0\}$ satisfies $|x|^N |f(x)| \to 0$ when x approaches the origin 0 in a fixed angle $\theta_0 < \arg(x) < \theta_1$.*
(3) (Logarithmic series expansion) *Every holomorphic solution f on the universal covering of $U \backslash \{0\}$ can be expressed as a linear combination of functions of the form*

$$x^\lambda \cdot \sum_{i=0}^{m-1} g_i(x) \cdot (\log x)^i$$

where $\lambda \in \mathbf{C}$ and $g_i(x)$ is a holomorphic function on U.

When these three equivalent conditions are satisfied then the origin $x = 0$ is said to be a *regular singular point* of I. For example, $x = 0$ is not a regular singular point of $I = D \cdot \{x^2 \partial + 1\}$; indeed, $\mathrm{in}_{(-1,1)}(I) = D$ and the unique solution $f = \exp(1/x)$ of I has an essential singularity at $x = 0$.

We shall next present a generalization of the classical existence theorem to $n > 1$ variables. Let I be a D-ideal and $\mathrm{ch}(I) \subset \mathbf{C}^{2n}$ its characteristic variety. We define the *singular locus* of I, denoted by $\mathrm{Sing}(I)$, to be the Zariski closure of the image of $\mathrm{ch}(I) \backslash \{\xi_1 = \cdots = \xi_n = 0\}$ under the coordinate projection $\mathbf{C}^{2n} \to \mathbf{C}^n$, $(x, \xi) \mapsto x$. (Note that this extends the above definition for $n = 1$.) Thus the singular locus is the zero set of the following ideal:

$$\left(\mathrm{in}_{(0,e)}(I) : \langle \xi_1, \ldots, \xi_n \rangle^\infty \right) \cap \mathbf{C}[x_1, \ldots, x_n]. \qquad (1.32)$$

The *saturation* and *elimination* in formula (1.32) take place in the commutative polynomial ring $\mathbf{C}[x, \xi]$. Thus the singular locus can be computed from the characteristic ideal $\mathrm{in}_{(0,e)}(I)$ by commutative Gröbner basis methods; the saturation can be computed, e.g., by the formula

$$\left(\mathrm{in}_{(0,e)}(I) : \langle \xi_1, \ldots, \xi_n \rangle^\infty \right) = \left(\langle \mathrm{in}_{(0,e)}(I), z - f \rangle : z^\infty \right) \cap \mathbf{k}[x, \xi]$$

where $f = \xi_1 + y\xi_2 + \cdots + y^{n-1}\xi_n$ and y, z are new indeterminates. See [27, Exercise §4.4–8] and [32, §15.10.4 and §15.10.6].

Consider the characteristic ideal in Example 1.4.4. The saturation in (1.32) erases the second-to-last primary component and preserves the other nine. The subsequent elimination step shows that (1.32) equals the principal radical ideal

$$\text{ideal of } \mathrm{Sing}(I) \quad = \quad \langle\, x_1 \cdot x_2 \cdot x_3 \cdot x_4 \cdot (x_1 x_4 - x_2 x_3) \,\rangle. \qquad (1.33)$$

The following theorem relates the holonomic rank of a D-ideal I to the solution space of I.

Theorem 1.4.19. *Let I be a holonomic ideal and U a simply connected domain in $\mathbf{C}^n \backslash \mathrm{Sing}(I)$. Consider the system of differential equations $I \bullet f = 0$, i.e.,*

$$\ell \bullet f = 0, \quad \ell \in I, \qquad \text{for holomorphic functions } f \text{ on } U.$$

The dimension of the complex vector space of solutions is equal to $\mathrm{rank}(I)$.

This is a special case of the *Cauchy-Kowalevskii-Kashiwara Theorem* (= CKKT). The CKKT concerns cohomological solution spaces. It first appeared in the celebrated Master's Thesis of Kashiwara in 1971. To prove the CKKT one needs a sheaf-theoretic discussion of $\mathcal{D}$-modules and solution sheaves. We refer to [62, p.44], [75, 3.5, 5.3], [76].

In the following discussion we shall prove a variant of Theorem 1.4.19 which is slightly weaker, namely, we shall replace $\text{Sing}(I)$ by a possibly larger hypersurface. This still implies that the holonomic rank equals the dimension of holomorphic solutions in a small neighborhood of a generic point in $\mathbf{C}^n$. Our construction uses Gröbner bases in R and the following result on Pfaffian equations. Let $A_1, A_2, \ldots, A_n$ be $m \times m$ matrices whose entries are holomorphic functions. Consider the following *Pfaffian system* of first-order differential equations for a vector-valued holomorphic function $F(x)$:

$$\frac{\partial F}{\partial x_i} = A_i(x) \cdot F \qquad (i = 1, 2, \ldots, n) \qquad (1.34)$$

If there exist m linearly independent solutions, then the matrices satisfy the following *compatibility conditions*

$$\frac{\partial A_i}{\partial x_j} + A_i A_j = \frac{\partial A_j}{\partial x_i} + A_j A_i \quad \text{for all } 1 \le i < j \le n. \qquad (1.35)$$

Let us give a proof. Let $\tilde{F}$ be the $m \times m$ matrix consisting of m linearly independent solutions. Since

$$\frac{\partial}{\partial x_j} \frac{\partial \tilde{F}}{\partial x_i} = \frac{\partial A_i}{\partial x_j} \tilde{F} + A_i \frac{\partial \tilde{F}}{\partial x_j} = \left(\frac{\partial A_i}{\partial x_j} + A_i A_j \right) \tilde{F}$$

and

$$\frac{\partial}{\partial x_j} \frac{\partial \tilde{F}}{\partial x_i} = \frac{\partial}{\partial x_i} \frac{\partial \tilde{F}}{\partial x_j},$$

we have

$$\left(\frac{\partial A_i}{\partial x_j} + A_i A_j \right) \tilde{F} = \left(\frac{\partial A_j}{\partial x_i} + A_j A_i \right) \tilde{F}.$$

Multiplying the inverse matrix $\tilde{F}^{-1}$ from the right, we obtain (1.35). The converse is true as follows.

Theorem 1.4.20. *Suppose that the entries of $A_1, \ldots, A_n$ are holomorphic around a point $a \in \mathbf{C}^n$. If (1.35) holds then the space of holomorphic solutions to the Pfaffian system (1.34) at $x = a$ has $\mathbf{C}$-dimension m.*

This is a special case of the Frobenius theorem on the existence of integral manifolds in differential geometry. A constructive proof of Theorem 1.4.20 was given by Yoshida and Takano [104].

We next explain how to transform a holonomic ideal I into a first order system of the Pfaffian form (1.34). Consider the left ideal RI in the ring of differential operators R. Compute a Gröbner basis $\mathcal{G}$ of RI with respect to any term order $\prec$ on R and determine the set of *standard monomials*,

$$\{\ \partial\text{-monomials not in } \mathrm{in}_\prec(RI)\ \} \ = \ \{\partial^{p(1)}, \partial^{p(2)}, \ldots, \partial^{p(m)}\}, \quad p(i) \in \mathbf{N}^n.$$

This set is finite and has cardinality $m = \mathrm{rank}(I)$ by Corollary 1.4.14. We may assume $\partial^{p(1)} = 1$. For each $i \in \{1, \ldots, n\}$ and $j \in \{1, \ldots, m\}$ there is a unique relation

$$\partial_i \cdot \partial^{p(j)} - \sum_{k=1}^{m} a_{ijk}(x) \cdot \partial^{p(k)} \ \in \ RI.$$

The coefficients $a_{ijk}(x)$ are rational functions which are obtained by Gröbner basis normal form with respect to $\mathcal{G}$. For $i = 1, \ldots, n$ we define $A_i(x)$ to be the $m \times m$-matrix whose entry in row j and column k equals $a_{ijk}(x)$.

Lemma 1.4.21. *The matrices $A_i(x)$ satisfy the compatibility conditions (1.35).*

Proof. We have $\partial_j \partial_i \partial^{p(c)} = \partial_i \partial_j \partial^{p(c)}$ in R. Therefore, we have

$$\partial_j \sum_k a_{ick}(x) \partial^{p(k)} = \partial_i \sum_k a_{jck}(x) \partial^{p(k)} \mod RI$$

This is equivalent to

$$\sum_k \frac{\partial a_{ick}(x)}{\partial x_j} \partial^{p(k)} + \sum_k a_{ick}(x) \partial_j \partial^{p(k)}$$

$$= \sum_k \frac{\partial a_{jck}(x)}{\partial x_i} \partial^{p(k)} + \sum_k a_{jck}(x) \partial_i \partial^{p(k)} \mod RI.$$

Since, $\partial_j \partial^{p(k)} = \sum a_{jke}(x) \partial^{p(e)} \mod RI$ and $\partial_i \partial^{p(k)} = \sum a_{ike}(x) \partial^{p(e)} \mod RI$, we obtain the conclusion from the fact that the standard monomials $\partial^{p(e)}$ are $\mathbf{k}(x)$-linearly independent modulo RI. $\square$

Let $V(x)$ be a common denominator of all the rational functions $a_{ijk}(x)$. The polynomial $V(x)$ defines a hypersurface in $\mathbf{C}^n$ which contains the singular locus $\mathrm{Sing}(I)$. This containment may be proper. Let U be any simply connected domain in $\{x \in \mathbf{C}^n : V(x) \neq 0\}$. Thus the matrices $A_i(x)$ are holomorphic in U. It follows from our construction of the matrices $A_i(x)$ that the holomorphic solutions of (1.34) on U are precisely the vectors

$$F(x) \ = \ \begin{pmatrix} \partial^{p(1)} \bullet f(x) \\ \partial^{p(2)} \bullet f(x) \\ \vdots \\ \partial^{p(m)} \bullet f(x) \end{pmatrix}$$

where $f(x)$ runs over the holomorphic solutions of I on U.

Theorem 1.4.20 and Lemma 1.4.21 imply the following theorem-algorithm.

Theorem 1.4.22. *Let I be a holonomic D-ideal. Then I can be transformed into a Pfaffian system (1.34) by Gröbner bases. The holonomic rank of I equals the dimension of holomorphic solutions in a neighborhood of any point at which the entries of the resulting matrices $A_i(x)$ are holomorphic.*

Example 1.4.23. Let $n = 4$ and let I be the D-ideal generated by the four operators in (1.22) for $a = 1/2, b = 1/2, c = 1$. This is the GKZ system associated with the period map in Proposition 1.3.6. We fix a term order on R with $\partial_1 \succ \partial_2 \succ \partial_3 \succ \partial_4$. The reduced Gröbner basis of RI equals

$$\mathcal{G} = \{ x_1\underline{\partial_1} - x_4\partial_4, \ 2x_2\underline{\partial_2} + 2x_4\partial_4 + 1, \ 2x_3\underline{\partial_3} + 2x_4\partial_4 + 1,$$
$$(4x_1x_4^2 - 4x_2x_3x_4) \cdot \underline{\partial_4^2} + (8x_1x_4 - 4x_2x_3) \cdot \partial_4 + x_1 \}.$$

The initial monomials are underlined. Hence the standard monomials are $\mathcal{B} = \{1, \partial_4\}$. The algorithm of Theorem 1.4.22 yields the following matrices:

$$A_1(x) = \begin{pmatrix} x_4/x_1 & x_4/(x_2x_3 - x_1x_4) \\ 0 & 1/(4x_2x_3 - 4x_1x_4) \end{pmatrix}$$

$$A_2(x) = \begin{pmatrix} -x_4/x_2 & (x_1x_4 + x_2x_3)/(2x_1x_2x_4 - 2x_2^2x_3) \\ -1/(2x_2) & x_1/(4x_1x_2x_4 - 4x_2^2x_3) \end{pmatrix}$$

$$A_3(x) = \begin{pmatrix} -x_4/x_3 & (x_1x_4 + x_2x_3)/(2x_1x_3x_4 - 2x_2x_3^2) \\ -1/(2x_3) & x_1/(4x_1x_3x_4 - 4x_2x_3^2) \end{pmatrix}$$

$$A_4(x) = \begin{pmatrix} 1 & (x_2x_3 - 2x_1x_4)/(x_1x_4^2 - x_2x_3x_4) \\ 0 & x_1/(4x_2x_3x_4 - 4x_1x_4^2) \end{pmatrix}$$

Hence hypergeometric system $I \bullet f = 0$ is equivalent to the Pfaffian system

$$\frac{\partial}{\partial x_i}\begin{pmatrix} f \\ \partial f/\partial x_4 \end{pmatrix} = A_i(x) \cdot \begin{pmatrix} f \\ \partial f/\partial x_4 \end{pmatrix} \qquad (i = 1, 2, 3, 4).$$

The four matrices have the common denominator $V(x) = x_1x_2x_3x_4(x_1x_4 - x_2x_3)$. This is precisely the equation of the singular hypersurface (1.33).

Note that Theorem 1.4.22 is weaker than Theorem 1.4.19 because it can happen that the common denominator $V(x)$ of the $A_i(x)$ vanishes at points outside the singular locus $\text{Sing}(I)$.

Let us see an example among the classical Appell hypergeometric functions F_i of two variables [5]. The discrepancy does not appear for the hypergeometric systems F_1, F_2 and F_3, but it appears for F_4.

Example 1.4.24. Consider the D- ideal I generated by

$$\partial_1(\theta_1 + c - 1) - (\theta_1 + \theta_2 + a)(\theta_1 + \theta_2 + b), \qquad (1.36)$$
$$\partial_2(\theta_2 + c' - 1) - (\theta_1 + \theta_2 + a)(\theta_1 + \theta_2 + b), \qquad (1.37)$$

where a, b, c, c' are generic complex numbers. This is the system for the Appell function F_4. The singular locus of I is $x_1 x_2 (x_1^2 + x_2^2 - 2x_1 x_2 - 2x_1 - 2x_2 + 1) = 0$. Let us transform it into a Pfaffian system by the weight vector $(10, 1)$. The reduced Gröbner basis contains an element of the form

$$x_1 (x_1^2 + x_2^2 - 2x_1 x_2 - 2x_1 - 2x_2 + 1)(px_1 + qx_2 + r)\partial_{x_2}^4 + \text{lower order terms}$$

where p, q, r are constants. There exists a factor $px_1 + qx_2 + r$, which persists as a denominator of the matrices $A_i(x)$. Therefore, the zero set of the common denominator $V(x)$ of the $A_i(x)$ is larger than $\text{Sing}(I)$ in this case.

We close this section with bibliographic notes. The algorithm (cf. (1.26)) of using the characteristic ideal to evaluate the rank is due to Oaku [76]. Algorithm 1.4.17 to evaluate the rank in R appears in several papers in the 1980's. Holonomic rank in Ore algebras is studied in the thesis of Chyzak [29]. He also provides a MAPLE package for Gröbner basis computation in the ring of differential operators and, more generally, in Ore algebras.

For a given ideal I in $\mathbf{k}[x]$, the Hilbert polynomial of I, the dimension of I, and the degree (multiplicity) of I can be computed by various computer systems for commutative rings. See the introduction of Chapter 4. Practical implementations of primary ideal decompositions are available in **Asir** (command **primadec**) [74] and **Singular** [43].

The Appell functions were introduced by Appell [5]. These are classical hypergeometric functions of two variables and can be regarded as GKZ systems via Gale duality, as in Section 3.3.

The singular locus of the GKZ system associated to a matrix A is the zero set of the *principal A-determinant* of A. The polynomial $x_1 x_2 x_3 x_4 (x_1 x_4 - x_2 x_3)$ in (1.33) is that for $A = \begin{pmatrix} 1 & 0 & 0 & -1 \\ 0 & 1 & 0 & 1 \\ 0 & 0 & 1 & 1 \end{pmatrix}$. The principal $\mathcal{A}$-determinant is a main topic of the book by Gel'fand, Kapranov and Zelevinsky [41].

1.5 Integrals of Products of Linear Forms

Let us consider the integral of a product of powers of linear forms

$$\Phi(\alpha; z) \quad = \quad \int_C \prod_{j=1}^n (z_{1j} + s \cdot z_{2j})^{\alpha_j} ds, \tag{1.38}$$

where α_i are parameters satisfying $\sum_{i=1}^n \alpha_i = -2$, $z = (z_{ij})$ is a variable $2 \times n$-matrix and C is a path in the complex plane with the coordinate s that connects two branch points $-z_{1i}/z_{2i}$ and $-z_{1j}/z_{2j}$ of the integrand. By virtue of the condition on the parameters, the integrand does not have a branch point at $s = \infty$. If we specialize $n = 4$, $\alpha_2 = b-1, \alpha_3 = c-b-1, \alpha_4 = -a$, and

$$z = \begin{pmatrix} 1 & 0 & 1 & 1 \\ 0 & 1 & -1 & -x \end{pmatrix} \text{ then (1.38) satisfies Gauss' hypergeometric equation}$$

in x.

Note that if we put $s = t_2/t_1$, then our integral is formally written as

$$\int \prod_{j=1}^{n} (t_1 z_{1j} + t_2 z_{2j})^{\alpha_j} (t_1 dt_2 - t_2 dt_1), \tag{1.39}$$

since $d(t_2/t_1) = (t_1 dt_2 - t_2 dt_1)/t_1^2$.

More generally, one can consider the following integral for $\sum_{i=1}^{n} \alpha_i = -k$:

$$\int \left(\prod_{j=1}^{n} (t_1 z_{1j} + \cdots + t_k z_{kj})^{\alpha_j} \right) \sum_{i=1}^{k} (-1)^{i-1} t_i dt_1 \wedge \cdots \wedge dt_{i-1} \wedge dt_{i+1} \wedge \cdots \wedge dt_k$$

Such integrals have been studied by many authors including Aomoto [3] and Gel'fand [36]. Our aim is to derive series expansions from Gröbner deformations of the differential equations satisfied by such integrals. For simplicity of exposition, we only consider the case $k = 2$ appearing in (1.39).

We fix the parameter vector $\alpha = (\alpha_1, \dots, \alpha_n)$, and we regard $\Phi(\alpha; z)$ as a (multi-valued) function on the space of matrices $\mathbf{C}^{2 \times n}$. The function $\Phi(\alpha; z)$ is equivariant with respect to the left action of $g \in GL(2, \mathbf{C})$ on $z \in \mathbf{C}^{2 \times n}$.

Lemma 1.5.1. ([36]) $\Phi(\alpha; gz) = \det(g)^{-1} \cdot \Phi(\alpha; z)$ for all $g \in GL(2, \mathbf{C})$.

Proof. The proof is purely formal. Put $g = \begin{pmatrix} g_{11} & g_{12} \\ g_{21} & g_{22} \end{pmatrix}$. Then, we have

$$\Phi(\alpha; gz) = \int_C \prod_{j=1}^{n} ((g_{11} z_{1j} + g_{12} z_{2j}) + s(g_{21} z_{1j} + g_{22} z_{2j}))^{\alpha_j} \, ds.$$

Let $\tilde{g} = \begin{pmatrix} \tilde{g}_{11} & \tilde{g}_{12} \\ \tilde{g}_{21} & \tilde{g}_{22} \end{pmatrix}$ be the inverse matrix of g. We introduce the change of coordinates

$$s = \frac{\tilde{g}_{22} t + \tilde{g}_{12}}{\tilde{g}_{21} t + \tilde{g}_{11}}.$$

Then, we have

$$\Phi(\alpha; gz) = \int_{\tilde{C}} (\tilde{g}_{21} t + \tilde{g}_{11})^{-\sum \alpha_i} \prod_{j=1}^{n} (z_{1j} + t z_{2j})^{\alpha_j - 1} \frac{\det(g)^{-1}}{(\tilde{g}_{21} t + \tilde{g}_{11})^2} \, dt,$$

because $ds = \det(g)^{-1} dt/(\tilde{g}_{21} t + \tilde{g}_{11})^2$. Since $-\sum \alpha_i = 2$, we are done. $\qquad\square$

Lemma 1.5.1 suggests that we consider Φ as a function on an affine chart of the Grassmann variety $GL(2, n) \backslash \mathbf{C}^{2 \times n}$. Set $m := n - 2$ and abbreviate

$\beta_1 := -\alpha_{m+1} - 1$ and $\beta_2 := -\alpha_{m+2} - 1$. Then $\sum_{j=1}^{m} \alpha_j = \beta_1 + \beta_2$. Define Ψ as the restriction of the function Φ to the subspace $\mathbf{C}^{2 \times m}$ given by $z_{m+1,1} = z_{m+2,2} = 1$ and $z_{m+1,2} = z_{m+2,1} = 0$. For example, when $n = 5$ we have

$$\Psi\left(\alpha, \beta; \begin{pmatrix} z_{11} & z_{12} & z_{13} \\ z_{21} & z_{22} & z_{23} \end{pmatrix}\right) = \Phi\left(\alpha; \begin{pmatrix} z_{11} & z_{12} & z_{13} & 1 & 0 \\ z_{21} & z_{22} & z_{23} & 0 & 1 \end{pmatrix}\right).$$

Let D be the Weyl algebra generated by $z_{1j}, z_{2j}, \partial_{1j}, \partial_{2j}$ for $j = 1, 2, \ldots, m$.

Theorem 1.5.2. *The function Ψ is annihilated by the D-ideal generated by*

$$\sum_{j=1}^{m} z_{1j}\partial_{1j} - \beta_1, \quad \sum_{j=1}^{m} z_{2j}\partial_{2j} - \beta_2, \tag{1.40}$$

$$z_{11}\partial_{11} + z_{21}\partial_{21} - \alpha_1, \; z_{12}\partial_{12} + z_{22}\partial_{22} - \alpha_2, \; \ldots, \; z_{1m}\partial_{1m} + z_{2m}\partial_{2m} - \alpha_m,$$

$$\partial_{1i}\partial_{2j} - \partial_{1j}\partial_{2i} \quad \text{for} \quad 1 \le i < j \le m.$$

Proof. The equations $\partial_{1i}\partial_{2j} - \partial_{1j}\partial_{2i}$ can be proved by differentiating under the integral sign. They are obviously satisfied by the integrand of the integral

$$\Psi(\alpha, \beta; z) = \int_C \prod_{j=1}^{m} (z_{1j} + s \cdot z_{2j})^{\alpha_j} \cdot s^{-\beta_2 - 1} ds \tag{1.41}$$

and hence they are satisfied by the integral itself. The second set of equations states that $\Psi(\alpha, \beta; z)$ is homogeneous of degree α_j with respect to the variables in the j-th column of z. This clearly holds for the kernel of the integral and hence it holds for the integral. To establish the first two Euler relations in (1.40), it now suffices to prove that $\Psi(\alpha, \beta; z)$ is homogeneous of degree β_2 with respect to the second row of z. This is done by performing the substitution $z_{2j} \mapsto t z_{2j}$ followed by the change of variables $s \mapsto \tilde{s}/t$. $\square$

The equations in Theorem 1.5.2 lie in two distinguished $2m$-dimensional commutative polynomial subrings of the Weyl algebra D. Let $\mathbf{C}[\partial]$ denote the polynomial ring in the variables ∂_{ij}, and let $\mathbf{C}[\theta]$ be the polynomial ring in the variables $\theta_{ij} = z_{ij}\partial_{ij}$. Let H be the ideal in $\mathbf{C}[\theta]$ generated by the $m+2$ linear forms $\sum_{j=1}^{m} \theta_{ij} - \beta_j$ for $i = 1, 2$ and $\sum_{i=1}^{2} \theta_{ij} - \alpha_j$ for $j = 1, \ldots, m$. Let I_A be the ideal in $\mathbf{C}[\partial]$ generated by all the 2×2-minors of the matrix

$$\begin{pmatrix} \partial_{11} & \partial_{12} & \partial_{13} & \cdots & \partial_{1m} \\ \partial_{21} & \partial_{22} & \partial_{23} & \cdots & \partial_{2m} \end{pmatrix}.$$

Theorem 1.5.2 states that Ψ is annihilated by the D-ideal $D \cdot I_A + D \cdot H$. This D-ideal is a typical *GKZ-hypergeometric system*. For $m = 2$ we get (1.22) under the correspondence $z_{11} = x_1$, $z_{12} = x_2$, $z_{21} = x_3$, $z_{22} = x_4$, $\beta_2 = -b$, $\alpha_1 = c - b - 1$, and $\alpha_2 = -a$.

Theorem 1.5.3. *The D-ideal $D \cdot I_A + D \cdot H$ is holonomic of rank m.*

This theorem is a special case of the results of Gel'fand, Kapranov and Zelevinsky [38]. A complete proof will be given in Chapter 4. In the next two paragraphs we informally discuss the ingredients of the proof.

The zero set of the determinantal ideal I_A consists of all $2 \times m$-matrices of rank ≤ 1. The projectivization of this affine variety is the *Segre embedding* of the product of projective spaces $\mathbf{P}^1 \times \mathbf{P}^{m-1}$ into $\mathbf{P}^{2m-1}$. The $\binom{m}{2}$ given generators of the ideal I_A form the reduced Gröbner basis with respect to any term order. This is proved in [96, Example 1.4]; see also Proposition 1.5.9 below. In particular, the ideal I_A has a square-free initial ideal (i.e., the radical of the initial ideal agrees with itself) and defines an irreducible variety (namely, $\mathbf{P}^1 \times \mathbf{P}^{m-1}$). These two facts imply that I_A is a prime ideal. Following [96, Section 5], we say that I_A is the *toric ideal* associated with the m-dimensional polytope $\Delta_1 \times \Delta_{m-1} = 1\text{-}simplex \times (m-1)\text{-}simplex$. In the case of $m = 3$, the product $\Delta_1 \times \Delta_2$ is a triangular prism which we can be represented as the convex hull of the columns of the matrix

$$A = \begin{pmatrix} 0 & 0 & 0 & 1 & 1 & 1 \\ 1 & 0 & 0 & 1 & 0 & 0 \\ 0 & 1 & 0 & 0 & 1 & 0 \\ 0 & 0 & 1 & 0 & 0 & 1 \end{pmatrix}.$$

The determinantal ideal I_A has an important algebraic property called "Cohen-Macaulay". We will define and discuss this property in Chapter 4. Here it suffices to say that, for a toric ideal, the Cohen-Macaulay property follows from the existence of a square-free initial monomial ideal.

We next argue that the determinantal ideal I_A has degree m. Together with the Cohen-Macaulay property, this will imply that the hypergeometric system $D \cdot I_A + D \cdot H$ has rank m for all parameter values α, β. Fix any term order w on $\mathbf{k}[\partial]$ which selects the antidiagonal monomial as the leading terms in each of the 2×2-minors. The initial monomial ideal of I_A equals

$$\mathrm{in}_w(I_A) \quad = \quad \langle\, \partial_{1j}\partial_{2i} \,:\, 1 \leq i < j \leq m \,\rangle. \tag{1.42}$$

This is the irredundant intersection of m prime ideals of codimension $m-1$:

$$\mathrm{in}_w(I_A) \quad = \quad \bigcap_{i=1}^{m} \langle \partial_{21}, \partial_{22}, \ldots \partial_{2,i-1}, \partial_{1,i+1}, \ldots, \partial_{1m} \rangle. \tag{1.43}$$

The reader is encouraged to compute this intersection in (1.43) and verify that the result equals (1.42). The formula (1.43) implies that the monomial ideal $\mathrm{in}_w(I_A)$ has degree m, has codimension $m-1$, and is Cohen-Macaulay. The same three properties are inherited by the determinantal ideal I_A.

We have used w, which is usually used to denote a weight vector, to denote a term order. Readers might think that this notation generates confusions, but it is known that for any term order $\prec$ and any ideal $I \subset \mathbf{k}[\partial_{11}, \ldots, \partial_{2m}]$, there exists a non-negative integer vector $w \in \mathbf{N}^{2m}$ such that $\mathrm{in}_w(I) = \mathrm{in}_\prec(I)$. See Proposition 2.1.5.

Let us now assume that α, β are sufficiently generic. We shall explicitly construct m linearly independent series solutions to $D \cdot I_A + D \cdot H$, using the technique of Gröbner deformations to be presented in greater generality later in this book. The first step is to replace the binomial operators $\partial_{1j}\partial_{2i} - \partial_{1i}\partial_{2j}$ by their leading monomials $\partial_{1j}\partial_{2i}$. In other words, we first consider the

$$\text{initial system:} \qquad D \cdot \text{in}_w(I_A) + D \cdot H. \tag{1.44}$$

The *initial system* defined above will be called the *fake initial ideal* in subsequent sections. The solution space of the initial system (1.44) is spanned by certain monomials

$$z^u \;=\; z_{11}^{u_{11}} z_{12}^{u_{12}} \cdots z_{1m}^{u_{1m}} z_{21}^{u_{21}} z_{22}^{u_{22}} \cdots z_{2m}^{u_{2m}}.$$

What does it mean for a monomial z^u to be annihilated by H and by $\text{in}_w(I_A)$?

Lemma 1.5.4. *A monomial z^u is annihilated by H if and only of its exponent matrix $u = (u_{ij})$ has row sums β_1, β_2 and has column sums $\alpha_1, \alpha_2, \ldots, \alpha_m$.*

Proof. These are precisely the homogeneity relations prescribed by the $m+2$ Euler operators which generate H. (See also the proof of Theorem 1.5.2).   □

Lemma 1.5.5. *A monomial z^u is annihilated by $\text{in}_w(I_A)$ if and only if*

$$z^u \;=\; z_{11}^{u_{11}} z_{12}^{u_{12}} \cdots z_{1i}^{u_{1i}} z_{2i}^{u_{2i}} z_{2,i+1}^{u_{2,i+1}} \cdots z_{2,m}^{u_{2,m}} \quad \text{for some } i \in \{1, 2, \ldots, m\}.$$

Proof. This is clear from the prime decomposition (1.43).   □

The previous two lemmas immediately imply the following result:

Proposition 1.5.6. *The solution space to the initial system (1.44) is the complex vector space spanned by the m monomials $z^{u^{(1)}}$, $z^{u^{(2)}}$, $\ldots$, $z^{u^{(m)}}$, where*

$$u^{(i)} \;=\; \begin{pmatrix} \alpha_1 & \cdots & \alpha_{i-1} & \beta_1 - \sum_{j=1}^{i-1} \alpha_j & 0 & \cdots & 0 \\ 0 & \cdots & 0 & \beta_2 - \sum_{j=1}^{i+1} \alpha_j & \alpha_{i+1} & \cdots & \alpha_m \end{pmatrix}$$

Proof. Among all matrices with fixed row and column sums (as in Lemma 1.5.4), only the above matrix $u^{(i)}$ has the support required in Lemma 1.5.5.   □

The next step in our Gröbner deformation method is to extend the solution basis of the initial system to a solution basis of the original system. To see what this extension process means geometrically, we represent the term order used above by an integral weight matrix, for instance,

$$w \;=\; \begin{pmatrix} 1 & 2 & 3 & \cdots & m-1 & m \\ m-1 & m-2 & m-3 & \cdots & 1 & 0 \end{pmatrix}.$$

This means that operator ∂_{1j} has weight w_{ij}, where $w_{1j} = j$ and $w_{2j} = m-j$. The weight matrix does indeed select the antidiagonal initial ideal (1.42).

Consider any solution $\Psi(z_{11}, z_{12}, \ldots, z_{2m})$ of the GKZ-hypergeometric system $D \cdot I_A + D \cdot H$, for instance the integral (1.41). Consider the new function

$$\Psi(t^w z) \quad = \quad \Psi(t^{w_{11}} z_{11}, t^{w_{12}} z_{12}, \ldots, t^{w_{2m}} z_{2m})$$

in one more variable, and assume that it admits a series expansion with respect to the new variable t. That series will start out as follows:

$$\Psi(t^w z) \quad = \quad \mathrm{in}_w(\Psi)(z) \cdot t^\alpha + \text{terms of higher degree in } t. \qquad (1.45)$$

The initial coefficient $\mathrm{in}_w(\Psi)$ defined as a coefficient of t^α is a function in z which is annihilated by the initial system (1.44). Precise statements of this observation can be found in Theorems 2.5.5 and 3.1.3. We may therefore assume that $\mathrm{in}_w(\Psi)(z) = z^{u^{(i)}}$ for some i, i.e., the initial coefficient in (1.45) equals one of the m monomials in Proposition 1.5.6. We can reconstruct the series from its initial monomial.

Theorem 1.5.7. *For each $i \in \{1, 2, \ldots, m\}$, the GKZ hypergeometric system $D \cdot I_A + D \cdot H$ has a unique series solution of the following form*

$$\Psi^{(i)} = z^{u^{(i)}} \cdot (1 + \text{Laurent monomials in } z \text{ with positive } w\text{-weight}) \quad (1.46)$$

The series $\Psi^{(1)}, \ldots, \Psi^{(m)}$ have a common domain of convergence in $\mathbf{C}^{2 \times m}$.

Theorem 1.5.7 is a special case of results to be proved in Section 3.4. Using the notation of that section, we write down the series $\Psi^{(i)}$ explicitly. Let e_{ij} denote the $2 \times n$ matrix whose entry in position (i, j) is 1 and whose other entries are zero. We introduce the following *additive monoid* (a set which is closed by addition and contains the unit element 0) isomorphic to $\mathbf{N}^{m-1}$:

$$M^{(i)} = \mathbf{N}(e_{21} + e_{1i} - e_{11} - e_{2i}) + \cdots + \mathbf{N}(e_{2,i-1} + e_{1i} - e_{1,i-1} - e_{2i})$$
$$+ \mathbf{N}(-e_{2,i+1} - e_{1i} + e_{1,i+1} + e_{2i}) + \cdots + \mathbf{N}(-e_{2m} - e_{1i} + e_{1,m} + e_{2i}).$$

Then the series in Theorem 1.5.7 equals

$$\Psi^{(i)}(z) \quad = \quad z^u \cdot \sum_{v \in M^{(i)}} \frac{[u]_{v_-}}{[v+u]_{v_+}} \cdot z^v, \qquad (1.47)$$

where $v = v_+ - v_-$, $(v_\pm \in \mathbf{N}^{2m})$ and

$$[a]_s := \prod_{ij} \prod_{p=0}^{s_{ij}-1} (a_{ij} - p).$$

The m series $\Psi^{(i)}(z)$ have a common domain of convergence in $\mathbf{C}^{2 \times m}$ because the weight vector w has strictly positive inner product with the elements in $M^{(i)} \backslash \{0\}$.

For analytic purposes it is sometimes desirable to rewrite the coefficients of the series (1.47) using the classical Γ-function. We have

$$\Psi^{(i)}(z) \;=\; \Gamma(u+1) \cdot \sum_{v \in M^{(i)}} \frac{1}{\Gamma(u+v+1)} \cdot z^{u+v}, \qquad (1.48)$$

where $\Gamma(u+v+1) := \prod_{i=1}^{2} \prod_{j=1}^{m} \Gamma(u_{ij}+v_{ij}+1)$. The equality of this series with (1.47) is derived from the relation $\Gamma(a+k+1) = \Gamma(a+1) \cdot [a+k]_k$. The formula (1.47) can be used to give a nice basis of solutions to get connection formulae among series solutions as in [90]. One application of connection formulae is to draw global graphs of hypergeometric functions.

Example 1.5.8. Let us consider an example with integral parameter vectors:

$$m = 3, \; \alpha_1 = -1, \alpha_2 = -1, \alpha_3 = -2, \; \beta_1 = -6, \; \beta_2 = 2.$$

The general form of integral representations (1.41) of solutions equals

$$\Psi(z) \;=\; \int_C \frac{s \cdot ds}{(z_{11} + sz_{21})(z_{12} + sz_{22})(z_{13} + sz_{23})^2}.$$

If we take C to be a sufficiently large positively oriented circle around the origin in the complex plane, then this integral equals the coefficient of s^{-1} in the Taylor series expansion of the kernel. We obtain the Laurent polynomial

$$3\frac{z_{23}^2}{z_{11}z_{12}z_{13}^4} + 2\frac{z_{22}z_{23}}{z_{11}z_{12}^2z_{13}^3} + \frac{z_{22}^2}{z_{11}z_{12}^3z_{13}^2} + 2\frac{z_{21}z_{23}}{z_{11}^2z_{12}z_{13}^3} + \frac{z_{21}z_{22}}{z_{11}^2z_{12}^2z_{13}^2} + \frac{z_{21}^2}{z_{11}^3z_{12}z_{13}^2}.$$

The first monomial is precisely the monomial $z^{u^{(3)}}$ in Proposition 1.5.6. Therefore the series $3 \cdot \Psi^{(3)}(z)$ terminates and is equal to the Laurent polynomial above. Note that the seven terms are listed in increasing w-order. The reader is invited to investigate the other two formal series solutions $\Psi^{(1)}(z)$ and $\Psi^{(2)}(z)$, whose starting monomials have the exponent matrices

$$u^{(1)} = \begin{pmatrix} -6 & 0 & 0 \\ 5 & -1 & -2 \end{pmatrix} \quad \text{and} \quad u^{(2)} = \begin{pmatrix} -1 & -5 & 0 \\ 0 & 4 & -2 \end{pmatrix}.$$

One of these series does not exist because α, β is not generic enough. We shall see in Section 3.5 on how to replace it with a series that involves logarithms.

Returning to our general discussion, we can ask if there exist other series solutions to $D \cdot I_A + D \cdot H$ with respect to different term orders. The answer is "no" because all initial monomial ideals of I_A are isomorphic to (1.42).

Proposition 1.5.9. [96, p.2, Example 1.4]
The initial ideal $\mathrm{in}_w(I_A)$ is generated by monomials if

$$w_{1i} + w_{2j} - w_{1j} - w_{2i} \neq 0, \; 1 \leq i \neq j \leq n.$$

There exist n! such monomial ideals; namely, they are

$$\langle \partial_{1\sigma(j)} \partial_{2\sigma(i)} , \ i < j \rangle,$$

where σ runs over all permutations of $\{1, 2, \ldots .n\}$.

This proposition implies that we have only $n!$ many choices of a set of starting monomials. Since the system of equations itself admits the action of S_n by $z_{ij} \mapsto z_{i\sigma(j)}$, we have only one set of solutions up to S_n symmetry.

For example, consider the case $m = 2$. The possible exponent matrices of starting monomials are

$$u^{[21]} = \begin{pmatrix} \alpha_1 & \beta_1 - \alpha_1 \\ 0 & \beta_2 \end{pmatrix}$$

$$u^{[12]} = \begin{pmatrix} \beta_1 & 0 \\ \alpha_1 - \beta_1 & \alpha_2 \end{pmatrix} \quad \text{for } \underline{\partial_{21}\partial_{12} - \partial_{11}\partial_{22}}$$

and

$$u^{[22]} = \begin{pmatrix} \beta_1 - \alpha_2 & \alpha_2 \\ \beta_2 & 0 \end{pmatrix}$$

$$u^{[11]} = \begin{pmatrix} 0 & \beta_1 \\ \alpha_1 & \alpha_2 - \beta_1 \end{pmatrix} \quad \text{for } \underline{\partial_{11}\partial_{22} - \partial_{12}\partial_{21}}.$$

Each of these matrices has row sums β_1, β_2 and column sums α_1, α_2. Hence we obtain four possible series solutions arising from Gröbner deformations:

$$\Psi^{[21]}(x) = z_{11}^{\alpha_1} z_{12}^{\beta_1 - \alpha_1} z_{22}^{\beta_2} \cdot F(\alpha_1, \beta_2, \alpha_1 - \beta_1 - 1; t)$$

$$\Psi^{[12]}(x) = z_{11}^{\beta_1} z_{21}^{\alpha_1 - \beta_1} z_{22}^{\alpha_2} \cdot F(\alpha_2, \beta_1, \beta_1 - \alpha_1 - 1; t) \quad \text{for } \underline{\partial_{21}\partial_{12} - \partial_{11}\partial_{22}}$$

and

$$\Psi_{[22]}(x) = z_{11}^{\beta_1 - \alpha_2} z_{12}^{\alpha_2} z_{21}^{\beta_2} \cdot F\left(\alpha_2, \beta_2, \alpha_2 - \beta_1 - 1; \frac{1}{t}\right)$$

$$\Psi_{[11]}(x) = z_{12}^{\beta_1} z_{21}^{\alpha_1} z_{22}^{\alpha_2 - \beta_1} \cdot F\left(\alpha_1, \beta_1, \beta_1 - \alpha_2 - 1; \frac{1}{t}\right) \quad \text{for } \underline{\partial_{11}\partial_{22} - \partial_{12}\partial_{21}}$$

where $t = -z_{12}z_{21}/z_{11}z_{22}$, and F denotes the Gauss series in Theorem 1.3.4.

In case of $m = 3$, $\Psi^{(i)}(z)$, $(i = 1, 3)$ can be expressed in terms of the Appell hypergeometric function F_1. In fact, for the initial ideal

$$\text{in}_w(I_A) = \langle \partial_{12}\partial_{21}, \partial_{13}\partial_{21}, \partial_{13}\partial_{22} \rangle,$$

the monomials $z^{u^{(i)}}$ are solutions of $D \cdot \text{in}_w(I_A) + D \cdot H$ where

$$u^{(1)} = \begin{pmatrix} \beta_1 & 0 & 0 \\ \alpha_1 - \beta_1 & \alpha_2 & \alpha_3 \end{pmatrix}, \quad u^{(2)} = \begin{pmatrix} \alpha_1 & \beta_1 - \alpha_1 & 0 \\ 0 & \beta_2 - \alpha_3 & \alpha_3 \end{pmatrix},$$

$$u^{(3)} = \begin{pmatrix} \alpha_1 & \alpha_2 & \alpha_3 - \beta_2 \\ 0 & 0 & \beta_2 \end{pmatrix}.$$

Assume that none of $\alpha_1, \alpha_2, \alpha_3, \beta_1, \beta_2, \beta_1 - \alpha_1, \beta_2 - \alpha_3$ is an integer. To express series solutions in terms of classical hypergeometric functions, we rearrange (1.47) slightly:

$$\Psi^{(1)}(z) = z_{11}^{\beta_1} z_{21}^{\alpha_1 - \beta_1} z_{22}^{\alpha_2} z_{23}^{\alpha_3}$$
$$\times F_1 \left(-\beta_1, -\alpha_2, -\alpha_3, \alpha_1 - \beta_1 + 1, \left(\frac{z_{21} z_{12}}{z_{11} z_{22}} \right), \left(\frac{z_{21} z_{13}}{z_{11} z_{23}} \right) \right),$$

$$\Psi^{(2)}(z) = z_{11}^{\alpha_1} z_{12}^{\beta_1 - \alpha_1} z_{22}^{\beta_2 - \alpha_3} z_{23}^{\alpha_3}$$
$$\times G_2 \left(-\alpha_1, -\alpha_3, \alpha_1 - \beta_1, \alpha_3 - \beta_2, \left(-\frac{z_{21} z_{12}}{z_{11} z_{22}} \right), \left(-\frac{z_{22} z_{13}}{z_{12} z_{23}} \right) \right),$$

$$\Psi^{(3)}(z) = z_{11}^{\alpha_1} z_{12}^{\alpha_2} z_{13}^{\alpha_3 - \beta_2} z_{23}^{\beta_2}$$
$$\times F_1 \left(-\beta_2, -\alpha_1, -\alpha_2, \alpha_3 - \beta_2 + 1, \left(\frac{z_{21} z_{13}}{z_{11} z_{23}} \right), \left(\frac{z_{22} z_{13}}{z_{12} z_{23}} \right) \right).$$

Here F_1 is the *Appell hypergeometric function* F_1 and G_2 is the hypergeometric function G_2 in *Horn's list* (cf. [33, Section 5.7.1]). The above three series have the open domain

$$\left| \frac{z_{21} z_{12}}{z_{11} z_{22}} \right|, \left| \frac{z_{22} z_{13}}{z_{12} z_{23}} \right| < 1$$

as a common domain of convergence. They are linearly independent and thus they form a fundamental system of solutions of the GKZ system on that domain. Let $I \subset D_2$ be the D-ideal generated by

$$\theta_1(\theta_1 + \theta_2 + c - 1) - x_1(\theta_1 + \theta_2 + a)(\theta_1 + b),$$
$$\theta_2(\theta_1 + \theta_2 + c - 1) - x_2(\theta_1 + \theta_2 + a)(\theta_2 + b'),$$
$$(x_1 - x_2)\partial_1 \partial_2 - b' \partial_1 + b \partial_2.$$

Here, a, b, b', c are constants. The Appell function

$$F_1(a, b, b', c; x_1, x_2) := \sum_{m,n=0}^{\infty} \frac{(a)_{m+n}(b)_m(b')_n}{(c)_{m+n}(1)_m(1)_n} x_1^m x_2^n$$

is annihilated by the ideal I which has holonomic rank 3. The system of differential equations $I \bullet f = 0$ is called *Appell's system of differential equations for F_1* [5]. The singular locus $\text{Sing}(I)$ is the curve with the equation $x_1 x_2 (x_1 - 1)(x_2 - 1)(x_1 - x_2) = 0$. Our fundamental set of solutions of the corresponding GKZ system gives $3! = 6$ linearly independent solutions of I by the action of S_3.

We next present the general definition of *GKZ systems*.

Definition 1.5.10. Let $A = (a_{ij})$ be an integer $d \times n$-matrix of rank d, and fix a vector of parameters $\beta = (\beta_1, \dots, \beta_d)^T \in \mathbf{k}^d$. The GKZ system

$H_A(\beta)$ is the following system of linear partial differential equations for an indeterminate function $f = f(x)$:

$$(\sum_{j=1}^{n} a_{ij}x_j\partial_j - \beta_i) \bullet f = 0, \quad \text{for} \quad i = 1, \ldots, d, \quad \text{and}$$

$$(\partial^u - \partial^v) \bullet f = 0 \quad \text{for all} \quad u, v \in \mathbf{N}^n \quad \text{with} \quad Au = Av.$$

We regard $H_A(\beta)$ as a left ideal in the Weyl algebra D. The second group of operators $\partial^u - \partial^v$ generate the *toric ideal* I_A of A. GKZ systems are also called GKZ hypergeometric system, GKZ hypergeometric differential equations, and A-hypergeometric differential equations.

Example 1.5.11. Let $A = \begin{pmatrix} 3 & 2 & 1 & 0 \\ 0 & 1 & 2 & 3 \end{pmatrix}$. The GKZ hypergeometric ideal $H_A(\beta)$ is generated by $3x_1\partial_1 + 2x_2\partial_2 + x_3\partial_3 - \beta_1$, $x_2\partial_2 + 2x_3\partial_3 + 3x_4\partial_4 - \beta_2$, $\partial_1\partial_3 - \partial_2^2$, $\partial_1\partial_4 - \partial_2\partial_3$, and $\partial_2\partial_4 - \partial_3^2$. We note that an algorithm for finding generators of I_A is given in [96, Algorithm 4.5, 12.3].

Example 1.5.12. Let A be the following $(m+1) \times 2m$ matrix:

$$A = \begin{pmatrix} 0 & 0 & \cdots & 0 & 1 & 1 & \cdots & 1 \\ 1 & 0 & \cdots & 0 & 1 & 0 & \cdots & 0 \\ 0 & 1 & \cdots & 0 & 0 & 1 & \cdots & 0 \\ 0 & 0 & \cdots & 0 & 0 & 0 & \cdots & 0 \\ 0 & 0 & \cdots & 0 & 0 & 0 & \cdots & 0 \\ 0 & 0 & \cdots & 1 & 0 & 0 & \cdots & 1 \end{pmatrix}. \tag{1.49}$$

Then A can be considered as the set of vertices of the product of simplices $\Delta_1 \times \Delta_{m-1}$. Index the variables in such a way those corresponding to the former m columns are z_{1j}, and those to the latter m columns are z_{2j} ($1 \le j \le m$). For $u = (u_{ij})_{i=1,2;\, 1\le j\le m}$, $v = (v_{ij})_{i=1,2;\, 1\le j\le m} \in \mathbf{N}^{2m}$, we have $Au = Av$ if and only if the sum of any column or any row of u equals the corresponding sum of v. The D-ideal (1.40) agrees with the GKZ system $H_A\begin{pmatrix} \beta_2 \\ \alpha \end{pmatrix}$.

Remark 1.5.13. As we have seen in the case of the two GKZ systems for Gauss' hypergeometric function (1.22) and (1.49) ($m = 2$), different A's may give the same GKZ system. In fact, let A, A' be two $d \times n$ integer matrices of rank d. If there exists an invertible matrix $G \in \mathrm{GL}_d(\mathbf{Q})$ such that $A' = GA$, then $I_A = I_{A'}$ and $H_A(\beta) = H_{A'}(G\beta)$.

We close this section with bibliographic notes and remarks. The GKZ system was first introduced and studied by Gel'fand, Kapranov and Zelevinsky [38]. They derived the Γ-series (1.48) but they used a somewhat different combinatorial description. Their construction is based on *regular triangulations* of the convex hull of A. For a geometric introduction to regular triangulations, we refer to the book of Gel'fand, Kapranov and Zelevinsky [41] or the book of Ziegler [105].

Their construction of Γ-series is implicit in our approach using Gröbner deformations. We wish to explain this using Example 1.5.12. Consider the zero set of the square-free monomial ideal $\mathrm{in}_w(I_A)$ in (1.42) and (1.43). It is the union of m coordinate linear subspaces of dimension m in projective $(2m-1)$-space $\mathbf{P}^{2m-1}$. We can interpret this arrangement of linear subspaces as a pure m-dimensional simplicial complex Δ_w having m facets. The ideal $\mathrm{in}_w(I_A)$ is called the *Stanley-Reisner ideal* of the simplicial complex Δ_w. It turns out that Δ_w is a regular triangulation of the polytope $A = \Delta_1 \times \Delta_{m-1}$. There is a general correspondence between initial ideals of toric ideals and regular triangulations of polytopes, which can be found in the book of Sturmfels [96, Section 8]. For an introduction to simplicial complexes and their Stanley-Reisner ideals we refer to the book of Stanley [94] or the book of Hibi [48].

As an example of the correspondence consider the case $m = 3$. Here the simplicial complex Δ_w associated with $\mathrm{in}_w(I_A)$ consists of three tetrahedra:

$$\Delta_w = \left\{ \begin{pmatrix} * & * & * \\ & & * \end{pmatrix}, \begin{pmatrix} * & * & \\ & * & * \end{pmatrix}, \begin{pmatrix} * & & \\ * & * & * \end{pmatrix} \right\}.$$

These three tetrahedra define a triangulation of the triangular prism $\Delta_1 \times \Delta_2$, whose vertices naturally correspond to the coordinates of a 2×3-matrix.

For an m-dimensional polytope P with vertices in an m-dimensional lattice $L \simeq \mathbf{Z}^m$, we define its *normalized volume* by the volume of P with the normalization that the volume of the convex hull of $0, e_1, \ldots, e_m$ is 1. Here, $\{e_i\}$ are the basis of the lattice. For example, the normalized volume of the 1-simplex $\times$ $(m-1)$-simplex is m, which agrees with the holonomic rank of the GKZ system (1.40). It is proved by Gel'fand, Kapranov and Zelevinsky that the holonomic rank of the GKZ system associated to A is equal to the normalized volume of the convex hull of A when the affine toric variety I_A is Cohen-Macaulay [38], [40] . We will discuss relations between the holonomic rank and the normalized volume in Chapter 4 in more detail.

We started this section with an integral representation. To make a precise discussion of domains of integration, which are called cycles, we need a theory of *twisted cycles*. The segment C in the integral (1.38) should be regarded as an element of the twisted homology group of locally finite chains H_1^{lf}. For generic α_j, the group is isomorphic to the twisted homology group H_1. See Section 5.4 and the book of Aomoto and Kita [4].

2. Solving Regular Holonomic Systems

A left ideal I in the Weyl algebra $D = \mathbf{C}\langle x_1, \ldots, x_n, \partial_1, \ldots, \partial_n \rangle$ represents a system of linear partial differential equations with polynomial coefficients. The holomorphic solutions to I around a generic point in $\mathbf{C}^n$ form a vector space. This space has finite dimension rank(I) if I is holonomic. In this chapter we are interested in convergent series solutions to I of the form

$$f = \sum_{a,b} c_{a,b} \cdot x_1^{a_1} x_2^{a_2} \cdots x_n^{a_n} \cdot \log(x_1)^{b_1} \cdots \log(x_n)^{b_n}, \qquad (2.1)$$

where the a_i are complex numbers and the b_j are nonnegative integers. In Chapter 3 we shall study series (2.1) which satisfy GKZ hypergeometric equations. The terms of the series f are partially ordered by means of a generic weight vector $w \in \mathbf{R}^n$. The sum of its smallest terms, $\mathrm{in}_w(f)$, is a solution to the *Gröbner deformation* $\mathrm{in}_{(-w,w)}(I)$. Our idea is to find $\mathrm{in}_w(f)$ first, by means of the techniques in Section 2.3, and to construct f from it. This approach is justified by the following result to be proved in Section 2.2:

$$\mathrm{rank}(\mathrm{in}_{(-w,w)}(I)) \leq \mathrm{rank}(I). \qquad (2.2)$$

This inequality can be strict, which means that not all solutions to I are expressible in the form (2.1). For instance, this happens for $n = 1$ if some solution has an essential singularity at 0 or at ∞. On the other hand, if the system I is *regular holonomic*, then equality holds in (2.2). An algorithmic proof for this theorem is given in Section 2.4. In Section 2.5 we introduce term orders on the *Nilsson ring* of series (2.2); for each such term order, the regular holonomic system I has a *canonical solution basis*. Section 2.6 gives an algorithm for computing this basis. The terms appearing in the canonical series are governed by the Gröbner cone containing w. The set of these Gröbner cones is finite and forms the *small Gröbner fan*. This is a piece of the *Gröbner fan* of I, which indexes the finite set of all possible initial ideals of I. We begin our discussion in Section 2.1 with these combinatorial objects.

2.1 The Gröbner Fan

An important issue in Gröbner bases theory is to describe all possible initial monomial ideals of a fixed ideal and to understand the natural adjacency relations among these ideals. In the familiar commutative case this is accomplished by the Gröbner fan [72] or, by its polar dual object, the state polytope. For a comprehensive introduction to these topics we refer to [96].

The purpose of this section is to extend the construction of the Gröbner fan to the Weyl algebra. This was also done by Assi, Castro-Jiménez and Granger in [7]. All definitions and statements in this section are self-contained; however, we will frequently refer to Sturmfels' book [96, §1,2] for details used in the proofs. As a main application of the Gröbner fan, we prove a semicontinuity theorem (Theorem 2.2.1) for the holonomic rank of a D-ideal.

Let us begin by reviewing some basic notions from polyhedral geometry (see, e.g., [105]). A *polyhedron* is a finite intersection of closed half-spaces in $\mathbf{R}^m$. Let P be any polyhedron in $\mathbf{R}^m$ and $w \in \mathbf{R}^m$. We define

$$\mathrm{face}_w(P) := \{u \in P \mid w \cdot u \ge w \cdot v \text{ for all } v \in P\}.$$

A subset F of P which has this form is called a *face* of P. Every polyhedron P has only finitely many faces. Each face is again a polyhedron. A face of P of dimension $\dim P - 1$ is called a *facet* of P. If F is a face of P then its *normal cone* is

$$N_P(F) = \{w \in \mathbf{R}^m \mid \mathrm{face}_w(P) = F\}.$$

Note that the set $N_P(F)$ is relatively open, i.e., it is an open subset in its linear span. We have the dimension formula $\dim\big(N_P(F)\big) + \dim(F) = m$.

A polyhedron of the form $\{\mathbf{x} \in \mathbf{R}^m \mid A \cdot \mathbf{x} \le 0\}$, where A is a matrix with m columns, is called a *(polyhedral) cone*. For a cone C, there exist vectors $c_1, \ldots, c_p \in \mathbf{R}^m$ such that

$$C = \{\lambda_1 c_1 + \cdots + \lambda_p c_p \mid \lambda_1, \ldots, \lambda_p \in \mathbf{R}_+\}.$$

When we can take the c_i's in $\mathbf{Q}^m$, the cone is called a *rational cone*. For a given polyhedral cone C, the *polar (dual)* C^* is defined by $C^* = \{y \in \mathbf{R}^m \mid y \cdot x \ge 0 \text{ for all } x \in C\}$. A collection Δ of polyhedral cones is called a *(polyhedral) fan* when the following two conditions are satisfied:

1. if $P \in \Delta$ and F is a face of P, then $F \in \Delta$;
2. if $P_1, P_2 \in \Delta$, then $P_1 \cap P_2$ is a face of P_1 and P_2.

The union of all the cones in the fan Δ is called the *support of the fan* Δ. If the support of the fan Δ agrees with the whole space $\mathbf{R}^m$, the fan Δ is called a *complete fan*. A fan Δ' is called a *refinement* of another fan Δ if for all $P' \in \Delta'$ there exists $P \in \Delta$ such that $P' \subseteq P$.

Let P be a polyhedron. The set of the closures of all normal cones $N_P(F)$, where F runs over all faces of P, is a fan. It is called the *normal fan* of P. The *Minkowski sum* of two polyhedra P_1 and P_2 is defined to be

$$P_1 + P_2 := \{p_1 + p_2 \mid p_1 \in P_1, \ p_2 \in P_2\}.$$

A polyhedron is called a *polytope* if it is a bounded subset of $\mathbf{R}^m$. The minimal polyhedron containing a finite point set $S \subset \mathbf{R}^m$ is a polytope. It is called the *convex hull* of S and is denoted by $\mathrm{conv}(S)$.

Fix a D-ideal I. Two weight vectors for the Weyl algebra D, say $(u, v) \in \mathbf{R}^{2n}$ and $(u', v') \in \mathbf{R}^{2n}$, are called *equivalent* for I if $\mathrm{in}_{(u,v)}(I) = \mathrm{in}_{(u',v')}(I)$. Two weight vectors $(t, u, v), (t', u', v') \in \mathbf{R}^{1+2n}$ for the homogenized Weyl algebra $D^{(h)}$ are *equivalent* for a fixed $D^{(h)}$-ideal I if $\mathrm{in}_{(t,u,v)}(I) = \mathrm{in}_{(t',u',v')}(I)$. The following result is due to Assi, Castro-Jiménez and Granger [7].

Theorem 2.1.1. *Fix a $D^{(h)}$-ideal I generated by homogeneous elements. Each equivalence class of weight vectors for $D^{(h)}$ is a relatively open rational polyhedral cone. The set of the closures of the open cones is finite and forms a polyhedral fan in*

$$\{(t, u, v) \in \mathbf{R}^{1+2n} \mid 2t \leq u_i + v_i\}.$$

The fan is called the *Gröbner fan* of I. This theorem generalizes the following result from commutative Gröbner bases.

Theorem 2.1.2. ([72], [96, §2]) *Fix a homogeneous ideal I in $\mathbf{k}[x]$. Each equivalence class of weight vectors $w \in \mathbf{R}^n$ is a relatively open rational polyhedral cone. The set of the closures of the open cones is finite and forms a polyhedral fan in $\mathbf{R}^n$.*

If the ideal $I \subset \mathbf{k}[x]$ is not homogeneous and the weight vector w has negative coordinates, then equivalence classes need not be polyhedral cones. For instance, if $n = 2$ and $I = \langle 1+x_1, 1+x_2 \rangle$, then the equivalence class of the weight vector $(-1, -1)$ is the non-convex open set $\mathbf{R}^2 \setminus \{w \mid w_1 \geq 0, w_2 \geq 0\}$. The same phenomenon happens in the Weyl algebra D for weight vectors of the form $(w, -w) \in \mathbf{R}^{2n}$; for instance, take the D-ideal generated by $1 + \partial_1$ and $1 + \partial_2$. However, the following slightly weaker result is true for D.

Theorem 2.1.3. *Fix an arbitrary D-ideal I in the Weyl algebra D.*

(1) *Each equivalence class of weight vectors $(u, v) \in \mathbf{R}^{2n}$ is a finite union of relatively open convex rational polyhedral cones.*

(2) *Consider only weight vectors $(u, v) \geq 0$. Then each equivalence class of weight vectors is a relatively open convex rational polyhedral cone and the collection of the closures of them is a fan in $\{(u, v) \in \mathbf{R}^{2n} \mid (u, v) \geq 0\}$.*

Proof. We derive (1) from Theorem 2.1.1 as follows. Let F^h be the homogenization of a set of generators of I. The Gröbner fan of $I^h = D^{(h)} \cdot \{F^h\}$ restricted to the hyperplane $\{t = 0\}$ is also a fan. Since $\mathrm{in}_{(0,u,v)}(I^h) = \mathrm{in}_{(0,u',v')}(I^h)$ implies $\mathrm{in}_{(u,v)}(I) = \mathrm{in}_{(u',v')}(I)$, we get the conclusion.

The statement (2) can be shown similarly to Theorem 2.1.1 since $\prec_{(u,v)}$ is a term order, and there is the reduced Gröbner basis with respect to $\prec_{(u,v)}$. See our proof of Theorem 2.1.1 below. $\qquad\square$

We define the *Gröbner fan* of a D-ideal I to be the restriction of the Gröbner fan of its homogenization $H(I) \subset D^{(h)}$ to the hyperplane $\{t = 0\}$ in $\mathbf{R}^{2n+1}$.

In what follows we outline the proof of Theorem 2.1.1. The proof is analogous to that in [96, Section 2]. In order to clarify the difference among the ring of polynomials $\mathbf{k}[x]$, the Weyl algebra D and the homogenized Weyl algebra $D^{(h)}$, we present a sequence of lemmas and propositions for these three rings. Let $\prec$ be a term order in $\mathbf{k}[x]$ or D or $D^{(h)}$. Recall that, in case of D and $D^{(h)}$, it must satisfy the condition $x_i\partial_i \succ 1$ or $x_i\partial_i \succ h^2$ respectively.

Proposition 2.1.4.

(1) *For every ideal $I \subset \mathbf{k}[x]$ and $w \in \mathbf{R}^n$, we have $\mathrm{in}_{\prec}(\mathrm{in}_w(I)) = \mathrm{in}_{\prec_w}(I)$.*
(2) *For every D-ideal I and $(u,v) \in \mathbf{R}^{2n}$ with $u_i + v_i \geq 0$, we have*
$$\mathrm{in}_{\prec}(\mathrm{in}_{(u,v)}(I)) = \mathrm{in}_{\prec_{(u,v)}}(I).$$
(3) *For every $D^{(h)}$-ideal I and $(t,u,v) \in \mathbf{R}^{1+2n}$ with $2t \leq u_i + v_i$, we have*
$$\mathrm{in}_{\prec}(\mathrm{in}_{(t,u,v)}(I)) = \mathrm{in}_{\prec_{(t,u,v)}}(I).$$

Proof. Statement (1) appears in [96, Proposition 1.8]. Statements (2) and (3) can be shown in an analogous way. For instance, let us prove (2). First note that $\mathrm{in}_{\prec_{(u,v)}}(f) = \mathrm{in}_{\prec}(\mathrm{in}_{(u,v)}(f))$ for each $f \in D$. Hence, monomials in $\mathrm{in}_{\prec_{(u,v)}}(I)$ are also in $\mathrm{in}_{\prec}(\mathrm{in}_{(u,v)}(I))$. Let $x^a\xi^b$ be a monomial in $\mathrm{in}_{\prec}(\mathrm{in}_{(u,v)}(I))$. It follows from Corollary 1.1.2 that there exists $f \in I$ with $x^a\xi^b = \mathrm{in}_{\prec_{(u,v)}}(f)$. Therefore the two monomial ideals $\mathrm{in}_{\prec}(\mathrm{in}_{(u,v)}(I))$ and $\mathrm{in}_{\prec_{(u,v)}}(I)$ contain the same set of monomials, and hence they are equal. $\square$

Proposition 2.1.5.

(1) *For any term order $\prec$ and any ideal $I \subset \mathbf{k}[x]$, there exists a positive weight vector $w > 0$ such that $\mathrm{in}_w(I) = \mathrm{in}_{\prec}(I)$.*
(2) *For any term order $\prec$ and any D-ideal $I \subset D$, there exists a positive weight vector $(u,v) > 0$ such that $\mathrm{in}_{(u,v)}(I) = \mathrm{in}_{\prec}(I)$.*
(3) *For any term order $\prec$ and any $D^{(h)}$-ideal $I \subset D^{(h)}$, there exists a positive weight vector $(t,u,v) > 0$ such that $2t < u_i + v_i$ and $\mathrm{in}_{(t,u,v)}(I) = \mathrm{in}_{\prec}(I)$.*

Proof. Part (1) appears in [96, Proposition 1.1]. Let us prove (3). Let $\mathcal{G} = \{g_1, \ldots, g_r\}$ be the reduced Gröbner basis of I with respect to $\prec$. Write
$$g_i = c_{i0}h^{\lambda_{i0}}x^{\alpha_{i0}}\partial^{\beta_{i0}} + \cdots + c_{i\nu_i}h^{\lambda_{i\nu_i}}x^{\alpha_{i\nu_i}}\partial^{\beta_{i\nu_i}}.$$

We define $\mathcal{C}_{I,\prec}$ to be the set of all non-negative vectors (t,u,v) in $\mathbf{R}^{1+2n}$ such that $\mathrm{in}_{(t,u,v)}(g_i) = c_{i0}h^{\lambda_{i0}}x^{\alpha_{i0}}\xi^{\beta_{i0}}$ and $0 < 2t < u_i + v_i$. Equivalently,
$$\mathcal{C}_{I,\prec} = \{(t,u,v) \,|\, t \cdot (\lambda_{i0} - \lambda_{i\ell}) + u \cdot (\alpha_{i0} - \alpha_{i\ell}) + v \cdot (\beta_{i0} - \beta_{i\ell}) > 0 \quad \text{for}$$
$$i = 1, \ldots, r, \ \ell = 1, \ldots, \nu_i \quad \text{and} \quad 0 < 2t < u_j + v_j \text{ for } j = 1, \ldots, n\}.$$

Suppose that $\mathcal{C}_{I,\prec}$ is empty. By Farkas' Lemma of linear programming [92, Section 7.8], there exist non-negative integers $r_{i\ell}, a_j$, not all zero, such that

$$\sum_{i=1}^{r}\sum_{\ell=1}^{\nu_i} r_{i\ell} \cdot (\lambda_{i0} - \lambda_{i\ell}, \alpha_{i0} - \alpha_{i\ell}, \beta_{i0} - \beta_{i\ell}) + \sum_{j=1}^{n} a_j(-2, e_j, e_j) \; \le \; (0,0,0).$$

By the multiplicativity property of our term order, this translates into

$$\prod_{i=1}^{r}\prod_{\ell=1}^{\nu_i} \left(h^{\lambda_{i0}}\right)^{r_{i\ell}} \left(x^{\alpha_{i0}}\right)^{r_{i\ell}} \left(\xi^{\beta_{i0}}\right)^{r_{i\ell}} \left(\prod_{j=1}^{n} x_j^{a_j}\xi_j^{a_j}\right) \tag{2.3}$$

$$\preceq \; \prod_{i=1}^{r}\prod_{\ell=1}^{\nu_i} \left(h^{\lambda_{i\ell}}\right)^{r_{i\ell}} \left(x^{\alpha_{i\ell}}\right)^{r_{i\ell}} \left(\xi^{\beta_{i\ell}}\right)^{r_{i\ell}} \prod_{j=1}^{n}(h^2)^{a_j}. \tag{2.4}$$

The hypothesis $\mathrm{in}_{\prec}(g_i) = c_{i0}h^{\lambda_{i0}}x^{\alpha_{i0}}\xi^{\beta_{i0}}$ implies $h^{\lambda_{i0}}x^{\alpha_{i0}}\partial^{\beta_{i0}} \succ h^{\lambda_{i\ell}}x^{\alpha_{i\ell}}\partial^{\beta_{i\ell}}$, and the definition of the multiplicative monomial order implies $x_j\xi_j \succ h^2$. By multiplying the left hand sides and right hand sides of these $\succ$-relations, we find that $(2.3) \succ (2.4)$. This is a contradiction, and we conclude that $\mathcal{C}_{I,\prec}$ is a non-empty open convex cone. Take $(t,u,v) \in \mathcal{C}_{I,\prec}$. Then, $\mathrm{in}_{(t,u,v)}(I) = \mathrm{in}_{\prec}(I)$, the proof of which is analogous to that in [96, page 5], and the entire argument for part (2) is analogous. $\square$

The next lemma is a key tool for the local structure of the Gröbner fan.

Lemma 2.1.6.

(1) Let $I \subset \mathbf{k}[x]$ be an ideal, $w, w' \in \mathbf{R}^n$ and $\varepsilon > 0$ sufficiently small. Then

$$\mathrm{in}_w(\mathrm{in}_{w'}(I)) = \mathrm{in}_{w'+\varepsilon w}(I).$$

(2) Let I be a D-ideal and $(u,v), (u',v') \in \mathbf{R}^{2n}$ weight vectors for the Weyl algebra. For sufficiently small $\varepsilon > 0$ we have

$$\mathrm{in}_{(u,v)}(\mathrm{in}_{(u',v')}(I)) = \mathrm{in}_{(u',v')+\varepsilon(u,v)}(I). \tag{2.5}$$

(3) Let I be a $D^{(h)}$-ideal and $(t,u,v), (t',u',v') \in \mathbf{R}^{1+2n}$ weight vectors for the homogenized Weyl algebra. For sufficiently small $\varepsilon > 0$ we have

$$\mathrm{in}_{(t,u,v)}(\mathrm{in}_{(t',u',v')}(I)) = \mathrm{in}_{(t',u',v')+\varepsilon(t,u,v)}(I).$$

Proof. We shall prove case (2). The other two cases are similar. Let $\prec_\varepsilon$ be the multiplicative monomial order on the Weyl algebra D defined as follows:

$$x^\alpha\partial^\beta \prec_\varepsilon x^a\partial^b$$

$$:\Longleftrightarrow \quad (u'+\varepsilon u)\alpha + (v'+\varepsilon v)\beta < (u'+\varepsilon u)a + (v'+\varepsilon v)b$$

$$\text{or } [(u'+\varepsilon u)\alpha + (v'+\varepsilon v)\beta = (u'+\varepsilon u)a + (v'+\varepsilon v)b \text{ and}$$

$$u'\alpha + v'\beta < u'a + v'b]$$

$$\text{or } [(u'+\varepsilon u)\alpha + (v'+\varepsilon v)\beta = (u'+\varepsilon u)a + (v'+\varepsilon v)b \text{ and}$$

$$u'\alpha + v'\beta = u'a + v'b \text{ and } x^\alpha\partial^\beta <_t x^a\partial^b]$$

where $<_t$ is any tie breaking term order. Fix a sufficiently large integer d, and choose $\varepsilon = \varepsilon(d)$ small enough for all of the following assertions to be true. Every element $g \in I$ of total degree less than d is a k-linear combination of monomials $x^a \partial^b$, $(|a| + |b| < d)$ and this sum breaks into four pieces:

$$g(x, \partial) \quad = \quad g_0(x, \partial) + g_1(x, \partial) + g_2(x, \partial) + g_3(x, \partial),$$

where $g_0 = \text{in}_{\prec_\varepsilon}(g)$, $g_0 + g_1 = \text{in}_{(u'+\varepsilon u, v'+\varepsilon v)}(g)$, and $g_0 + g_1 + g_2 = \text{in}_{(u',v')}(g)$. Furthermore, we have

$$\text{in}_{(u,v)}\big(g_0 + g_1 + g_2\big) \quad = \quad g_0 + g_1. \tag{2.6}$$

Let F be any finite set of generators of I. Let F^h be the homogenization of F and let $\prec'_\varepsilon$ be the induced order in the homogenized Weyl algebra. Let G^h be the output of the Buchberger algorithm for F^h and $\prec'_\varepsilon$. This output consists of elements of degree less than d, and the computation is independent of the choice of ε. The dehomogenization $\mathcal{G} := G^h_{|h=1}$ is a Gröbner basis of I with respect to $\prec_\varepsilon$. From this Gröbner basis we derive the following three auxiliary Gröbner bases in the commutative polynomial ring $\mathbf{k}[x, \xi]$, in the ring $\text{gr}_{(u'+\varepsilon u, v'+\varepsilon v)}(D)$, and in the ring $\text{gr}_{(u',v')}(D)$ respectively:

(a) The monomials $g_0(x, \xi)$ for $g \in \mathcal{G}$ are a Gröbner basis for the monomial ideal $\text{in}_{\prec_\varepsilon}(I)$.
(b) The initial forms $g_0 + g_1$ for $g \in \mathcal{G}$ are a Gröbner basis for the initial ideal $\text{in}_{(u'+\varepsilon u, v'+\varepsilon v)}(I)$ with respect to $\prec_\varepsilon$.
(c) The initial forms $g_0 + g_1 + g_2$ for $g \in \mathcal{G}$ are a Gröbner basis for the initial ideal $\text{in}_{(u',v')}(I)$ with respect to $\prec_\varepsilon$.

Property (c) and the definition of $\prec_\varepsilon$ shows that the polynomials $g_0 + g_1 + g_2$ for $g \in \mathcal{G}$ are a Gröbner basis for $\text{in}_{(u',v')}(I)$ with respect to the weights (u, v). In view of (2.6), the polynomials $g_0 + g_1$ for $g \in \mathcal{G}$ generate the ideal $\text{in}_{(u,v)}(\text{in}_{(u',v')}(I))$. Property (b) now implies (2.5). $\qquad \square$

We fix a $D^{(h)}$-ideal I generated by homogeneous elements. We denote by $C_I[(t, u, v)]$ the equivalence class of the weight vector $(t, u, v) \in \mathbf{R}^{1+2n}$ with respect to I. The following proposition expresses this equivalence class as the solution set of a system of linear equations and strict linear inequalities. This implies that $C_I[(t, u, v)]$ is a relatively open convex polyhedral cone.

Proposition 2.1.7. *Let $\mathcal{G}$ be the reduced Gröbner basis of I with respect to $\prec_{(t,u,v)}$ where $\prec$ is a term order. The equivalence class $C_I[(t, u, v)]$ equals*

$$\{(t', u', v') \in \mathbf{R}^{1+2n} \mid \text{in}_{(t,u,v)}(g) = \text{in}_{(t',u',v')}(g) \text{ for all } g \in \mathcal{G}, \ 2t' \leq u'_i + v'_i\}.$$

Proof. See [96, p.12, Proposition 2.3]. $\qquad \square$

Let us now construct the Gröbner fan, starting with the easy case of principal ideals. Consider a normally ordered expression,

$$f = \sum_{(\alpha,\beta)\in E} c_{\alpha\beta} x^{\alpha} \partial^{\beta} \in D,$$

where $c_{\alpha\beta} \neq 0$ for $(\alpha,\beta) \in E$. The convex hull of the set E in $\mathbf{R}^{2n}$ is denoted by $\text{New}(f)$ and called the *Newton polytope* of f. Similarly, we define $\text{New}(f)$ for any element f in the homogenized Weyl algebra $D^{(h)}$ by taking the convex hull of all exponent vectors appearing in the normally ordered expression for f. The following lemma gives the Newton polytope of an initial form of f.

Lemma 2.1.8.

(1) *For $f \in \mathbf{k}[x]$ and $w \in \mathbf{R}^n$, we have $\text{face}_w(\text{New}(f)) = \text{New}(\text{in}_w(f))$.*
(2) *For $f \in D$ and $(u,v) \in \mathbf{R}^{2n}$, $\text{face}_{(u,v)}(\text{New}(f)) = \text{New}(\text{in}_{(u,v)}(f))$.*
(3) *For $f \in D^{(h)}$, $(t,u,v) \in \mathbf{R}^{1+2n}$, $\text{face}_{(t,u,v)}(\text{New}(f)) = \text{New}(\text{in}_{(t,u,v)}(f))$.*

Proof. Straightforward from the definitions; see e.g. [96, page 12]. □

For the principal ideal generated by a homogeneous element f^h in $D^{(h)}$, the normal fan of the Newton polytope $\text{New}(f^h)$ is the Gröbner fan of the $D^{(h)}$-ideal $D^{(h)} \cdot \{f^h\}$. Figure 2.1 is a picture of the Newton polytope and the Gröbner fan for $f = x^3\partial^2 + (x + x^2)\partial + 1$ restricted to $t = 0$.

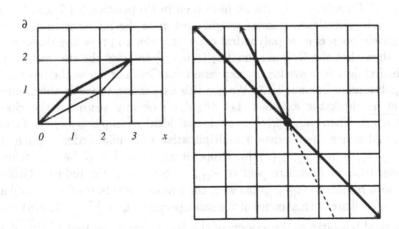

Fig. 2.1. The Newton polytope and the Gröbner fan of $x^3\partial^2 + (x + x^2)\partial + 1$.

We fix a $D^{(h)}$-ideal I generated by homogeneous elements. Let $\mathcal{G}$ be the reduced Gröbner basis of I with respect to the order $\prec_{(t,u,v)}$ where $\prec$ is a term order. In Proposition 2.1.7 we characterized the equivalence class of (t,u,v) by a set of linear inequalities. Here is a different characterization as a normal cone of a polytope defined as the Minkowski sum of Newton polytopes.

Lemma 2.1.9. *Let* $Q := \sum_{g \in \mathcal{G}} \text{New}(g)$ *and* $W = \{(t, u, v) \mid 2t \leq u_i + v_i\}$. *Then*

$$C_I[(t, u, v)] = N_Q \left(\text{face}_{(t,u,v)}(Q) \right) \cap W.$$

Proof. The normal cone at a face of a Minkowski sum $P = P_1 + \cdots + P_m$ satisfies the following general identity [44, Lemma 2.1.5]:

$$N_P(\text{face}_w(P)) = \bigcap_{i=1}^m N_{P_i} \left(\text{face}_w(P_i) \right).$$

In the particular situation under consideration, this translates into

$$N_Q(\text{face}_{(t,u,v)}(Q)) = \bigcap_{g \in \mathcal{G}} N_{\text{New}(g)} \left(\text{face}_{(t,u,v)}(\text{New}(g)) \right).$$

On the other hand, Lemma 2.1.8 (3) implies

$$\{(t', u', v') \in W \mid \text{in}_{(t',u',v')}(g) = \text{in}_{(t,u,v)}(g)\}$$
$$= N_{\text{New}(g)} \left(\text{face}_{(t,u,v)}(\text{New}(g)) \right) \cap W.$$

Take the intersection over $g \in \mathcal{G}$ on both sides and apply Proposition 2.1.7. □

Proof (of Theorem 2.1.1). As we have seen in Proposition 2.1.7 and Lemma 2.1.9, each equivalence class of weight vectors is the intersection of W and a relatively open convex polyhedral cone. In order to prove the theorem, we must show that the Gröbner fan $\{\overline{C_I[(t, u, v)]}\}$ satisfies the two axioms of a polyhedral fan. We have locally expressed our Gröbner fan as the normal fan of a polytope in Lemma 2.1.9. We use this expression to verify the axioms.

Let us check the axiom 1. Let (t', u', v') be any vector in the closure $\overline{C_I[(t, u, v)]}$. Then, $\text{in}_{(t,u,v)}(I)$ is an initial ideal of $\text{in}_{(t',u',v')}(I)$ by Lemma 2.1.6, and hence there exists a multiplicative monomial order $\prec$ such that $\text{in}_{\prec_{(t,u,v)}}(I) = \text{in}_{\prec_{(t',u',v')}}(I)$ by Proposition 2.1.4. Let $\mathcal{G}$ be the reduced Gröbner basis of I with respect to $\prec_{(t,u,v)}$. Since $\mathcal{G}$ is the reduced Gröbner basis with respect to $\prec_{(t',u',v')}$ as well, the equivalence classes $C_I[(t, u, v)]$ and $C_I[(t', u', v')]$ are normal cones of the same polytope $Q = \sum_{g \in \mathcal{G}} \text{New}(g)$. Since the normal fan satisfies the axioms of the fan, we conclude that $\overline{C_I[(t', u', v')]}$ is a face of $\overline{C_I[(t, u, v)]}$.

We verify the axiom 2 for being a fan in a similar manner to the above. See [96, page 13; Proposition 2.4]. □

A concept closely related to the Gröbner fan is the universal Gröbner basis. A subset G of an ideal I in $\mathbf{k}[x]$, D or $D^{(h)}$ is called a *universal Gröbner basis* if G is a Gröbner basis for I with respect to any weight vector. Similarly, $G \subset D$ is called a *small universal Gröbner basis* if G is a Gröbner basis for I with respect to any weight vector of the form (u, v), $u + v = 0$. The existence of a finite universal Gröbner basis follows from the finiteness of the Gröbner fan (Theorem 2.1.1 and 2.1.3). For algorithms for transforming the Gröbner fan into a universal Gröbner basis and vice versa, see [96, §3].

Corollary 2.1.10. *Every left ideal in* $\mathbf{k}[x]$, D *or* $D^{(h)}$ *has a finite universal Gröbner basis.*

Example 2.1.11. Let $n = 2$ and consider the following $D^{(h)}$-ideal

$$I = D^{(h)} \cdot \left\{ 2x_1\partial_1 + 3x_2\partial_2 + 6h^2 , \; 2hx_2\partial_1 + 3x_1^2\partial_2 \right\}.$$

Its dehomogenization $I_{|_{h=1}} \subset D$ is the annihilating D-ideal of the rational function $(x_1^3 - x_2^2)^{-1}$ (see Example 5.3.14). By means of an exhaustive computer search, we found that the following seven elements form a universal Gröbner basis for I:

$$2x_1\partial_1 + 3x_2\partial_2 + 6h^2,$$
$$3x_1^2\partial_2 + 2x_2\partial_1 h,$$
$$-x_1^3\partial_2 + x_2^2\partial_2 h + 2x_2 h^3,$$
$$x_1^3\partial_1 - x_2^2\partial_1 h + 3x_1^2 h^2,$$
$$9x_1x_2\partial_2^2 - 4x_2\partial_1^2 h + 15x_1\partial_2 h^2,$$
$$-9x_1^2\partial_2^2 + 4x_1\partial_1^2 h + 10\partial_1 h^3,$$
$$27x_2^2\partial_2^3 + 8x_2\partial_1^3 h + 135x_2\partial_2^2 h^2 + 105\partial_2 h^4.$$

The weight vector for $x_1, x_2, \partial_1, \partial_2, h$ is denoted by (u_1, u_2, v_1, v_2, t). First note that the ideal I is homogeneous with respect to the weight vectors $(1, 1, 1, 1, 1)$ and $(2, 3, -2, -3, 0)$. Therefore the Gröbner fan of I is a 3-dimensional fan times the 2-dimensional subspace spanned by these two vectors. We remove this subspace by working in the 3-dimensional subspace $\{u_1 = 0, t = 0\}$.

We shall describe the Gröbner fan of I in (v_1, p_2, p_3)-space. The facet hyperplanes of this fan are given by the following seven linear forms:

$$v_1,$$
$$p_2 := u_2 + v_2,$$
$$v_1 - p_2,$$
$$p_3 := 3u_1 - 2u_2,$$
$$h_1 := v_1 - p_2 - p_3,$$
$$h_2 := 2v_1 - 2p_2 - p_3,$$
$$h_3 := 3v_1 - 3p_2 - p_3.$$

The Gröbner fan has nine maximal cones. We list the defining inequalities for each. The inequalities $v_1, p_2 > 0$ are assumed tacitly; they characterize weight vectors for the homogenized Weyl algebra. For each cone we also list a representative weight vector and the corresponding initial ideal of I. A diagram of the Gröbner fan is given in Figure 2.2.

initial ideals	inequalities	weight vectors
(1) $\langle x_1\xi_1, x_2\xi_1 h, x_1^3\xi_2 \rangle$	$p_3 > 0, h_1 > 0$	$(0, -3, 8, 4, 0)$
(2) $\langle x_1\xi_1, x_1^2\xi_2, x_2\xi_1^2 h \rangle$	$p_3 > 0, h_1 < 0, h_2 > 0$	$(0, -3, 5, 4, 0)$
(3) $\langle x_1\xi_1, x_1^2\xi_2, x_1 x_2\xi_2^2, x_2\xi_1^3 h \rangle$	$p_3 > 0, h_2 < 0, h_3 > 0$	$(0, -6, 6, 7, 0)$
(4) $\langle x_1\xi_1, x_1^2\xi_2, x_1 x_2\xi_2^2, x_2^2\xi_2^3 \rangle$	$p_3 > 0, h_3 < 0, v_1 - p_2 > 0$	$(0, -6, 2, 7, 0)$
(5) $\langle x_2\xi_2, x_1^2\xi_2, x_1^3\xi_1 \rangle$	$p_3 > 0, v_1 - p_2 < 0$	$(0, -3, 1, 6, 0)$
(6) $\langle x_2\xi_2, x_1^2\xi_2, x_2^2\xi_1 h \rangle$	$p_3 < 0, h_1 < 0$	$(0, 3, 1, 5, 0)$
(7) $\langle x_2\xi_2, x_2\xi_1 h, x_1^2\xi_2^2 \rangle$	$p_3 < 0, h_1 > 0, h_2 < 0$	$(0, 3, 1, 2, 0)$
(8) $\langle x_2\xi_2, x_2\xi_1 h, x_1\xi_1^2 h \rangle$	$p_3 < 0, h_2 > 0, v_1 - p_2 < 0$	$(0, 6, 2, 1, 0)$
(9) $\langle x_1\xi_1, x_2\xi_1 h, x_2^2\xi_2 h \rangle$	$p_3 < 0, v_1 - p_2 > 0$	$(0, 3, 2, -2, 0)$

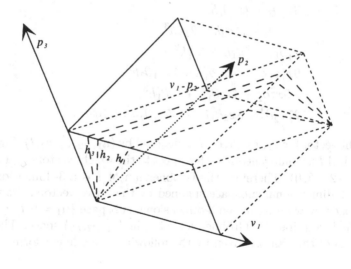

Fig. 2.2. The Gröbner fan in $0 \le v_1, p_2, p_3 \le 1$.

The Gröbner fan of a homogeneous $D^{(h)}$-ideal I has as its support the closed cone $W = \{(t, u, v) \in \mathbf{R}^{1+2n} \mid 2t \le u_i + v_i \text{ for all } i\}$. The cone W contains the linear space $L = \{(t, u, v) \in \mathbf{R}^{1+2n} \mid 2t = u_i + v_i \text{ for all } i\}$. This is a linear subspace of dimension $n + 1$. The restriction of the Gröbner fan to L is a complete fan of dimension $n + 1$; we call it the *small Gröbner fan* of I. The *small Gröbner fan* of a D-ideal I is defined as the small Gröbner fan of its homogenization $I^{(h)}$ restricted to the hyperplane $\{t = 0\}$. Thus the small Gröbner fan of I is an n-dimensional fan; its cells are in many-to-one correspondence with the initial ideals of the form $\mathrm{in}_{(-w,w)}(I)$. The importance of the small Gröbner fan will become clear when we discuss the torus action on the Weyl algebra D. We close this section with two examples.

Example 2.1.12. Let I be the D-ideal of all operators which annihilate $f = 1/(x_1^3 - x_2^2)$. This is the dehomogenization of the $D^{(h)}$-ideal in Example 2.1.11. The small Gröbner fan of I covers the 2-dimensional space with coordinates $(u_1, u_2, -u_1, -u_2, 0)$. It has three cells given by $p_3 = 3u_1 - 2u_2 > 0$, $p_3 = 0$, and $p_3 < 0$. The two open cells $p_3 > 0$ and $p_3 < 0$ correspond to choosing one of the two terms in the denominator polynomial $x_1^3 - x_2^2$ to be the leading term. Thus the cells of the small Gröbner fan are in bijection with the possible Gröbner deformations of the rational function f. In Theorems 2.5.5 and 2.5.12, the correspondence will be formulated precisely.

Example 2.1.13. ($\Delta_1 \times \Delta_2$-GKZ system, Appell function F_1)
Consider the following GKZ hypergeometric system for $\Delta_1 \times \Delta_2$:

$$
\begin{pmatrix}
0 & 0 & 0 & 1 & 1 & 1 \\
1 & 0 & 0 & 1 & 0 & 0 \\
0 & 1 & 0 & 0 & 1 & 0 \\
0 & 0 & 1 & 0 & 0 & 1
\end{pmatrix}
\begin{pmatrix}
\theta_{11} \\ \theta_{12} \\ \theta_{13} \\ \theta_{21} \\ \theta_{22} \\ \theta_{23}
\end{pmatrix}
-
\begin{pmatrix}
a_2 \\ b_1 \\ b_2 \\ b_3
\end{pmatrix}, \quad a_i, b_j \in \mathbf{N},
$$

$$\partial_{1i}\partial_{2j} - \partial_{1j}\partial_{2i}, \quad 1 \le i < j \le 3.$$

Let us describe the small Gröbner fan for the D-ideal generated by these operators. We assume that the values of the parameters a_i, b_j are generic. We can prove that the set of homogenizations of the differential operators above is already a small universal Gröbner basis. Moreover, up to S_3-symmetry, there is only one $\mathrm{in}_{(-w,w)}(I)$ for generic w. See Proposition 1.5.9 and (1.44).

The small Gröbner fan is two-dimensional, after dividing out the four-dimensional space of homogeneities of the given equations. We denote by u_{ij} the weight for x_{ij} and by $-u_{ij}$ the weight for ∂_{ij}. Put $p = u_{11} + u_{22} - u_{12} - u_{21}$, $q = u_{12} + u_{23} - u_{13} - u_{22}$. These two coordinates suffice to describe the small Gröbner fan. By examining linear inequalities for the weight vector w derived from the small universal Gröbner basis, we see that the facet hyperplanes are given by $p = 0$, $q = 0$, and $p + q = 0$. There are six maximal dimensional cones. See Figure 2.3.

This fan has the following combinatorial meaning. Consider the set of all non-negative integer 2×3-matrices with fixed row and column sums:

$$F = \{(p_{ij}) \in \mathbf{N}^{2\times 3} \mid p_{1i} + p_{2i} = b_i, \ p_{21} + p_{22} + p_{23} = a_2\}.$$

This gives rise to the hypergeometric polynomial (cf. Lemma 3.4.10 or [89])

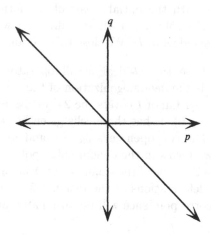

Fig. 2.3. Small Gröbner fan for Example 2.1.13.

$$\Phi(x) = \sum_{p \in F} \frac{x_{11}^{p_{11}} x_{12}^{p_{12}} x_{13}^{p_{13}} x_{21}^{p_{21}} x_{22}^{p_{22}} x_{23}^{p_{23}}}{p_{11}! p_{12}! p_{13}! p_{21}! p_{22}! p_{23}!}.$$

The space of polynomial solutions to the GKZ system above is spanned by $\Phi(x)$. The small Gröbner fan is a refinement of the normal fan of the Newton polytope of $\Phi(x)$. For suitable choices of a_2, b_1, b_2, b_3 both fans are equal.

2.2 Semi-Continuity of the Holonomic Rank

In this section we present some algebraic applications of the Gröbner fan. The main result is the following inequality for the holonomic rank.

Theorem 2.2.1. *Let I be a holonomic D-ideal and $w \in \mathbf{R}^n$. Then the initial D-ideal $\mathrm{in}_{(-w,w)}(I)$ is also holonomic and*

$$\mathrm{rank}\big(\mathrm{in}_{(-w,w)}(I)\big) \quad \leq \quad \mathrm{rank}(I) \tag{2.7}$$

Our proof of Theorem 2.2.1 is based on a technique known as the *Gröbner walk* [25]. This refers to a walk jumping from cell to cell in the Gröbner fan, which was constructed in the previous section. For our purposes, the following description is sufficient. Consider the closed cone $W = \{ (u, v) : u + v \geq 0 \}$ in $\mathbf{R}^{2n}$. For each vector (u, v) in that W, we have an initial ideal $\mathrm{in}_{(u,v)}(I)$ in the associated graded ring $\mathrm{gr}_{(u,v)}(D)$. The *Gröbner fan* of I is a finite polyhedral fan with support W such that $\mathrm{in}_{(u,v)}(I)$ is constant as (u, v) ranges over any of its cones. The restriction of the Gröbner fan to its linear subspace $\{(u, v) : u + v = 0\} \simeq \mathbf{R}^n$ is the small Gröbner fan. Its cones correspond to the initial D-ideals $\text{in}_{(-w,w)}(I)$ as in Theorem 2.1.3 (1).

We start with two lemmas in elementary commutative algebra.

Lemma 2.2.2. *Let J be any ideal in $\mathbf{k}[x, \xi]$ and $(u, v) \in \mathbf{R}^{2n}$.*
(1) $\dim(\text{in}_{(u,v)}(J)) \leq \dim(J)$.
(2) *If (u, v) is a non-negative vector then $\dim(\text{in}_{(u,v)}(J)) = \dim(J)$.*

Proof. We shall use the following characterization of the *Krull dimension* $\dim(J)$ of a polynomial ideal $J \subset \mathbf{k}[x, \xi]$. Let $\mathcal{M}$ be any set of monomials $x^a \xi^b$ which forms a $\mathbf{k}$-vector space basis for the residue ring $\mathbf{k}[x, \xi]/J$. Let $\mathcal{M}_{\leq d}$ denote the (finite) subset of monomials in $\mathcal{M}$ of degree at most d. Then the numerical function $H_{\mathcal{M}} : \mathbf{N} \to \mathbf{N}$, $d \mapsto \#\mathcal{M}_{\leq d}$ is a polynomial for $d \gg 0$, and $\dim(J)$ is the degree of that polynomial. Note that the function $H_{\mathcal{M}}$ itself is not an invariant of J but depends on our choice of $\mathcal{M}$, unless J is a homogeneous ideal; only the asymptotic order of growth of $H_{\mathcal{M}}$ is an invariant of J. See Section 9.3 in the undergraduate text book [27] for details.

Let $\prec$ be any term order on $\mathbf{k}[x, \xi]$ and let $\prec_{(u,v)}$ denote the multiplicative monomial order on the set of monomials in $\mathbf{k}[x, \xi]$ which is defined as follows:

$$x^a \xi^b \prec_{(u,v)} x^{a'} \xi^{b'} : \Longleftrightarrow$$
$$au + bv < a'u + b'v \text{ or } (au + bv = a'u + b'v \text{ and } x^a \xi^b \prec x^{a'} \xi^{b'}).$$

Note that $\prec_{(u,v)}$ is a term order on $\mathbf{k}[x, \xi]$ if and only if the vector (u, v) is non-negative; otherwise the order $\prec_{(u,v)}$ is not Noetherian, i.e., there exists an infinite strictly decreasing chain $f_1 \succ_{(u,v)} f_2 \succ_{(u,v)} \cdots$.

Let $\mathcal{M}'$ be the set of all monomials of $\mathbf{k}[x, \xi]$ which do not lie in the monomial ideal

$$\text{in}_{\prec_{(u,v)}}(J) = \text{in}_{\prec}(\text{in}_{(u,v)}(J)).$$

The set $\mathcal{M}'$ is a $\mathbf{k}$-vector space basis for $\mathbf{k}[x, \xi]/\text{in}_{(u,v)}(J)$, since $\prec$ is a term order. Therefore $\dim(\text{in}_{(u,v)}(J))$ is the degree of the polynomial $H_{\mathcal{M}'}(d)$ for $d \gg 0$.

The set $\mathcal{M}'$ is $\mathbf{k}$-linearly independent in $\mathbf{k}[x, \xi]/J$, and it is a $\mathbf{k}$-basis for $\mathbf{k}[x, \xi]/J$ when $\prec_{(u,v)}$ is a term order, i.e., when (u, v) is non-negative. Hence $\dim(J)$ is always greater than or equal to $\dim(\text{in}_{(u,v)}(J))$, and equality holds when (u, v) is non-negative. $\qquad\qquad \square$

For any ideal J in $\mathbf{k}[x, \xi]$ we abbreviate

$$\rho(J) := \dim_{\mathbf{k}(x)}(\mathbf{k}(x)[\xi]/\mathbf{k}(x)[\xi] \cdot J). \qquad (2.8)$$

Here we allow $\rho(J) = \infty$. For a holonomic D-ideal I we have $\text{rank}(I) = \rho(\text{in}_{(0,e)}(I)) < \infty$.

Lemma 2.2.3. *Let J be any ideal in $\mathbf{k}[x, \xi]$ and $(u, v) \in \mathbf{R}^{2n}$.*

(1) *If $\dim(J) \leq n$ then $\rho\big(\mathrm{in}_{(u,v)}(J)\big) \leq \rho(J) < \infty$.*

(2) *If $\dim(J) \leq n$ and v is a positive vector in $\mathbf{R}^n$ then $\rho\big(\mathrm{in}_{(0,v)}(J)\big) = \rho(J)$.*

Proof. Let $\mathcal{M}$ be any set of monomials $x^a \xi^b$ which is $\mathbf{k}$-linearly independent in $\mathbf{k}[x, \xi]/J$. Let $\mathcal{N}$ be the set of all monomials ξ^b in $\mathbf{k}[\xi]$ such that $x^a \xi^b \in \mathcal{M}$ for all $a \in \mathbf{N}^n$. Then $\mathcal{N}$ is also $\mathbf{k}(x)$-linearly independent in $\mathbf{k}(x)[\xi]/\mathbf{k}(x)[\xi] \cdot J$. Note, however, that if $\mathcal{M}$ is a $\mathbf{k}$-basis of $\mathbf{k}[x, \xi]/J$ then $\mathcal{N}$ need not be a $\mathbf{k}(x)$-basis of $\mathbf{k}(x)[\xi]/\mathbf{k}(x)[\xi] \cdot J$.

Let $\prec$ be an *elimination term order* of the form $\xi \succ x$ on $\mathbf{k}[x, \xi]$, and suppose that $\mathcal{M}$ is the set of monomials not in $\mathrm{in}_\prec(J)$. Clearly, $\mathcal{M}$ is a $\mathbf{k}$-basis of $\mathbf{k}[x, \xi]/J$. We claim that the corresponding set $\mathcal{N}$ is also a $\mathbf{k}(x)$-basis of $\mathbf{k}(x)[\xi]/\mathbf{k}(x)[\xi] \cdot J$. To see that $\mathcal{N}$ spans this $\mathbf{k}(x)$-vector space, consider any monomial ξ^b not in $\mathcal{N}$. There exists $a \in \mathbf{N}^n$ such that $x^a \xi^b \notin \mathcal{M}$. Modulo J we can replace $x^a \xi^b$ by a $\mathbf{k}$-linear combination of monomials $x^c \xi^d \in \mathcal{M}$. Then, ξ^b is expressed as a $\mathbf{k}(x)$-linear combination of monomials ξ^d such that $\xi^d \prec \xi^b$ modulo $\mathbf{k}(x)[\xi]J$. (Here we are using the elimination property of $\prec$). For occurring monomials ξ^d belonging to $\mathcal{N}$, we keep those terms. For other occurring monomials, we repeat the rewriting process: some x-monomial multiple of ξ^d is the $\prec$-initial monomial of an element in J etc... Since $\prec$ is Noetherian, this process terminates and the final result is a $\mathbf{k}(x)$-linear combination of monomials in $\mathcal{N}$ expressing ξ^b.

For $d \gg 0$ the cardinality of $\mathcal{M}_{\leq d}$ has order at least $\frac{1}{n!} d^n$ times $\#\mathcal{N}$. By the argument in the previous paragraph we have $\rho(J) = \#\mathcal{N}$. This number must be finite if $\dim(J) \leq n$. Therefore $\rho(J) < \infty$.

Let $\prec$ be an elimination term order as above, let $\mathcal{M}'$ be the set of monomials not in $\mathrm{in}_\prec(\mathrm{in}_{(u,v)}(J))$, and let $\mathcal{N}'$ be the corresponding set of monomials in $\mathbf{k}[\xi]$. Then $\rho\big(\mathrm{in}_{(u,v)}(J)\big) = \#\mathcal{N}'$. Since $\mathcal{M}'$ is $\mathbf{k}$-linearly independent modulo J, the set $\mathcal{N}'$ is $\mathbf{k}(x)$-linearly independent modulo $\mathbf{k}(x)[\xi] \cdot J$. Therefore $\rho(J) \geq \#\mathcal{N}'$, which proves part (1).

For $u = 0$ and v positive, the term order $\prec_{(u,v)}$ defined in the proof of Lemma 2.2.2 is also an elimination order of the form $\xi \succ x$. In this case, $\mathcal{M}'$ is a $\mathbf{k}$-basis modulo J, and, moreover, the set $\mathcal{N}'$ is a $\mathbf{k}(x)$-basis modulo $\mathbf{k}(x)[\xi] \cdot J$. This proves part (2).   □

Proof (of Theorem 2.2.1). For any real number t in the half-open interval $(0, 1]$ consider the ideal $J_t := \mathrm{in}_{(1-t)(-w,w)+t(0,e)}(I)$ in the commutative polynomial ring $\mathbf{k}[x, \xi]$. Clearly, J_1 equals the characteristic ideal of I, and, by Lemma 2.1.6, there exists a number t_1 such that J_ε equals the characteristic ideal of $\mathrm{in}_{(-w,w)}(I)$, for any $\varepsilon \in (0, t_1)$. Thus we are considering a homotopy or "Gröbner walk" between the two characteristic ideals of interest. Our task is to show the inequalities

$$\dim(J_\varepsilon) \ \leq \ \dim(J_1) \quad \text{and} \quad \rho(J_\varepsilon) \ \leq \ \rho(J_1). \tag{2.9}$$

The segment from $(-w, w)$ to $(0, e)$ intersects finitely many distinct walls of the Gröbner fan. Thus there are real numbers $0 = t_0 < t_1 < t_2 < \cdots < t_r = 1$ such that J_t remains unchanged as the parameter t ranges inside the open interval (t_{i-1}, t_i). We use the notation $J_{(t_{i-1}, t_i)} := J_t$ for this ideal. By Lemma 2.1.6, it is expressed as follows:

$$\text{in}_{(0,e)}(J_{t_{i-1}}) = J_{(t_{i-1}, t_i)} = \text{in}_{(0,-e)}(J_{t_i}). \tag{2.10}$$

By applying Lemmas 2.2.2 (2) and 2.2.3 (2) to the first equation we obtain

$$\dim(J_{t_{i-1}}) = \dim(J_{(t_{i-1}, t_i)}) \quad \text{and} \quad \rho(J_{t_{i-1}}) = \rho(J_{(t_{i-1}, t_i)}). \tag{2.11}$$

By applying Lemmas 2.2.2 (1) and 2.2.3 (1) to the second equation we obtain

$$\dim(J_{(t_{i-1}, t_i)}) \le \dim(J_{t_i}) \quad \text{and} \quad \rho(J_{(t_{i-1}, t_i)}) \le \rho(J_{t_i}). \tag{2.12}$$

Concatenating the inequalities (2.11) and (2.12) for all i, we obtain the inequalities (2.9). □

The Gröbner walk in the above proof can also be used to give more algorithmic robustness to the concept of holonomicity. Recall that we defined a D-ideal I to be *holonomic* if the characteristic ideal $\text{in}_{(0,e)}(I)$ is an n-dimensional ideal in $\mathbf{k}[x, \xi]$. As we stated in Theorem 1.4.12, for the definition of holonomicity, we can replace the specific ideal $\text{in}_{(0,e)}(I)$ by any other commutative initial ideal $\text{in}_{(u,v)}(I) \subset \mathbf{k}[x, \xi]$ of I, provided (u, v) is non-negative and $u + v > 0$.

Proof (of Theorem 1.4.12). Let $(u, v) > 0$ and $(u', v') > 0$ be positive weight vectors. We claim that

$$\dim \text{in}_{(u,v)}(I) = \dim \text{in}_{(u',v')}(I). \tag{2.13}$$

Since each equivalence class of weights is a convex cone, we have $\text{in}_{(u',v')}(I) = \text{in}_{(\lambda u', \lambda v')}(I)$ for any $\lambda > 0$, and we can choose λ such that $(u, v) - \lambda(u', v') > 0$. Hence, we may assume $(u, v) - (u', v') > 0$. We consider a Gröbner walk on $(u', v') + t(u - u', v - v')$, $t \in [0, 1]$. It follows from Lemma 2.2.2 that

$$\dim \text{in}_{(u,v)}(I) \ge \dim \text{in}_{(u',v')}(I).$$

This implies (2.13) since everything is symmetric with respect to (u, v) and (u', v').

Let us prove that $\dim \text{in}_{(u,v)}(I) = \dim \text{in}_{(u',v')}(I)$ for all $u + v > 0$, $u' + v' > 0$, $(u, v) \ge 0$, $(u', v') \ge 0$. Note that $\text{in}_{(u,v)}(I)$ is an ideal in the commutative polynomial ring $\mathbf{k}[x, \xi]$. Let $\prec$ be a term order. Then $\prec_{(u,v)}$ is a term order. From Proposition 2.1.5, there exists a positive weight vector (u'', v'') such that $\text{in}_{(u'',v'')}(I) = \text{in}_{\prec_{(u,v)}}(I)$. Since the Krull dimension of a polynomial ideal remains fixed under passing to the initial monomial ideal with respect to a term order, we find that

$$\dim\mathrm{in}_{(u'',v'')}(I) \;=\; \dim\mathrm{in}_{\prec_{(u,v)}}(I) \;=\; \dim\mathrm{in}_{\prec}(\mathrm{in}_{(u,v)}(I)) \;=\; \dim\mathrm{in}_{(u,v)}(I).$$

From this the conclusion follows, because $(u'', v'') > 0$.   □

Corollary 2.2.4 (Weak FTAA; the Bernstein inequality). *For any* $(u,v) \geq$ *0,* $u + v > 0$ *and proper D-ideal I, we have*

$$\dim\mathrm{in}_{(u,v)}(I) \geq n.$$

Proof. The inequality can be shown in an elementary way when $(u,v) = (e,e)$. The method of the proof is known as Joseph's proof. See, e.g., [26, Chapter 9]. We then derive the conclusion from Theorem 1.4.12.   □

We do not know at present whether the strong FTAA (Fundamental Theorem of Algebraic Analysis) holds for all initial ideals $\mathrm{in}_{(u,v)}(I)$ not just $(u,v) = (0,e)$. Recent work of Greg Smith suggests an affirmative answer.

As an immediate application of Theorem 1.4.12, we obtain the following algorithm to determine if a given D-ideal I is holonomic or not.

Algorithm 2.2.5 (Is I holonomic or not?)
(1) Choose a term order $\prec$.
(2) Compute a Gröbner basis of I with respect to $\prec$.
(3) Compute the Krull dimension of the initial monomial ideal $\mathrm{in}_{\prec}(I)$. (This is a monomial ideal in the commutative polynomial ring $\mathbf{k}[x,\xi]$.)
(4) If the Krull dimension is n, then I is holonomic, otherwise it is not.

2.3 Torus Action and Frobenius Ideals

Let D denote the Weyl algebra of dimension n over a field $\mathbf{k}$ of characteristic zero. The multiplicative group $T = (\mathbf{k}^*)^n$, called the n-dimensional algebraic *torus* over $\mathbf{k}$, acts on the Weyl algebra D as follows:

$$T \times D \to D, \quad (t, \partial_i) \mapsto t_i \cdot \partial_i, \quad (t, x_i) \mapsto t_i^{-1} \cdot x_i. \qquad (2.14)$$

Here $t = (t_1, \ldots, t_n)$. Given $t \in T$ and $f \in D$, we write $t \circ f$ for the image under this action. In this section, we first reformulate the Gröbner basis theory in D for the weight vector $(-w, w)$ in terms of the torus action.

Our motivation for studying torus actions on D-ideals is the problem of solving systems of linear partial differential equations with polynomial coefficients. Suppose we are given a "nice" function $f(x)$ which is a solution of I. Then the new function $f(t^w y) = f(t^{w_1} y_1, \ldots, t^{w_n} y_n)$ in one more variable is also "nice", which means that it admits a series expansion with respect to the new variable t:

$$f(t^w y) \;=\; f_\alpha(y) \cdot t^\alpha + \text{terms of higher order in } t.$$

The crucial fact is that $f_\alpha(y)$ is annihilated by

$$I' \quad := \quad \mathrm{in}_{(-w,w)}(I)|_{x_i \mapsto y_i,\, \partial_i \mapsto \partial_{y_i}}.$$

In fact, let us introduce the following change of variables

$$x_i \;=\; t^{w_i} y_i \tag{2.15}$$

where t is regarded as a parameter. Then, we have

$$\frac{\partial}{\partial x_i} \;=\; t^{-w_i} \frac{\partial}{\partial y_i}. \tag{2.16}$$

From the definition of $\mathrm{in}_{(-w,w)}(I)$, the D-ideal I' is generated by the leading coefficients of the expansion of $\ell \in I$ in terms of t under the change of variables (2.15) and (2.16) where $1 \succ t \succ t^2 \succ \cdots$. It implies that I' annihilates $f_\alpha(y)$. This will be made precise in Section 2.5. In Section 2.6 we shall reverse this process by showing that the function f can be reconstructed from $\mathrm{in}_w(f) = f_\alpha$. For $n = 1$ this is the classical *Frobenius method* for constructing series solutions to linear ordinary differential equations, as we have seen in (1.16) for the Gauss hypergeometric differential equation. In that case, we choose $(-w, w) = (-1, 1)$ and $\mathrm{in}_w(f)$ is x^s where s is a root of the indicial equation $s(s+1-c) = 0$; x^s is a solution of $\theta(\theta+1-c)$ where $\mathrm{in}_{(-1,1)}(xP) = \theta(\theta+1-c)$.

If $t \in T$ then $t \circ I$ denotes the D-ideal consisting of all elements $t \circ f$ where $f \in I$. A D-ideal I which satisfies $I = t \circ I$ for all $t \in T$ is called *torus-fixed* or *torus-invariant*. We abbreviate $\theta_i = x_i \partial_i$ and $\mathbf{k}[\theta] = \mathbf{k}[\theta_1, \ldots, \theta_n]$. Since $\theta_i \theta_j = \theta_j \theta_i$, $\mathbf{k}[\theta]$ is a commutative polynomial subring of D. Note that $\mathbf{k}[\theta]$ consists precisely of those elements of D which are fixed under T.

Lemma 2.3.1. *A D-ideal I is torus-fixed if and only if I is generated by elements of the form $x^a \cdot p(\theta) \cdot \partial^b$ where $a, b \in \mathbf{N}^n$ and $p(\theta) \in \mathbf{k}[\theta]$.*

Proof. The if-direction is obvious since $t \circ \left(x^a \cdot p(\theta) \cdot \partial^b \right) = t^{b-a} \cdot x^a \cdot p(\theta) \cdot \partial^b$. For the converse suppose $I = t \circ I$. Every element f of I can be written as a finite sum $f = \sum_{a,b} c_{ab} \cdot x^a \cdot p_{ab}(\theta) \cdot \partial^b$, where $c_{ab} \in \mathbf{k}$, and $a, b \in \mathbf{N}^n$ with disjoint support. The element $t \circ f = \sum_{a,b} c_{ab} \cdot t^{b-a} \cdot x^a \cdot p_{ab}(\theta) \cdot \partial^b$ lies also in I, for all $t \in T$. By taking $\mathbf{k}$-linear combinations for suitable elements $t \in T$, we conclude that $x^a \cdot p_{ab}(\theta) \cdot \partial^b \in I$ for all a, b with $c_{ab} \neq 0$. Therefore I is generated by elements of the required form. $\qquad\square$

We now study the action of one-parameter subgroups of T on D. Let us recall the definition of initial forms and some properties discussed in Section 1.1 in terms of torus action. The w-weight (or w-degree) of a T-eigenvector $x^a \cdot p(\theta) \cdot \partial^b$ is the inner product $w \cdot (b - a)$. Let $f \in D$ and write $f = \sum_{a,b} c_{ab} \cdot x^a \cdot p_{ab}(\theta) \cdot \partial^b$ as above. The *initial form* $\mathrm{in}_{(-w,w)}(f)$ is the subsum of all non-zero summands $c_{ab} \cdot x^a \cdot p_{ab}(\theta) \cdot \partial^b$ which have maximum w-weight. This can be restated as follows. Consider $t \circ f$ for $t = (s^{w_1}, \ldots, s^{w_n})$. The initial form $\mathrm{in}_{(-w,w)}(f)$ is the leading coefficient of the expansion of $t \circ f$ in terms of s and the order $1 \prec s \prec s^2 \prec \cdots$.

The next lemma follows from the Leibnitz formula in Theorem 1.1.1.

Lemma 2.3.2. *For all $f, g \in D$ and $w \in \mathbf{R}^n$ we have*

$$\mathrm{in}_{(-w,w)}(g \cdot f) \quad = \quad \mathrm{in}_{(-w,w)}(g) \cdot \mathrm{in}_{(-w,w)}(f). \tag{2.17}$$

Let I be any D-ideal and $w \in \mathbf{R}^n$. We define its *initial ideal* to be the **k**-vector space

$$\mathrm{in}_{(-w,w)}(I) \quad := \quad \mathbf{k} \cdot \big\{ \mathrm{in}_{(-w,w)}(f) : f \in I \big\}.$$

As we have seen in Corollary 1.1.2, this is a D-ideal.

We now fix a D-ideal I and we vary the weights; we consider the small Gröbner fan of I. A weight vector $w \in \mathbf{R}^n$ is said to be *generic for I* if w lies in an open cone of the small Gröbner fan. The cones in the small Gröbner fan being rational means that each equivalence class contains an integer vector $w \in \mathbf{Z}^n$. Geometrically and informally speaking, for integral weights, we can interpret the initial ideal $\mathrm{in}_{(-w,w)}(I)$ as the limit, in a suitable Hilbert scheme, of I under a one-parameter subgroup:

$$\mathrm{in}_{(-w,w)}(I) \quad = \quad \lim_{\tau \to \infty} (\tau^{w_1}, \ldots, \tau^{w_n}) \circ I. \tag{2.18}$$

An invariant D-ideal by the action of the torus is characterized as follows.

Theorem 2.3.3. *Let I be a D-ideal.*
(1) *I is torus-fixed if and only if $\mathrm{in}_{(-w,w)}(I) = I$ for all $w \in \mathbf{R}^n$.*
(2) *If w is generic for I then the initial ideal $\mathrm{in}_{(-w,w)}(I)$ is torus-fixed.*

Proof. Part (1) follows from the observation that I is torus-fixed if and only if $f \in I$ implies $\mathrm{in}_{(-w,w)}(f) \in I$ for any $w \in \mathbf{R}^n$. Part (2) is an immediate consequence of (1) and Lemma 2.1.6. $\square$

Every D-ideal I gives rise to an ideal $\widetilde{I}$ in the commutative polynomial ring $\mathbf{k}[\theta]$ as follows. Recall that R is the ring of differential operators with rational function coefficients. We define $\widetilde{I} := RI \cap \mathbf{k}[\theta]$. The operation $I \mapsto \widetilde{I}$ is particularly interesting when I is torus-invariant. In this case we call $\widetilde{I}$ the *distraction* of I, and we have the following explicit description of the generators of $\widetilde{I}$. For $b = (b_1, \ldots, b_n) \in \mathbf{N}^n$ we abbreviate

$$[\theta]_b \quad := \quad \prod_{i=1}^{n} \prod_{j=0}^{b_i - 1} (\theta_i - j).$$

This is our multivariate notation for *falling factorials*.

Theorem 2.3.4. *The distraction $\widetilde{I}$ of a torus-fixed D-ideal I is the $\mathbf{k}[\theta]$-ideal generated by $[\theta]_b \cdot p(\theta - b)$ where $x^a \cdot p(\theta) \cdot \partial^b$ runs over a generating set of I.*

Proof. The following identity shows that the proposed vectors do lie in $\tilde{I}$:

$$[\theta]_b \cdot p(\theta - b) \quad = \quad x^{b-a} \cdot x^a \cdot p(\theta) \cdot \partial^b \quad \in RI. \tag{2.19}$$

Let J be a torus-fixed ideal generated by a finite set $\{x^a p_{ab}(\theta)\partial^b \mid (a,b) \in C\}$. Let $q(\theta)$ be an element of $\mathbf{k}(x)J \cap \mathbf{k}[\theta_1, \ldots, \theta_n]$. It can be expressed as

$$q(\theta) \quad = \quad \sum_{(a,b)\in C} \frac{c_{ab}(x,\partial)}{d_{ab}(x)} p_{ab}(\theta - b)[\theta]_b,$$

where $c_{ab}(x,\partial) \in D$ and d_{ab} is a polynomial in x. Choose generic integers $w_i > 0$ and replace x_i by $t^{w_i}x_i$ and ∂_i by $t^{-w_i}x_i$. Expanding both sides as a Laurent series in t, we have

$$\sum_{k=-K}^{\infty} t^k \left(\sum_{(a,b)\in C} e_{kab}(x,\partial)p_{ab}(\theta - b)[\theta]_b \right) \quad = \quad q(\theta), \ e_{kab}(x,\partial) \in D.$$

Hence, we have

$$\sum_{(a,b)\in C} e_{0ab}(x,\partial)p_{ab}(\theta - b)[\theta]_b \quad = \quad q(\theta).$$

Since $p_{ab}(\theta - b)[\theta]_b$ is torus-fixed and $e_{0ab}(x,\partial)$ can be expressed as a polynomial in θ, we conclude that $q(\theta)$ can be expressed as a $\mathbf{k}[\theta_1, \ldots, \theta_n]$-linear combination of $p_{ab}(\theta - b)[\theta]_b$. □

Corollary 2.3.5. *If J is a torus fixed D-ideal, then*

$$\mathbf{k}[x_1^{\pm}, \ldots, x_n^{\pm}] J \cap \mathbf{k}[\theta] \quad = \quad \mathbf{k}(x_1, \ldots, x_n) J \cap \mathbf{k}[\theta].$$

Here, $\mathbf{k}[x_1^{\pm}, \ldots, x_n^{\pm}] := \mathbf{k}[x_1^{-1}, \ldots, x_n^{-1}, x_1, \ldots, x_n]$.

The term *distraction* comes from algebraic geometry, namely, from Hartshorne's work on the Hilbert scheme [46]. If I is a D-ideal generated by monomials ∂^b, then $\tilde{I}$ corresponds to the distraction of I as defined in [46]. The structure of these distractions will be characterized in Theorem 3.2.2.

In order to find w-lowest terms of series solutions to I, we must solve the differential equations given by $\mathrm{in}_{(-w,w)}(I)$. Assuming w to be generic, it follows from Lemma 2.3.1 and Theorem 2.3.3 (2) that $\mathrm{in}_{(-w,w)}(I)$ is generated by elements of the form $x^a \cdot p(\theta) \cdot \partial^b$, where $a, b \in \mathbf{N}^n$ and $p \in \mathbf{k}[\theta]$. Since we want classical solutions ("nice functions"), we may replace $x^a \cdot p(\theta) \cdot \partial^b$ by $x^b \cdot p(\theta) \cdot \partial^b \in \mathbf{k}[\theta]$ and assume that the generators themselves are torus-fixed. This motivates the following definition.

A D-ideal F which is generated by elements in $\mathbf{k}[\theta]$ is called a *Frobenius ideal*. Hence every Frobenius ideal is torus-invariant and can be written as $F = D \cdot I$ where I is an ideal in the commutative polynomial ring $\mathbf{k}[\theta]$. We call the ideal I *Artinian* if the residue ring $\mathbf{k}[\theta]/I$ is finite-dimensional as a $\mathbf{k}$-vector space. In this case we set $\mathrm{rank}(I) := \dim_{\mathbf{k}}(\mathbf{k}[\theta]/I)$.

Proposition 2.3.6. *A Frobenius ideal $F = D \cdot I$ is holonomic if and only if the underlying commutative ideal I is Artinian. In this case, $\mathrm{rank}(I)$ equals the holonomic rank of F.*

We prove this by examining the interplay between the Buchberger algorithm in the Weyl algebra D and in the commutative polynomial ring $\mathbf{k}[\theta]$.

Lemma 2.3.7. *Let $\prec$ be any term order on the Weyl algebra D. Let f and g be elements in $\mathbf{k}[\theta]$ and $\mathrm{sp}(f,g)$ their S-pair taken in D. Then $\mathrm{sp}(f,g)$ lies in $\mathbf{k}[\theta]$.*

Proof. To form the S-pair in the Weyl algebra D, we must rewrite the input as follows:

$$
\begin{aligned}
f(\theta) &= a \cdot x^\alpha \partial^\alpha + \text{lower terms in } \prec, \\
g(\theta) &= b \cdot x^\beta \partial^\beta + \text{lower terms in } \prec.
\end{aligned}
$$

Let x^γ be the least common multiple of the monomials x^α and x^β. Then

$$
\mathrm{sp}(f,g) = x^{\gamma-\alpha}\partial^{\gamma-\alpha} \cdot f - \frac{a}{b} \cdot x^{\gamma-\beta}\partial^{\gamma-\beta} \cdot g = [\theta]_{(\gamma-\alpha)} \cdot f - \frac{a}{b} \cdot [\theta]_{(\gamma-\beta)} \cdot g
$$

is an identity in $\mathbf{k}[\theta]$. Thus the restriction of the Buchberger algorithm from D to $\mathbf{k}[\theta]$ can be viewed as the distraction of the Buchberger algorithm for polynomials. They differ but give the same result for input in $\mathbf{k}[\theta]$.   □

Corollary 2.3.8. *Let $\prec$ be any term order on D and $\prec'$ its restriction to $\mathbf{k}[\theta]$ defined by $\theta^\alpha \prec' \theta^\beta :\iff x^\alpha \partial^\alpha \prec x^\beta \partial^\beta$. A finite subset $\mathcal{G}$ of $\mathbf{k}[\theta]$ is a Gröbner basis with respect to $\prec'$ if and only if $\mathcal{G}$ is a Gröbner basis in D with respect to $\prec$.*

Proof (of Proposition 2.3.6). We use the definition of holonomicity and holonomic rank given in Definition 1.4.8. Let $\prec$ be any term order on the Weyl algebra D which refines the weight $(0,\ldots,0;1,\ldots,1)$. By Lemma 2.3.7, the reduced Gröbner basis $\mathcal{G}$ of $F = D \cdot I$ is a subset of $\mathbf{k}[\theta]$. By Corollary 2.3.8, $\mathcal{G}$ is a Gröbner basis for a degree-compatible term order $\prec_e$ on $\mathbf{k}[\theta]$. Let $g_1(\theta),\ldots,g_r(\theta)$ be the initial (leading) forms (with respect to total degree) of the elements in $\mathcal{G}$. The characteristic ideal of F is the ideal in $\mathbf{k}[x_1,\ldots,x_n,\xi_1,\ldots,\xi_n]$ generated by the polynomials $g_i(x_1\xi_1,\ldots,x_n\xi_n)$, $i = 1,\ldots,r$. This ideal has Krull dimension n if and only if F is holonomic. Clearly, this happens if and only if $\langle g_1(\theta),\ldots,g_r(\theta)\rangle$ is Artinian if and only if I is Artinian. In this case the rank of I coincides with the integer in (1.26).   □

A Frobenius ideal naturally appears by distracting the initial ideal of a given holonomic ideal I. Namely, for any generic weight vector $w \in \mathbf{R}^n$, we can form the *indicial ideal*

$$
\mathrm{ind}_w(I) := R \cdot \mathrm{in}_{(-w,w)}(I) \cap \mathbf{k}[\theta].
$$

Theorem 2.3.9. *Let I be a holonomic ideal and $w \in \mathbf{R}^n$ generic. Then $\mathrm{ind}_w(I)$ is a holonomic Frobenius ideal whose rank equals $\mathrm{rank}(\mathrm{in}_{(-w,w)}(I))$.*

Proof. From Theorem 2.2.1, $\mathrm{in}_{(-w,w)}(I)$ is holonomic. Hence, $R \cdot \mathrm{in}_{(-w,w)}(I)$ is a zero-dimensional ideal in R by Corollary 1.4.14. By Theorem 2.3.4, the zero-dimensional ideal is generated by elements in $\mathbf{k}[\theta]$, which together with Corollary 2.3.8 implies the rank equality and that $R \cdot \mathrm{in}_{(-w,w)}(I) \cap \mathbf{k}[\theta]$ is a zero-dimensional ideal in $\mathbf{k}[\theta]$. $\square$

The indicial ideal $\mathrm{ind}_w(I)$ will reappear in our constructions of series in Section 2.5 and in Chapters 3 and 4. In the remainder of this section we give an algorithm to find all classical solutions of holonomic Frobenius ideals. Later on, this will be applied to solve indicial ideals. In case of ordinary differential equations, the indicial ideal is the classical indicial polynomial.

Example 2.3.10. For $n = 1$ and $I = D \cdot \{\sum_{k=0}^m x^k f_k(\theta)\}$ where $f_k(\theta)$ are polynomials in θ, the initial ideal $\mathrm{in}_{(-1,1)}(I)$ is generated by $f_0(\theta)$, the *indicial polynomial* of the given operator. We factor this polynomial into linear factors,

$$f_0(\theta) = \prod_{i=1}^p (\theta - \lambda_i)^{m_i},$$

over an algebraic extension of $\mathbf{k}$. Then the functions

$$x^{\lambda_i} (\log x)^j, \quad i = 1, \ldots, p, \ 0 \le j \le m_i - 1$$

span the classical solution space of the ordinary differential equation $f_0(\theta) \bullet h = 0$. The holonomic rank of the initial ideal is

$$\dim_\mathbf{k} \mathbf{k}[\theta]/\langle f_0(\theta) \rangle = \sum_{i=1}^p m_i.$$

We next consider the general case $n > 1$. We take $\mathbf{k}$ to be the field $\mathbf{C}$ of complex numbers, to simplify the presentation. Let I be an Artinian ideal in $\mathbf{C}[\theta]$ and consider the holonomic Frobenius ideal $F = D \cdot I$. The variety defined by I is a finite subset $V(I) \subset \mathbf{C}^n$. The primary decomposition of I looks like

$$I = \bigcap_{p \in V(I)} Q_p(\theta - p), \tag{2.20}$$

where Q_p is an Artinian ideal, and $Q_p(\theta - p)$ denotes the ideal obtained from Q_p by replacing $\theta_i \mapsto \theta_i - p_i$ for $i = 1, \ldots, n$. Thus $Q_p(\theta - p)$ is primary to the maximal ideal of the point p, i.e., for any $f \in Q_p(\theta - p)$, there exists m such that $f \in \langle \theta_1 - p_1, \ldots, \theta_n - p_n \rangle^m$. In a computer system like Macaulay 2, the primary component can be computed as an iterated ideal quotient:

$$Q_p(\theta - p) = \left(I : \left(I : \langle \theta_1 - p_1, \ldots, \theta_n - p_n \rangle^\infty \right) \right) \tag{2.21}$$

The ideal Q_p is primary to the homogeneous maximal ideal $\langle \theta_1, \ldots, \theta_n \rangle$. We define its *orthogonal complement*:

$$Q_p^\perp := \{ g \in \mathbf{C}[x_1, \ldots, x_n] \mid f(\partial_1, \ldots, \partial_n) \bullet g(x_1, \ldots, x_n) = 0$$
$$\text{for all } f = f(\theta_1, \ldots, \theta_n) \in Q_p \}. \quad (2.22)$$

This is a finite-dimensional **k**-vector space. Its dimension equals $\text{rank}(Q_p)$.

Let $\mathbf{U}$ be a small open ball disjoint from the union of the coordinate hyperplanes in $\mathbf{C}^n$ and $\mathcal{O}^{\text{an}}(\mathbf{U})$ the complex vector space of holomorphic functions on $\mathbf{U}$. We fix a branch of the logarithm, so that $\log(x_i)$ is a well-defined element of $\mathcal{O}^{\text{an}}(\mathbf{U})$. Likewise, for any complex number α, the monomial $x_i^\alpha = \exp(\alpha \cdot \log(x_i))$ is a well-defined element of $\mathcal{O}^{\text{an}}(\mathbf{U})$.

We denote by $Hom_D(D/I, \mathcal{O}^{\text{an}}(\mathbf{U}))$ the space of solutions of a D-ideal I in $\mathcal{O}^{\text{an}}(\mathbf{U})$. This is the standard way in the theory of D-modules to denote the space of solutions. In fact, take φ in $Hom_D(D/I, \mathcal{O}^{\text{an}}(\mathbf{U}))$. Then, $\varphi(1)$ is an element of $\mathcal{O}^{\text{an}}(\mathbf{U})$, and, for any operator $\ell \in I$, the function $\varphi(1)$ satisfies

$$\ell \bullet \varphi(1) = \varphi(\ell \cdot 1) = \varphi(0) = 0,$$

which means that $\varphi(1)$ is a solution of I. Conversely, if we are given a solution $f \in \mathcal{O}^{\text{an}}(\mathbf{U})$, then the solution f determines an element φ of $Hom_D(D/I, \mathcal{O}^{\text{an}}(\mathbf{U}))$ by $\varphi(m) = m \bullet f$ for all $m \in D/I$.

Theorem 2.3.11. *Let $F = D \cdot I$ be a holonomic Frobenius ideal. Its solution space $Hom_D(D/F, \mathcal{O}^{\text{an}}(\mathbf{U}))$ has dimension $\text{rank}(I)$ and is spanned by the holomorphic functions $x_1^{p_1} x_2^{p_2} \cdots x_n^{p_n} \cdot g(\log(x_1), \ldots, \log(x_n))$, where $p = (p_1, \ldots, p_n) \in V(I)$ and $g \in Q_p^\perp$.*

Proof. Let $\mathcal{E}_I$ denote the **C**-vector space of all entire functions $P(z_1, \ldots, z_n)$ on $\mathbf{C}^n$ which are annihilated by the operators $f(\partial_{z_1}, \ldots, \partial_{z_n})$ for all $f = f(\theta_1, \ldots, \theta_n) \in I$. A classical theorem states that $\mathcal{E}_I$ has dimension $\text{rank}(I)$ and is spanned by the entire functions

$$\exp(p_1 z_1 + \cdots + p_n z_n) \cdot g(z_1, \ldots, z_n),$$
$$\text{where } p = (p_1, \ldots, p_n) \in V(I) \text{ and } g \in Q_p^\perp.$$

This theorem was rediscovered many times and appears in many places (see, e.g., the recent work of Oberst [83, Algorithm (61)]).

If $P(z_1, \ldots, z_n)$ is a non-zero element of $\mathcal{E}_I$ then $P(\log(x_1), \ldots, \log(x_n))$ defines a non-zero element of $Hom_D(D/F, \mathcal{O}^{\text{an}}(\mathbf{U}))$. Thus the functions in the statement of Theorem 2.3.11 span a subspace of dimension $\text{rank}(I)$ in $Hom_D(D/F, \mathcal{O}^{\text{an}}(\mathbf{U}))$. But this subspace must be the whole space, since F is holonomic of rank $\text{rank}(I)$, by Proposition 2.3.6. $\square$

In the study of hypergeometric functions we shall encounter holonomic Frobenius ideals which depend on certain parameters β_i. We present two examples, both to illustrate Theorem 2.3.11, and to show how the holonomic rank of a parametric Frobenius ideal may depend on the parameter value.

Example 2.3.12. Let $n = 2$. Our first example is the Frobenius ideal

$$F_1 \;\;=\;\; D \cdot \{\, \theta_1(\theta_1 + \beta),\; \theta_1(\theta_2 + \beta),\; \theta_2(\theta_1 + \beta),\; \theta_2(\theta_2 + \beta) \,\}.$$

The rank of F_1 equals 2 for $\beta \neq 0$ and it equals 3 for $\beta = 0$. Next consider

$$F_2 \;\;=\;\; D \cdot \{\, \theta_1\theta_2 - 1,\; \theta_1 - \beta\theta_2 - \beta + 1 \,\}.$$

The rank of F_1 equals 2 for $\beta \neq 0$ and it equals 1 for $\beta = 0$. The reader is invited to write down a basis, as in the previous theorem, for the solution space $Hom_D\big(D/F_i, \mathcal{O}^{\mathrm{an}}(\mathbf{U})\big)$ for $i = 1, 2$.

The proof of Theorem 2.3.11 shows that the problem of finding the solution space of a holonomic Frobenius ideal is equivalent to solving a holonomic system of linear partial differential equations with constant coefficients. In the remainder of this section we show how to do this using Gröbner bases. Mora [71] refers to Proposition 2.3.13 as *Gröbner duality*; see also [83].

Given an Artinian ideal I in $\mathbf{k}[\partial_1, \ldots, \partial_n]$, the first step is to compute its primary decomposition over the algebraic closure of $\mathbf{k}$. Algorithms for computing primary decomposition can be found in [12]. We assume that this has been done, so that I is primary to some maximal ideal of the form $\langle \partial_1 - p_1, \ldots, \partial_n - p_n \rangle$. Translating the point $(p_1, \ldots, p_n)$ to the origin, we may thus assume that I is an Artinian ideal primary to $\langle \partial_1, \ldots, \partial_n \rangle$ in $\mathbf{k}[\partial_1, \ldots, \partial_n]$.

Fix any term order $\prec$ and compute the reduced Gröbner basis $\mathcal{G}$ of the ideal I with respect to $\prec$. Our goal is to compute a canonical $\mathbf{k}$-basis for the solution space $I^{\perp}$. Let $\mathcal{S}_{\prec}(I)$ be the (finite) set of monomials in $\mathbf{k}[x_1, \ldots, x_n]$ which are annihilated by $\mathrm{in}_{\prec}(I)$. These are the *standard monomials* of I with respect to $\prec$. Clearly, $\mathcal{S}_{\prec}(I)$ is a canonical $\mathbf{k}$-basis of $(\mathrm{in}_{\prec}(I))^{\perp}$. Let $\mathcal{N}_{\prec}(I)$ denote the set of monomials in $\mathbf{k}[x_1, \ldots, x_n] \backslash \mathcal{S}_{\prec}(I)$.

For every non-standard monomial ∂^{α} there is a unique polynomial

$$\partial^{\alpha} \;-\; \sum_{x^{\beta} \in \mathcal{S}_{\prec}(I)} c_{\alpha,\beta} \cdot \partial^{\beta} \qquad \text{in the ideal } I,$$

which is gotten by taking the normal form modulo $\mathcal{G}$. Here $c_{\alpha,\beta} \in \mathbf{k}$.

Abbreviate $\beta! := \beta_1!\beta_2! \cdots \beta_n!$. For a standard monomial x^{β}, define

$$f_{\beta}(x) \;\;=\;\; x^{\beta} + \sum_{x^{\alpha} \in \mathcal{N}_{\prec}(I)} c_{\alpha,\beta} \frac{\beta!}{\alpha!} x^{\alpha}. \qquad (2.23)$$

This sum is finite because I is $\langle \partial_1, \ldots, \partial_n \rangle$-primary, i.e., if $|\alpha| \gg 0$, then $\partial^{\alpha} \in I$ and hence $c_{\alpha,\beta} = 0$. We can also write it as a sum over all $\alpha \in \mathbf{N}^n$:

$$f_{\beta}(x) \;\;=\;\; \sum_{\alpha} c_{\alpha,\beta} \frac{\beta!}{\alpha!} x^{\alpha}.$$

Proposition 2.3.13. *The polynomials f_β, where x^β runs over the set $S_\prec(I)$ of standard monomials, forms a $\mathbf{k}$-basis for the orthogonal complement $I^\perp$.*

Proof. The polynomials f_β are $\mathbf{k}$-linearly independent. Therefore, it suffices to show $g(\partial) \bullet f_\beta(x) = 0$ for $g(\partial) = \sum_u C_u \partial^u \in I$.

$$g(\partial) \bullet f_\beta(x) = \sum_\alpha \sum_u c_{\alpha,\beta} C_u \frac{\beta!}{\alpha!} (\partial^u \bullet x^\alpha)$$

$$= \sum_\alpha \sum_{u \le \alpha} c_{\alpha,\beta} C_u \frac{\beta!}{(\alpha - u)!} x^{\alpha - u}$$

$$= \sum_v \left(\sum_u c_{u+v,\beta} C_u \frac{\beta!}{v!} \right) x^v \quad \text{where } v = \alpha - u$$

$$= \beta! \sum_v \frac{1}{v!} \left(\sum_u c_{u+v,\beta} C_u \right) x^v.$$

The expression $\sum_u c_{u+v,\beta} C_u$ is the coefficient of ∂^β in the $\prec$-normal form of $\partial^v g(\partial)$. It is zero since $\partial^v g(\partial) \in I$.   □

If I is homogeneous, then we can write

$$f_\beta \;=\; x^\beta \;+\; \sum_{x^\alpha \in \mathcal{N}_\prec(I)_d} c_{\alpha,\beta} \cdot \frac{\beta!}{\alpha!} \cdot x^\alpha \tag{2.24}$$

where the degree of x^β is d and $\mathcal{N}_\prec(I)_d$ denotes the degree d elements in the set $\mathcal{N}_\prec(I)$ of non-standard monomials.

We summarize our algorithm for solving Frobenius ideals.

Algorithm 2.3.14
Input: A Frobenius ideal I which is primary to the maximal ideal $\langle \theta_1, \ldots, \theta_n \rangle$.
Output: A basis for the solutions to I.

1. Compute the reduced Gröbner basis of I for a term order $\prec$.
2. Let $\mathcal{S}_\prec(I)$ be the set of standard monomials for I.
3. Output $f_\beta(\log x_1, \ldots, \log x_n)$ for f_β in (2.23), for all $\theta^\beta \in \mathcal{S}_\prec(I)$.

Algorithm 2.3.15
Input: A holonomic Frobenius ideal $I = \langle f_1(\theta), \ldots, f_m(\theta) \rangle$
Output: A basis for the solutions to I.

1. `Sols` $:= \{ \ \}$.
2. Find the common zeros of $f_1(x) = \cdots = f_m(x) = 0$ in $\mathbf{C}^n$. We denote the set of common zeros by $V(I)$.
3. Repeat the following steps for each $p \in V(I)$:
   a) Compute the multiplicity of the root p
   b) If the multiplicity is one, then `Sols` $:=$ `Sols` $\cup \{x^p\}$.

c) If the multiplicity is more than one, then find the primary component $Q_p(\theta - p)$ and call Algorithm 2.3.14 with input Q_p. Let $\{f_\beta(\log x) \mid \beta \in S\}$ be the output. Put

$$\text{Sols} := \text{Sols} \cup \{x^p f_\beta(\log x) \mid \beta \in S\}.$$

4. Output Sols.

This algorithm requires three procedures for the ring of polynomials: solving a system of algebraic equations, computing the local multiplicity, and computing the primary component. For the reader who is not familiar with these, we note three introductory references to these topics.

1. See, e.g., [27, pp.112–119] for an elementary introduction to solving systems of algebraic equations by Gröbner bases.
2. Computing the multiplicity of I at p: Change coordinates so that p is the origin. Compute a Gröbner basis of I with respect to $\prec_{-e}$, $-e = (-1, \ldots, -1)$ and count the number of standard monomials. The number gives the multiplicity. See, e.g., [28, p.169] for details.
3. As to algorithms to get a primary decomposition, see, e.g., [12]. A quick way of extracting primary components is the formula (2.21) above.

Example 2.3.16. Let us solve the Frobenius ideal I generated by

$$\theta_1 + \theta_2 + \theta_3 + \theta_4 + \theta_5, \theta_1 + \theta_2 - \theta_4, \theta_2 + \theta_3 - \theta_4, \theta_1\theta_3, \theta_2\theta_4.$$

This ideal is primary to the maximal ideal $\langle \theta_1, \ldots, \theta_5 \rangle \subset \mathbf{k}[\theta]$, so we can apply Algorithm 2.3.14 to solve it. The reduced Gröbner basis for the graded reverse lexicographic order is

$$\underline{\theta_1} + 2\theta_4 + \theta_5, \underline{\theta_2} - 3\theta_4 - \theta_5, \underline{\theta_3} + 2\theta_4 + \theta_5, \underline{\theta_4^2} - (1/8)\theta_5^2, \underline{\theta_4\theta_5} + (3/8)\theta_5^2, \underline{\theta_5^3}.$$

There are four standard monomials $\{1, \theta_4, \theta_5, \theta_5^2\}$. Let us, for example, find a solution of degree two. The monomial θ_5^2 is the only one standard monomial of degree two. We need to compute all the normal forms of $\theta_i\theta_j$ to find the solution. The normal form of θ_2^2 is 0, the normal form of $\theta_1\theta_2$ is $\theta_5^2/8$, and so on. From these normal form computations, we get the solution $f(\log x_1, \ldots, \log x_5)$ corresponding to θ_5^2 where

$$f(y_1, \ldots, y_5) = (1/8)(y_2 + y_4 - 2y_5)(2y_1 - y_2 + 2y_3 + y_4 - 4y_5).$$

Similarly we obtain the other solutions

$$1, \ \log(x_1 x_3)/(x_2 x_5), \ \log(x_2 x_4 / x_5^2).$$

The following special case deserves particular attention. A homogeneous Artinian ideal I is called *Gorenstein* if there is a homogeneous polynomial $V(x)$ such that $I = \{p \in \mathbf{k}[\partial] : p(\partial)V(x) = 0\}$. In this case $I^\perp$ consists precisely of all polynomials which are gotten by taking successive partial derivatives of $V(x)$. For example, the ideal I generated by the elementary symmetric polynomials is Gorenstein. Here $V(x) = \prod_{1 \le i < j \le n}(x_i - x_j)$, the *discriminant*, and $I^\perp$ is the space of *harmonic polynomials*.

Suppose we wish to decide whether or not a given Frobenius ideal I is Gorenstein. We first compute a Gröbner basis $\mathcal{G}$ of I with respect to some term order $\prec$. A necessary condition is that there exists a unique standard monomial x^β of maximum degree, say t. For every monomial x^α of degree t there exists a unique constant $c_\alpha \in \mathbf{k}$ such that $x^\alpha - c_\alpha \cdot x^\beta \in I$. We can find the c_α's by normal form reduction modulo $\mathcal{G}$. Define $V := \sum_{\alpha:|\alpha|=t}(c_\alpha/\alpha!) \cdot x^\alpha$, and let $\partial^* V$ be the $\mathbf{k}$-vector space spanned by the polynomials

$$\partial^u V = \sum_{\alpha:|\alpha|=t-|u|}(c_{\alpha+u}/\alpha!) \cdot x^\alpha, \qquad (2.25)$$

where ∂^u runs over all monomials of degree at most t.

Proposition 2.3.17. *The ideal I is Gorenstein if and only if $\partial^* V = I^\perp$ if and only if $\dim_\mathbf{k}(\partial^* V)$ equals the number of standard monomials.*

Proof. Clear from the definitions. □

The previous two propositions provide a practical method for solving linear systems with constant coefficients. We illustrate this in a small example.

Example 2.3.18. For $n = 5$ consider the homogeneous ideal

$$I = \langle \partial_1\partial_3, \partial_1\partial_4, \partial_2\partial_4, \partial_2\partial_5, \partial_3\partial_5, \partial_1 + \partial_2 - \partial_4, \partial_2 + \partial_3 - \partial_5 \rangle.$$

Let $\prec$ be any term order with $\partial_5 \prec \partial_4 \prec \partial_3 \prec \partial_2 \prec \partial_1$. The reduced Gröbner basis of I with respect to $\prec$ equals

$$\mathcal{G} = \{\underline{\partial_1} - \partial_3 - \partial_4 + \partial_5, \underline{\partial_2} + \partial_3 - \partial_5, \underline{\partial_3^2} + \partial_4\partial_5, \underline{\partial_3\partial_5}, \underline{\partial_4^2}, \underline{\partial_3\partial_4} - \partial_4\partial_5, \underline{\partial_5^2}\}.$$

The underlined monomials generate the initial ideal $\mathrm{in}_\prec(I)$. The space of polynomials annihilated by $\mathrm{in}_\prec(I)$ is spanned by the standard monomials

$$\mathcal{S}_\prec(I) = \{1, x_3, x_4, x_5, x_4x_5\}.$$

There exists a unique standard monomial of maximum degree $t = 2$, so it makes sense to check whether I is Gorenstein. For any quadratic monomial x_ix_j, the normal form of x_ix_j with respect to $\mathcal{G}$ equals $c_{ij} \cdot x_4x_5$ for some constant $c_{ij} \in \mathbf{k}$. We collect these constants in the quadratic form

$$V = \frac{1}{2} \sum_{i=1}^{5} c_{ii} x_i^2 + \sum_{1 \le i < j \le 5} c_{ij} x_i x_j$$

$$= \underline{x_4 x_5} + x_1 x_5 + x_3 x_4 + x_2 x_3 + x_1 x_2 - \frac{1}{2} x_3^2 - \frac{1}{2} x_2^2 - \frac{1}{2} x_1^2.$$

This polynomial is annihilated by I, and its initial monomial is annihilated by $\mathrm{in}_{\prec}(I)$. We next compute the **k**-vector space $\partial^* V$ of all partial derivatives of V. It turns out that this space is five-dimensional. Using Proposition 2.3.17 we conclude that I is Gorenstein and its solution space $I^\perp$ equals $\partial^* V$.

Gorenstein indicial ideals such as the one in Example 2.3.18 will play an important role in our study of resonant hypergeometric series in Section 3.6.

2.4 Regular Holonomic Systems

The notion of regular holonomic $\mathcal{D}$-modules and D-ideals is a generalization of ordinary differential equation with regular singularities on $\mathbf{P}^1$. The definition itself is complicated, as we shall see below, but the set of regular holonomic $\mathcal{D}$-modules is a class of $\mathcal{D}$-modules which behaves nicely under functorial operations ([17], [60], [69]). In addition, the classical solutions to a regular holonomic system also admit a nice description. For instance, for an ordinary differential equation with regular singularities on $\mathbf{P}^1$, the space of classical solutions at $x = c$ is locally spanned by functions that belong to

$$\mathbf{k}[[x - c]] \left[(x - c)^{\alpha^1}, \ldots, (x - c)^{\alpha^r}, \log(x - c) \right].$$

These formal solutions can be constructed by classical algorithms. As for software, try the MAPLE V (release 5) command DEtools[formal_sol], which constructs the formal solutions of a given ordinary differential equation at any singular point, including irregular singular points.

A similar description is known in the several variable case, along with constructive methods for finding series solutions, under the assumption that an explicit resolution of singularities is available for the singular locus of the given regular holonomic system. Majima's research monograph [66] studies series solutions of Pfaffian systems at a normally crossing singular point, including the irregular case. Especially, see, pp. 19–32 and pp. 49–110 in [66].

Although the existence of a resolution of singularities was proved by Hironaka in the 1960's, no practical construction algorithm is known in general. So, we will take a different approach to constructing series solutions for regular holonomic systems. We do not use resolution of singularities at all. We restrict our discussion to series expansions at certain "infinities". The geometric meaning of the "infinities" will be understood via the toric compactification given by the small Gröbner fan of the given D-ideal. From this we shall derive a practical algorithm for constructing series solutions at "infinities".

In this section we shall prove that the space of classical solutions of a regular holonomic system at such an "infinity" is locally spanned by products of power functions, logarithmic functions, and convergent power series.

Recall that an operator ℓ in the one-dimensional Weyl algebra D_1 is called *Fuchsian* if all singular points of ℓ on the Riemann sphere $\mathbf{P}^1 = \overline{\mathbf{C}}$ are regular singularities. In this case, the D-ideal $I = D \cdot \{\ell\}$ is called *regular holonomic*.

The higher-dimensional definition of *regular holonomicity* is complicated, and several equivalent definitions are known. The definition used in this book (Definition 2.4.1 below) is based on a reduction to the one-dimensional case. It can be found in [17, p. 302, Definition 11.3.(ii)], [54, p.99], and [69]. Let $\mathcal{D}_X = \mathcal{O}_X \langle \partial_1, \ldots, \partial_n \rangle$ denote the sheaf of algebraic differential operators on $X = \mathbf{C}^n$. Any element of the germ $\mathcal{D}_p$ at $x = p$ is written as $\sum \frac{c_\beta(x)}{d_\beta(x)} \partial^\beta$ where $c_\beta(x), d_\beta(x)$ are polynomials and $d_\beta(p) \neq 0$. We use calligraphic fonts such as $\mathcal{D}$ to denote sheaves on algebraic varieties with the Zariski topology.

Definition 2.4.1. Let I be a holonomic ideal in the Weyl algebra D. Let C be a smooth curve in $X = \mathbf{C}^n$ and $j : C \to \mathbf{C}^n$ an embedding. A holonomic $\mathcal{D}_X$-module $\mathcal{D}_X/\mathcal{D}_X I$ is called *regular holonomic* when $L^k j^*(\mathcal{D}_X/\mathcal{D}_X I)$ is regular holonomic on a smooth compactification $\overline{C}$ for any such curve C and for all $k = 0, -1, \ldots, -n + 1$. When $\mathcal{D}_X/\mathcal{D}_X I$ is regular holonomic, we call I regular holonomic. Here, $L^k j^*$ is the k-th derived functor of j^*, as defined later.

Before explaining the meaning of this definition, we present two propositions that are necessary to prove the main results in this section: Theorem 2.4.12 (local solutions at normally crossing singularities), Corollary 2.4.14 (a degree bound for log), and Corollary 2.4.16 (local solutions on a cone).

Proposition 2.4.2. *Let I be a regular holonomic D-ideal. Take any point $x = c = (c_1, \ldots, c_n) \in \mathbf{C}^n$ and let f be a holomorphic solution of I around $x = c$. Then, the function $f(c_1, \ldots, c_{i-1}, x_i, c_{i+1}, \ldots, c_n)$ in the variable x_i satisfies a Fuchsian ordinary differential equation.*

Assume that we are given an $n \times n$ matrix (v_{ij}) with integer entries satisfying $\det(v_{ij}) = \pm 1$. Consider the multiplicative change of coordinates

$$x_j = t_j(y) := y_1^{v_{1j}} \cdots y_n^{v_{nj}}, \quad j = 1, \ldots, n. \tag{2.26}$$

It induces a biregular isomorphism of the complex torus $(\mathbf{C}^*)^n$.

Proposition 2.4.3. *Let I be a regular holonomic D-ideal and assume that its singular locus is contained in the zero set of a polynomial $d(x)$. Let $f(x)$ be a classical solution of I. Then, there exists a regular holonomic ideal*

$$J \subset \mathbf{C}\langle y_1, \ldots, y_n, \partial_{y_1}, \ldots, \partial_{y_n} \rangle$$

such that $J \bullet f(t_1(y), \ldots, t_n(y)) = 0$ and

$$\mathrm{Sing}(J) \cap (\mathbf{C}^*)^n \subseteq \{y \in (\mathbf{C}^*)^n \mid d(t_1(y), \ldots, t_n(y)) = 0\}.$$

Let us explain the definition of regular holonomic systems and give proofs to the two propositions. Near the end of the proof of Proposition 2.4.3 and the proof sketch of Theorem 2.4.9 (the GKZ system for homogeneous A is regular holonomic), we assume that readers are familiar with sheaf theory, cohomology theory in algebraic geometry (e.g., Chapters 2 and 3 of Hartshorne's text book [47]) and $\mathcal{D}$-module functors such as inverse images (restrictions) and direct images (integrations) (e.g., pp.232–248 of Borel's text book [17] or Chapter 1 of Hotta-Tanisaki's text book [54]). The reader not familiar with these can skip the details. Once he or she accepts Propositions 2.4.2, 2.4.3 and Theorem 2.4.9, then what follows should be understandable again.

As we stated in the beginning of this book, the elements of the Weyl algebra D act on the polynomial ring $\mathbf{k}[x]$. Thus we regard D as the subalgebra of $\mathrm{End}_{\mathbf{k}}\,\mathbf{k}[x]$ generated by elements of $\mathbf{k}[x]$ as multiplication operators and $\partial_1, \ldots, \partial_n$. We can define the sheaf of differential operators $\mathcal{D}_Y$ on any smooth algebraic variety Y with this point of view; namely, $\mathcal{D}_Y$ is the sheaf of subalgebras of $\mathcal{E}nd_{\mathbf{C}}\,\mathcal{O}_Y$ generated by sections of $\mathcal{O}_Y$ and the sheaf of (algebraic) vector fields Θ_Y (see, e.g., [17, pp.207–208], [54, pp.168–171]).

The $\mathcal{D}_X$-module $\mathcal{D}_X/\mathcal{D}_X I$ also has the structure of an $\mathcal{O}_X$-module, which is not necessarily coherent. Let C be a smooth curve in X as in Definition 2.4.1. We consider the $\mathcal{O}_C$-module

$$j^*(\mathcal{D}_X/\mathcal{D}_X I) \;\;=\;\; \mathcal{O}_C \otimes_{j^{-1}\mathcal{O}_X} j^{-1}(\mathcal{D}_X/\mathcal{D}_X I),$$

which is called the inverse image under j of the $\mathcal{O}_X$-module $\mathcal{D}_X/\mathcal{D}_X I$ in algebraic geometry (see, e.g., [47, p.110]). The k-th left derived functor of j^* is denoted by $L^k j^*$. Note $L^0 j^* = j^*$. It is known that the inverse image $\mathcal{N} := L^k j^*(\mathcal{D}_X/\mathcal{D}_X I)$ admits a structure of a left coherent $\mathcal{D}_C$-module. Moreover, it is a holonomic $\mathcal{D}_C$-module. The inverse image is called the *restriction* in the theory of $\mathcal{D}$-modules (see, e.g. [17, pp.233–234], [54, p.23], [59]).

Consider the special case $C = \{(x_1, \ldots, x_n) \,|\, x_1 = \cdots = x_{n-1} = 0\}$. It is known that, for a D-ideal I, the restriction $j^*(\mathcal{D}_X/\mathcal{D}_X I)$ is isomorphic to $\mathcal{D}_C \otimes_{D_C} D/(I + x_1 D + \cdots + x_{n-1}D)$ where $D_C = \mathbf{C}\langle x_n, \partial_n \rangle$ (see, e.g., [80]). The left D_C-module $D/(I + x_1 D + \cdots + x_{n-1}D)$ is called the restriction (in the Weyl algebra) of I to C. The isomorphism assures that we may study the restriction only in the Weyl algebra. We will give an algorithm for computing the restriction in the Weyl algebra in Section 5.2.

Theorem 2.4.4. *Let $\mathcal{N}$ be a left holonomic $\mathcal{D}_C$-module. There exists a Zariski open set U in C such that $\mathcal{N}_{|U}$ is a free $\mathcal{O}_U$-module of finite rank.*

For a proof of this theorem, see, e.g., [17, p.275], [54, p.89]. In case of a cyclic module $\mathcal{N} = \mathcal{D}_C/\mathcal{D}_C I$, this theorem is a consequence of Theorem 1.4.22. The general case can be proved essentially in the same way.

Let $u_1, \ldots, u_r$ be a set of free generators for the stalk at z in U. Then, there exist regular functions a_{ij} around z such that $\partial u_i = \sum a_{ij} u_j$ in $\mathcal{N}$. Here, we denote by x a local coordinate around z and by ∂ a vector field such that $\partial \bullet x = 1$. Consider the solution sheaf of the left holonomic $\mathcal{D}_C$-module $\mathcal{N}$, denoted by $\mathcal{S} := \mathcal{H}om_{\mathcal{D}_U}(\mathcal{N}_{|_U}, \mathcal{O}_C^{\mathrm{an}})$, which is locally an r-dimensional $\mathbf{C}$-vector space. Here, $\mathcal{O}_C^{\mathrm{an}}$ is the sheaf of holomorphic functions on C. The correspondence between $\mathcal{S}$ and the solution sheaf is as follows. Consider the ordinary differential equations

$$
\frac{\partial}{\partial x}
\begin{pmatrix} f_1 \\ \cdot \\ \cdot \\ f_r \end{pmatrix}
=
(a_{ij})
\begin{pmatrix} f_1 \\ \cdot \\ \cdot \\ f_r \end{pmatrix}.
$$

For a solution $(f_1, \ldots, f_r)^T$, define an element φ of $\mathcal{S}$ by $\varphi(u_i) = f_i \in \mathcal{O}_C^{\mathrm{an}}$. This gives a one-to-one correspondence between elements of $\mathcal{S}$ and the solutions of the ordinary differential equation, because

$$
\partial \bullet \varphi(u_i) = \varphi(\partial u_i) = \varphi\left(\sum a_{ij} u_i\right) = \sum a_{ij} \varphi(u_i).
$$

Let $\overline{C}$ be a smooth complete curve that is a compactification of $U \subseteq C$. We denote by $j_0 : U \to \overline{C}$ the open embedding. Note that U can be regarded as a Zariski open set in $\overline{C}$ by j_0. For a point $z \in \overline{C}$, we denote by B_z a sufficiently small open ball with the center z in the topology of $\overline{C}$ as a complex manifold. We call $\mathcal{N}$ *regular holonomic* if for all $z \in \overline{C}$ and for all $\varphi \in \mathcal{S}$ on $B_z \cap U \backslash \{z\}$, all $\varphi(u_1), \ldots, \varphi(u_r)$ grow polynomially at z, i.e., the condition (2) in Theorem 1.4.18 is satisfied. It is known that the definition does not depend on the choice of U. The polynomial growth condition also does not depend on the choice of completion, since birational transformations send a polynomially growing function to a polynomially growing function.

For $n = 1$, a D-ideal I is regular holonomic if and only if there exists a Zariski open set U in $\mathbf{C}$ such that all solutions on U grow polynomially on $\mathbf{P}^1$. We present two simple examples.

Example 2.4.5. Put $P = \theta_x^2 - x(\theta_x + 1/2)^2$ and $U = \overline{C} \backslash \{0, 1, \infty\}$, $\overline{C} = \mathbf{P}^1$. Then, $\mathcal{N} = \mathcal{D}_U / \mathcal{D}_U \cdot \{P\}$ is a coherent $\mathcal{O}_U$-module and 1 and θ_x are free generators of $\mathcal{N}$ as a $\mathbf{C}[x, 1/x, 1/(1-x)]$-module. In fact, we have

$$
\theta_x \begin{pmatrix} 1 \\ \theta_x \end{pmatrix} = \begin{pmatrix} 0 & 1 \\ \frac{x}{4(1-x)} & \frac{x}{1-x} \end{pmatrix} \begin{pmatrix} 1 \\ \theta_x \end{pmatrix}.
$$

By applying Theorem 1.4.18 to the points $0, 1, \infty$ and showing all solutions grow polynomially at any point on $\overline{C}$, we conclude that $D \cdot \{P\}$ is regular holonomic.

Example 2.4.6. Let $V = \{v_1, \ldots, v_m\} \subset \mathbf{C}$. Then $I = D \cdot \{\prod_{i=1}^m (x - v_i)\}$ is regular holonomic, since $(\mathcal{D}/\mathcal{D}I)_{|_{\mathbf{C} \backslash V}} = 0$ and the solution sheaf is zero on $\mathbf{C} \backslash V$.

We wish to emphasize that Definition 2.4.1 requires the polynomial growth condition to hold for all curves $C \subseteq X$ and all derived functors $L^k j^* (\mathcal{D}/\mathcal{D}I)$.

Example 2.4.7. Put $I = D \cdot \{x_1, \partial_2 - 1\}$, $(n = 2)$. For any curve C which is transversal to $x_1 = 0$, the polynomial growth condition is satisfied, because $L^k j^*(\mathcal{D}/\mathcal{D}I) = 0$ on $x_1 \neq 0$. Consider $C = \{(x_1, x_2) \mid x_1 = 0\}$. Then, $L^0 j^*(\mathcal{D}/\mathcal{D}I) = 0$, but $\mathcal{N} := L^{-1} j^*(\mathcal{D}/\mathcal{D}I) = \mathcal{D}_C/\mathcal{D}_C(\partial_2 - 1)$. This can be shown, for example, by the restriction algorithm in Chapter 5 and its generalization to the derived category in [80] (see Example 5.2.10). The solution for $\mathcal{N}$ is e^{x_2} on C. We conclude that I is not regular holonomic.   ·

The set of regular holonomic $\mathcal{D}$-modules is stable under several fundamental operations. We list some stability facts that are needed for our proofs of Proposition 2.4.2 and 2.4.3 (see, e.g., [17, p.303, p.308, p.316], [54, p.99]).

Theorem 2.4.8.

(1) *The category of regular holonomic $\mathcal{D}$-modules is an Abelian category. In particular, submodules of regular holonomic $\mathcal{D}$-modules are also regular.*
(2) *Regular holonomicity is preserved under the inverse image (restriction) and direct image (integral) functors associated to morphisms among smooth algebraic varieties.*

Now, we are ready to prove Propositions 2.4.2 and 2.4.3.

Proof (of Proposition 2.4.2). We may assume $i = 1$. Consider the curve

$$C = \{(x_1, \ldots, x_n) \in \mathbf{C}^n \mid x_2, \ldots, x_n \text{ have fixed values } c_2, \ldots, c_n.\}$$

and the inclusion $j : C \to \mathbf{C}^n$. Then, we have $j^*(\mathcal{D}/\mathcal{D}I) \simeq \mathcal{D}_C \otimes_{D_1} D/(I + (x_2 - c_2)D + \cdots + (x_n - c_n)D)$, which is regular holonomic by the definition. Let f be a holomorphic solution of I around the point $x = c_1$. By Proposition 5.2.4, the function $f(x_1, c_2, \ldots, c_n)$ is annihilated by $J := (I + (x_2 - c_2)D + \cdots + (x_n - c_n)D) \cap \mathbf{C}\langle x_1, \partial_1 \rangle$. Since $\mathcal{D}_C/(\mathcal{D}_C J)$ is a $\mathcal{D}_C$-submodule of $\mathcal{D}_C \otimes_{D_1} D/(I + (x_2 - c_2)D + \cdots + (x_n - c_n)D)$, the D_1-ideal J is regular holonomic by Theorem 2.4.8. Hence all classical solutions are polynomially growing functions. This implies $\mathbf{C}(x_1)\langle \partial_1 \rangle \cdot J$ is generated by a Fuchsian ordinary differential operator, by Theorem 1.4.18.   □

Proof (of Proposition 2.4.3). The map $x_j = y^{v_j}$ in (2.26) is a biregular map from $T = (\mathbf{C}^*)^n$ to $T_X := \{x \in (\mathbf{C}^*)^n\}$. The inverse map is written as $y_j = x^{u_j} = x_1^{u_{1j}} \cdots x_n^{u_{nj}}$. The biregular map induces an isomorphism of the ring of differential operators on the complex torus, which is given by $\partial_i \mapsto y^{-v_i} \sum_{j=1}^n u_{ij} y_j \partial_{y_j}$, $x_i \mapsto y^{v_i}$. Let $\ell_1, \ldots, \ell_m$ be generators of I. By applying the induced map to ℓ_i and multiplying with a suitable power of y, we obtain an element $\tilde{\ell}_i \in D_Y := \mathbf{C}\langle y_1, \ldots, y_n, \partial_{y_1}, \ldots, \partial_{y_n} \rangle$. Let I' be the ideal generated by these $\tilde{\ell}_i$. Since $x_j = y^{v_j}$ is biregular on the torus and $\mathcal{D}_{T_X}/\mathcal{D}_{T_X} I$ is regular holonomic on T_X, the $\mathcal{D}_T$-module $\mathcal{D}_T/\mathcal{D}_T I'$ is regular holonomic on T. Let $\mathcal{M}'$ be the direct image of $\mathcal{D}_T/\mathcal{D}_T I'$ under the inclusion $j_0 : T \to Y := \{y \in \mathbf{C}^n\}$. From the definition of the direct image by j_0 and Theorem 2.4.8 (2), $\mathcal{M}'$ is regular holonomic on Y and is equal to $\mathcal{D}_Y \otimes_{D_Y} M'$ where

$$M' \quad := \quad \mathbf{C}[y] \left[\frac{1}{y_1}, \ldots, \frac{1}{y_n}\right] \otimes_{\mathbf{C}[y]} D_Y/D_Y I'$$

(see, e.g., [54, p.31], [82]). Note that M' can be constructed from I in an algorithmic way by [82]. Consider the cyclic D_Y-submodule $D_Y/D_Y J$ generated by $1 \otimes 1 \in M'$. It is also regular holonomic by Theorem 2.4.8 (1) and J annihilates the function $f(t_1(y), \ldots, t_n(y))$. Since $J \supseteq I'$, our J satisfies the conditions of the proposition. □

We have introduced the notion of regular holonomic systems and proved two properties. However, at present we know no algebraic algorithm for deciding whether a given D-ideal is regular holonomic. There are several equivalent definitions of regular holonomic system, and regular holonomic systems satisfy a lot of nice properties. We will utilize some of these in this book. In fact, GKZ-hypergeometric systems, one of our main topics, associated with projective toric varieties are regular holonomic. Here is the precise statement.

Theorem 2.4.9. (Hotta, [53, p. 27]) *Let $H_A(\beta)$ be the hypergeometric D-ideal defined in Definition 1.5.10. If the vector $(1, 1, \ldots, 1)$ is in the $\mathbf{Q}$-span of the row vectors of A, then $H_A(\beta)$ is regular holonomic for all $\beta \in \mathbf{k}^d$*

The above hypothesis on A, stated again in (3.3), is equivalent to saying that the toric ideal I_A is generated by homogeneous binomials (see [96, §4]).

Proof (sketch). The algebra automorphism $D \ni f \mapsto \hat{f} \in D$ defined by $\hat{x}_j = -\partial_j$ and $\hat{\partial}_j = x_j$ $(j = 1, \ldots, n)$ is called *Fourier transform* of D. Since $(1, 1, \ldots, 1)$ is in the row vector space of A, the D-module $D/H_A(\beta)$ is *monodromic*, i.e., the action of the *Euler operator* $\theta_1 + \theta_2 + \cdots + \theta_n$ on $D/H_A(\beta)$ is locally finite: for any $p \in D/H_A(\beta)$, the set $\{(\theta_1 + \cdots + \theta_n)^m p \mid m \in \mathbf{N}\}$ spans a finite dimensional vector space over $\mathbf{C}$. Brylinski [19, Theorem 7.24] showed that a monodromic D-module is regular holonomic if and only if its Fourier transform is regular holonomic. The Fourier transform $D/\hat{H}_A(\beta)$ admits a natural action of the d-dimensional algebraic torus via the matrix A. Its support is the union of finitely many orbits. Such a D-module is known to be regular holonomic ([17, Theorem VII 12.11], [53, Corollary in §5]). □

Here is a GKZ-hypergeometric system which is not regular holonomic.

Example 2.4.10. Let $n = 3, d = 2$ and $A = \begin{pmatrix} 1 & 1 & 0 \\ 0 & -1 & 1 \end{pmatrix}$. The D-ideal

$$H_A(\beta) \quad = \quad D \cdot \{\partial_1 - \partial_2\partial_3, \; x_1\partial_1 + x_2\partial_2 - \beta_1, \; -x_2\partial_2 + x_3\partial_3 - \beta_2\}$$

is holonomic of rank 2 for all $\beta_1, \beta_2 \in \mathbf{k}$. But it is not regular holonomic. The initial ideal $\text{in}_{(-w,w)}(H_A(\beta))$ for $w = (1, 0, 0)$ is holonomic of rank 1, which is a contradiction to Theorem 2.5.1.

This example is the first instance of the family of hypergeometric systems associated with classical root systems which were studied by Gel'fand, Graev and Postnikov [37]. Any of these GKZ-systems is not regular holonomic for a generic parameter, since the toric ideal defined by the positive roots is not homogeneous. In fact, the following converse to Theorem 2.4.9 holds.

Theorem 2.4.11. *Let $A \in \mathbf{Z}^{d \times n}$ be a rank d matrix and $\beta \in \mathbf{k}^d$ generic. Then, the D-ideal $H_A(\beta)$ is regular holonomic if and only if the toric ideal I_A is homogeneous with respect to the usual grading $\deg(x_i) = 1$, $i = 1, \ldots, n$.*

Our proof of this theorem requires combinatorial techniques from Chapter 3. It will be given at the end of Section 3.2.

Now let us discuss local solutions of regular holonomic D-ideals. Let I be a regular holonomic D-ideal. We assume that the singular locus of I is contained in an algebraic hypersurface, whose defining polynomial $d \in \mathbf{k}[x]$ can be calculated using Gröbner bases as in (1.32). By Theorem 1.4.19, we see that any local holomorphic solution of I around a point in the open set $\{d(x) \neq 0\}$ can be analytically continued to any other point in $\{d(x) \neq 0\}$.

Let us assume first that the singular hypersurface is a normal crossing divisor locally at the origin. Algebraically this means that the polynomial $d(x)$ reduces to a monomial in the local ring at the origin. In other words,

$$d(x) \quad = \quad x^a \cdot (1 + \sum_u c_u x^u) \tag{2.27}$$

where $c_u \in \mathbf{k}^*$ and each x^u is a monomial of positive degree. Under this assumption, there exists a nice series representation for all the solutions of I in a neighborhood of the origin. The following theorem is well-known to the experts; for instance, it is stated in [60, p.862].

Theorem 2.4.12. *If the hypothesis (2.27) holds, then there exist vectors $\alpha_1, \ldots, \alpha_m$ in $\mathbf{C}^n$ such that any multivalued holomorphic solution of I on $\{\prod_{i=1}^n x_i \neq 0\}$ near the origin has a series expression in the ring*

$$\mathbf{C}[[x_1, \ldots, x_n]][x^{\alpha_1}, \ldots, x^{\alpha_m}, \log(x_1), \ldots, \log(x_n)]$$

These series converge around the origin.

To prove the theorem we need the following lemma.

Lemma 2.4.13. *Let $f(x)$ be a holomorphic function on the product of punctured disks $D' := D'_{\varepsilon_1} \times \cdots \times D'_{\varepsilon_n}$, $D'_{\varepsilon_i} := \{x_i \in \mathbf{C} \mid 0 < |x_i| < \varepsilon_i\}$. If $f(x)$ has a pole of finite order as a function of x_i for fixed values $x_1, \ldots, x_{i-1}, x_{i+1}, \ldots, x_n$ for all i, then $f(x)$ is meromorphic at $x = 0$.*

Proof. Fix a point $(x_2, \ldots, x_n)$ in $D'_{\varepsilon_2} \times \cdots \times D'_{\varepsilon_n}$ and consider the Laurent expansion of f in terms of x_1:

$$f(x) = \sum_{p \in \mathbf{Z}} f_p(x_2, \ldots, x_n) x_1^p.$$

From the assumption, $f(x)$ has a pole of finite order at x_1. However, the order may depend on $x_2, \ldots, x_n$. We will prove that the order is bounded. Consider the following set

$$U_q = \{(x_2, \ldots, x_n) \in D'_{\varepsilon_2} \times \cdots \times D'_{\varepsilon_n} \mid f_p = 0, \ \text{for all } p \le q\}.$$

The set U_q is closed in the classical topology and $\cup_{q \in \mathbb{Z}} U_q = D'_{\varepsilon_2} \times \cdots \times D'_{\varepsilon_n}$. By Baire's category theorem, there exists q_1 such that U_{q_1} contains an open ball in the classical topology. Then, $x_1^{q_1} f(x)$ is holomorphic in $D_{\varepsilon_1} \times U_{q_1}$ where D_{ε_1} is the open disk with radius ε_1. Since $D_{\varepsilon_1} \times U_{q_1}$ is open, $x_1^{q_1} f(x)$ can be extended to the holomorphic function $F(x_1, \ldots, x_n) :=$ $\frac{1}{2\pi\sqrt{-1}} \int_C \frac{\xi_1^{q_1} f(\xi_1, x_2, \ldots, x_n)}{\xi_1 - x_1} d\xi_1$ on $D_{\varepsilon_1} \times D'_{\varepsilon_2} \times \cdots \times D'_{\varepsilon_n}$. Here, C is a small circle with the center at the origin. By the same discussions for $x_2, \ldots, x_n$, the function $x_1^{q_1} \cdots x_n^{q_n} f(x)$ can be extended to a holomorphic function outside the codimension two set $\cup_{i \ne j}\{x \mid x_i = x_j = 0\}$. Since any codimension two singularity can be removed (see, e.g., [45, p.22]), we conclude that $x^q f(x)$ is holomorphic at $x = 0$. □

Proof (of Theorem 2.4.12). Let $\{f_1, \ldots, f_r\}$ be a basis of holomorphic solutions on a simply connected domain outside the singular locus near the origin. Suppose that the basis of R/RI as $\mathbf{C}(x)$-vector space is $\{1, \partial^{\alpha(1)}, \ldots, \partial^{\alpha(r-1)}\}$. Put

$$\Phi(x) \quad = \quad \begin{pmatrix} f_1 & \cdots & f_r \\ \partial^{\alpha(1)} f_1 & \cdots & \partial^{\alpha(1)} f_r \\ \vdots & \vdots & \vdots \\ \partial^{\alpha(r-1)} f_1 & \cdots & \partial^{\alpha(r-1)} f_r \end{pmatrix}.$$

Then, $\det(\Phi) \not\equiv 0$ and Φ is a (matrix valued) multi-valued holomorphic function on $D' = D'_{\varepsilon_1} \times \cdots \times D'_{\varepsilon_n}$ where $D'_{\varepsilon_i} = \{x_i \in \mathbf{C} \mid 0 < |x_i| < \varepsilon_i\}$. The fundamental group $\pi_1(D') = \pi_1(D'_{\varepsilon_1}) \times \cdots \times \pi_1(D'_{\varepsilon_n})$ is a free Abelian group generated by n elements γ_i which encircle $x_i = 0$. Consider the analytic continuation $\gamma_i^* \Phi$ of Φ along the path γ_i. Since the first row of $\gamma_i^* \Phi$ is again a basis of solutions, there exists a matrix M_i, which is called the *monodromy matrix*, satisfying $\gamma_i^* \Phi = \Phi M_i$. Since $\pi_1(D')$ is Abelian, the matrices M_i commute with one another and are invertible. Then, there exists a commuting family of matrices L_i such that $e^{2\pi\sqrt{-1}L_i} = M_i$. Now, consider

$$s(x) := \Phi(x) x_1^{-L_1} \cdots x_n^{-L_n}.$$

The monodromy of $\Phi(x)$ is killed by $x_1^{-L_1} \cdots x_n^{-L_n}$, $(\gamma_i^* x_i^{-L_i} = M_i^{-1} x_i^{-L_i})$. Hence $s(x)$ is a single valued function on the product of punctured disks D'.

We prove that $s(x)$ is a meromorphic function with poles on $x_1 \cdots x_n = 0$. Consider the curve

$$C = \{(x_1, \ldots, x_n) \in \mathbf{C}^n \mid x_2, \ldots, x_n \text{ have fixed values.}\}$$

Applying Proposition 2.4.2, we see that each component of $\Phi(x)$ satisfies a Fuchsian ordinary differential equation when it is restricted to C. Hence it grows polynomially when x_1 approaches 0 in C. Therefore, each component of $s(x)$ has polynomial growth and then $s(x)$ has a pole of finite order at x_1. The same holds for all other x_i. By applying Lemma 2.4.13, we conclude that there exists q such that $x^q s(x)$ is holomorphic at $x = 0$. Note that the first row of $s(x)x_1^{L_1} \cdots x_n^{L_n}$ gives a basis of solutions of I.   □

Corollary 2.4.14. *In the series representation of Theorem 2.4.12, each holomorphic solution to I is a polynomial in $\log(x_i)$ of degree at most* $\operatorname{rank}(I) - 1$.

Proof. The series $x_i^{L_i} = \exp(\log(x_i) \cdot L_i)$ is a polynomial in $\log(x_i)$ whose degree is less than the size of L_i. The same holds for $s(x)x_1^{L_1} \cdots x_n^{L_n}$.   □

Example 2.4.15. We consider the system of two differential operators

$$P_1 = \theta_x^2 - x(\theta_x + \theta_y + a)(\theta_x + b),$$
$$P_2 = \theta_y^2 - y(\theta_x + \theta_y + a)(\theta_y + b'),$$

where a, b, b' are parameters in $\mathbf{k}$. These are the classical differential equations for Appell's function F_2. They form a Gröbner basis with respect to $(-w, w)$ for any positive vector $w \in \mathbf{R}_+^2$. The underlined leading forms show that the dimension of the solution space is 4. The singular locus satisfies the hypothesis (2.27); it is given by $d(x, y) = xy(1 - x)(1 - y)(1 - x - y)$.

We will derive series solutions. Let e_1 and e_2 be indeterminate exponents and consider the following series with indeterminate coefficients:

$$f = \sum_{(i,j) \in \mathbf{N}^2} c_{ij} \cdot x^{e_1 + i} y^{e_2 + j}.$$

The differential equations $P_1 f = P_2 f = 0$ translate into linear recurrence relations for the coefficients c_{ij}:

$$(e_1 + i + 1)^2 c_{i+1,j} - (e_1 + i + e_2 + j + a)(e_1 + i + b)c_{i,j} = 0,$$
$$(e_2 + j + 1)^2 c_{i,j+1} - (e_1 + i + e_2 + j + a)(e_2 + j + b')c_{i,j} = 0.$$

In this case they can be solved explicitly:

$$c_{mn} = c_{00} \frac{(a + e_1 + e_2)_{m+n}(b + e_1)_m(b' + e_2)_n}{(e_1 + 1)_m^2 \cdot (e_2 + 1)_n^2}.$$

The resulting series f is a function of (e_1, e_2) which is holomorphic at $e_1 = 0$ and $e_2 = 0$. We choose $c_{00} = 1$. Then, we have

$$P_1 f = e_1^2 \cdot \sum_{j=0}^{\infty} c_{0j} x^{e_1} y^{e_2 + j} \quad \text{and} \quad P_2 f = e_2^2 \cdot \sum_{i=0}^{\infty} c_{i0} x^{e_1 + i} y^{e_2}.$$

We can force $P_1 f$ to vanish by two operations: either set $e_1 = 0$, or apply $\partial/\partial e_1$ and then set $e_1 = 0$. Likewise we can force $P_2 f$ to vanish by two operations: either set $e_2 = 0$, or apply $\partial/\partial e_2$ and then set $e_2 = 0$. These operations commute with P_1 and P_2 and with each other. By applying all four possible combinations to f, we obtain four linearly independent logarithmic series which are annihilated by P_1 and P_2. For example, we get the solution

$$\lim_{e_1 \to 0, e_2 \to 0} \frac{\partial^2 f}{\partial e_1 \partial e_2} = (\log x)(\log y)\left[1 + \frac{ab}{1^2}x + \frac{ab'}{1^2}y + \cdots\right]$$
$$+ (\log y)x(ab - 2a - 2b) + (\log x)y(ab' - 2a - 2b') + \cdots.$$

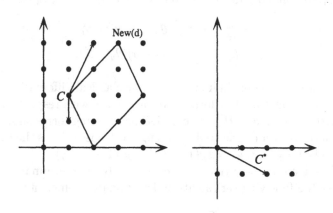

Fig. 2.4. The Newton polytope of d, a unimodular cone C and the polar cone

Our next goal is to remove the hypothesis (2.27) in Theorem 2.4.12. Consider an arbitrary regular holonomic D-ideal I. Let $d(x) = \sum_a c_a x^a$ be the defining polynomial of a singular hypersurface that contains $\mathrm{Sing}(I)$. We denote by $\mathrm{New}(d)$ the *Newton polytope* of $d(x)$. Thus $\mathrm{New}(d)$ is the convex hull in $\mathbf{R}^n$ of all points a with $c_a \neq 0$. We take a vertex q of $\mathrm{New}(d)$ and a unimodular cone C which has its vertex at q and contains $\mathrm{New}(d)$ (see Figure 2.4). An n-dimensional cone C is called *unimodular* if there exist n-vectors $u^1, \ldots, u^n \in \mathbf{Z}^n$ such that these n-vectors generate C, i.e., $C = \mathbf{R}_+ u^1 + \cdots + \mathbf{R}_+ u^n$, and the determinant of $(u^1, \ldots, u^n)$ is ± 1. Here $\mathbf{R}_+$ denotes the nonnegative real numbers. The $n \times n$-matrix $U = (u^1, \ldots, u^n)$ is invertible over $\mathbf{Z}$. Let $V = (v_{ij})$ be the inverse matrix of U. The row vectors $(v_{i1}, \ldots, v_{in})$ of V span the *polar cone* C^*, which consists of all vectors $w \in \mathbf{R}^n$ such that $w \cdot u^i \geq 0$ for $i = 1, \ldots, n$. Note that

$$\mathrm{in}_{-w}(d) \ = \ x^q \quad \text{for all vectors } w \text{ in the interior of } C^*. \tag{2.28}$$

We now apply the following multiplicative change of coordinates:

$$x_j \ = \ t_j(y) \ = \ y_1^{v_{1j}} y_2^{v_{2j}} \cdots y_n^{v_{nj}} \quad \text{for } j = 1, \ldots, n. \tag{2.29}$$

This transforms the defining equation of the singular locus as follows:

$$d(x_1, \ldots, x_n) \ = \ y^{Vq} \cdot \left(1 + \sum_u c_u y^u\right), \tag{2.30}$$

where $c_u \in \mathbf{k}^*$ and each appearing $u \in \mathbf{Z}^n$ is non-negative and non-zero. Let J be the new regular holonomic D-ideal constructed in Proposition 2.4.3. From (2.30) we see that the singular locus of the new system J is a normal crossing divisor around the origin.

By applying Theorem 2.4.12 to J, we obtain the following corollary.

Corollary 2.4.16. *The regular holonomic D-ideal I has a fundamental set of solutions on $0 < |x^{u^i}| \ll 1$ each of which is represented by a series in*

$$N \ = \ \mathbf{C}[[x^{u^1}, \ldots, x^{u^n}]][x^{\beta_1}, \ldots, x^{\beta_p}, \log(x_1), \ldots, \log(x_n)] \tag{2.31}$$

where $\beta_1, \ldots, \beta_p$ are suitable vectors in $\mathbf{C}^n$.

The condition $0 < |x^{u^i}| \ll 1$ means that

$$u^i \cdot (-\log|x_1|, \ldots, -\log|x_n|) \gg 0.$$

Therefore, each series in N which is a formal solution to I converges when $(-\log|x_1|, \ldots, -\log|x_n|)$ lies in a suitable translate of the polar cone C^*, in other words, when $(-\log|x_1|, \ldots, -\log|x_n|)$ lies in $p + C^*$ for some point p, the formal solutions converge. Note that C^* is a subcone of the normal cone of $\mathrm{New}(d)$ at the vertex q.

Example 2.4.17. Let I be the GKZ system associated to $\Delta_1 \times \Delta_2$ introduced in Theorem 1.5.2 ($m = 3$). It is easy to see that

$$\mathrm{in}_{(0,e)}(I) \subseteq \langle z_{11}\xi_{11} + z_{21}\xi_{21}, z_{12}\xi_{12} + z_{22}\xi_{22}, z_{13}\xi_{13} + z_{23}\xi_{23},$$
$$z_{11}\xi_{11} + z_{12}\xi_{12} + z_{13}\xi_{13}, z_{21}\xi_{21} + z_{22}\xi_{22} + z_{23}\xi_{23},$$
$$\xi_{11}\xi_{22} - \xi_{12}\xi_{21}, \xi_{12}\xi_{23} - \xi_{13}\xi_{22}, \xi_{11}\xi_{23} - \xi_{13}\xi_{21}\rangle.$$

In Theorem 4.3.8, we will prove that the right hand side agrees with the left hand side in this situation. By applying the saturation method (1.32) to get the singular locus, we can see that $\mathrm{Sing}(I) = \langle d(z)\rangle$, where

$$d(z) = \left(\prod z_{ij}\right)(z_{11}z_{22} - z_{12}z_{21})(z_{12}z_{23} - z_{13}z_{22})(z_{11}z_{23} - z_{13}z_{21}).$$

The Newton polytope of $d(z)$ is a hexagon which lies in $\mathbf{R}^6$. If we set $x = z_{12}z_{21}/z_{11}z_{22}$, $y = z_{13}z_{21}/z_{11}z_{23}$, then $d(z) = ($ a monomial in $z_{ij}) \cdot (1-x)(x-y)(1-y)$. See Figure 2.5 for the Newton polytope and polar cones. The series $\Psi^{(i)}$ ($i = 1, 2, 3$) in Theorem 1.5.7 converge when $(-\log|x|, -\log|y|)$ lies in a suitable translate of a polar cone C_j^* in Figure 2.5.

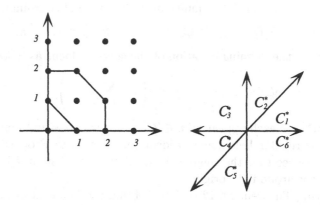

Fig. 2.5. The Newton polytope and polar cones

2.5 Canonical Series Solutions

In this section we prove that the rank of a holonomic D-ideal is preserved under Gröbner deformations if that D-ideal is regular holonomic, by introducing the notion of canonical series solutions and the Nilsson ring.

Theorem 2.5.1. *Let I be a regular holonomic D-ideal and w any weight vector in $\mathbf{R}^n$. Then*

$$\operatorname{rank}(I) = \operatorname{rank}(\operatorname{in}_{(-w,w)}(I)). \tag{2.32}$$

In view of Theorem 2.2.1 and Lemma 2.1.6 (2), it suffices to prove $\operatorname{rank}(I) \leq \operatorname{rank}(\operatorname{in}_{(-w,w)}(I))$ for a generic weight vector w. Our proof of this inequality is based on the representation (2.31) of the solutions to a regular holonomic system by formal power series with logarithmic terms.

We derive some basic facts about the set N in (2.31). First, it is not closed under differentiation in general, so we consider the ring $N[1/x] := N[1/x_1, \ldots, 1/x_n]$ instead of N. Let $d(x)$ be the defining polynomial of a singular hypersurface that contains $\operatorname{Sing}(I)$. Take a vertex q of $\operatorname{New}(d)$ and a unimodular cone C with basis $\{u^1, \ldots, u^n\}$ of $\mathbf{Z}^n$ so that w lies in the interior of the polar cone C^*, and that $C + q \supset \operatorname{New}(d)$. Then $w \cdot u^i > 0$ for $i = 1, \ldots, n$. We define a *monomial* in $N[1/x]$ to be an element of the form $x^a \log(x)^b = x_1^{a_1} x_2^{a_2} \cdots x_n^{a_n} \cdot \log(x_1)^{b_1} \cdots \log(x_n)^{b_n}$, where $b \in \mathbf{N}^n$ and a lies in the additive submonoid M of $\mathbf{C}^n$ which is generated by the $2n + p$ vectors $u^1, \ldots, u^n, \beta_1, \ldots, \beta_p, -e_1, \ldots, -e_n$ where $e_i = (0, \ldots, 0, 1, 0, \ldots, 0)$. Any element $f \in N[1/x]$ can be uniquely written as a formal sum of monomials. We easily see that the unique expression has the following property.

Proposition 2.5.2. *Take* $f \in N[1/x]$ *and let* $\sum c_{ab} x^a (\log x)^b$ *be the expression of* f *as a formal sum of monomials in* $N[1/x]$. *For any number* k, *the number of the non-zero coefficients* $c_{ab} \neq 0$ *such that* $\mathrm{Re}\,(w \cdot a) \leq k$ *is finite.*

Here $\mathrm{Re}(w \cdot a)$ denotes the real part of the complex number $w \cdot a$. We call the real number $\mathrm{Re}(w \cdot a)$ the *w-weight* (or *w-degree*) of the monomial $x^a \log(x)^b$. The proposition above states that the ring structure on $N[1/x]$ is well-defined. This fact is not trivial. For example, the product

$$\left(\sum_{i=-\infty}^{\infty} a_i x^i \right) \left(\sum_{j=-\infty}^{\infty} b_j x^j \right) = \sum_{k=-\infty}^{\infty} \left(\sum_{i+j=k} a_i b_j \right) x^k,$$

among elements in $\mathbf{k}[[x, 1/x]]$ is not well-defined, because $\sum a_i b_j$ does not converge in general. In case of $N[1/x]$, the product is well-defined since each coefficient of the product is a finite sum of products of coefficients of each factor by Proposition 2.5.2.

Proposition 2.5.3. *For any* k, *the monomials satisfying* $\mathrm{Re}\,(w \cdot a) \leq k$ *form a* $\mathbf{C}$-*vector space basis for* $N[1/x]/N_k$ *where*

$$N_k \quad := \quad \left\{ \sum_{\mathrm{Re}\,(w \cdot a) > k} c_{ab} x^a (\log x)^b \in N[1/x] \right\}.$$

A *partial term order* on N is a partial order $\preceq$ on the set of monomials such that $x^a \log(x)^b \preceq x^c \log(x)^d$ implies $x^{a+e} \log(x)^{b+f} \preceq x^{c+e} \log(x)^{d+f}$ for all $a, c, e \in M$ and $b, d, f \in \mathbf{N}^n$. A *term order* is a partial term order which is a total order on the set of monomials such that $1 \prec \log x_i$. For instance, the restriction of the *lexicographic order* on $\mathbf{C}^n \oplus \mathbf{N}^n \simeq \mathbf{R}^{2n} \oplus \mathbf{N}^n$ to $M \oplus \mathbf{N}^n$ defines a term order on N. We similarly consider partial term orders on $N[1/x]$.

The weight vector $w \in \mathbf{R}^n$ defines a partial term order $\leq$ on N as follows:

$$x^a \log(x)^b \leq x^c \log(x)^d \quad \Longleftrightarrow \quad \mathrm{Re}(w \cdot a) \leq \mathrm{Re}(w \cdot c). \qquad (2.33)$$

This partial term order on N induces an order in $N[1/x]$ by (2.33).

For any element $f \in N$ we define the *initial series* $\mathrm{in}_w(f)$ as follows.

Definition 2.5.4. Let $\min_w(f)$ denote the minimum w-weight of any monomial occurring in the monomial expansion $f = \sum_{a,b} c_{ab} x^a \log(x)^b$, which exists and is unique by Proposition 2.5.3. The minimum $\min_w(f)$ is a well-defined real number since $w \cdot u^i > 0$ for $i = 1, \ldots, n$. The *initial series* $\mathrm{in}_w(f)$ is the subsum of all terms $c_{ab} x^a \log(x)^b$ having w-weight equal to $\min_w(f)$. This subsum is finite by Proposition 2.5.2.

Theorem 2.5.5. *If* $f \in N$ *is a solution to* I *then* $\mathrm{in}_w(f)$ *is a solution to* $\mathrm{in}_{(-w,w)}(I)$.

Proof. If $\mathcal{U}$ is an open subset of $\mathbf{C}^n$ on which the series $f = \sum_{a,b} c_{ab} x^a \log(x)^b$ converges and represents a holomorphic function, then also the initial series $\mathrm{in}_w(f)$ converges and represents a holomorphic function on $\mathcal{U}$. It therefore suffices to assume that f is a formal solution to I and to show that $\mathrm{in}_w(f)$ is a formal solution to $\mathrm{in}_{(-w,w)}(I)$.

The result of applying a monomial $x^u \partial^v$ in D to a monomial $x^a \log(x)^b$ in $N[1/x]$ equals $x^{a+u-v} \cdot g(\log(x))$, where $g(\log(x))$ denotes a polynomial in $\log(x_1), \ldots, \log(x_n)$ which has degree $\leq b_i$ in $\log(x_i)$. Thus the expression $x^u \cdot \frac{\partial^v}{\partial x^v}(x^a \log(x)^b)$ is 0 or a finite sum of monomials in $N[1/x]$ whose w-degree is the w-degree of $x^a \log(x)^b$ minus the w-degree of $x^u \partial^v$.

Let $p = \sum_{(u,v) \in \tau} \gamma_{u,v} x^u \partial^v$ be any element of I. Here τ is a finite subset of $\mathbf{N}^{2n}$. Let τ' be the subset of τ such that $\mathrm{in}_{(-w,w)}(p) = \sum_{(u,v) \in \tau'} \gamma_{u,v} x^u \partial^v$, and set $\max_w(p) := w \cdot (v - u)$ for $(u,v) \in \tau'$. Since f is a formal solution of I, it is annihilated by p:

$$0 \;=\; p \bullet f \;=\; \sum_{(u,v) \in \tau} \sum_{(a,b)} \gamma_{u,v} \cdot c_{ab} \cdot x^u \cdot \frac{\partial^v}{\partial x^v}(x^a \log(x)^b).$$

Every summand in this double sum has w-degree at least $\min_w(f) - \max_w(p)$. The subsum of summands having w-degree precisely $\min_w(f) - \max_w(p)$ equals the result of applying the initial form $\mathrm{in}_{(-w,w)}(p)$ to the initial series $\mathrm{in}_w(f)$. This subsum must be zero, by Proposition 2.5.3. We conclude that $\mathrm{in}_w(f)$ is annihilated by $\mathrm{in}_{(-w,w)}(p)$ and hence by $\mathrm{in}_{(-w,w)}(I)$.   □

Let $\prec_w$ be the refinement of the partial term order (2.33) by the lexicographic term order. Every element g of N has a unique *initial monomial* $\mathrm{in}_{\prec_w}(g)$ with respect to the (total) term order $\prec_w$. The initial monomial satisfies

$$\mathrm{in}_{\prec_w}(g) \;=\; \mathrm{in}_{\mathrm{lex}}(\mathrm{in}_w(g)) \;=\; \mathrm{in}_{\prec_w}(\mathrm{in}_w(g)). \qquad (2.34)$$

The following lemma is a consequence of the definitions and Proposition 2.5.3.

Lemma 2.5.6. *Let* $g_1, \ldots, g_m \in N$.

(1) *If the initial monomials* $\mathrm{in}_{\prec_w}(g_1), \ldots, \mathrm{in}_{\prec_w}(g_m)$ *are distinct, then the initial series* $\mathrm{in}_w(g_1), \ldots, \mathrm{in}_w(g_m)$ *are* **C**-*linearly independent.*

(2) *If* $\mathrm{in}_w(g_1), \ldots, \mathrm{in}_w(g_m)$ *are* **C**-*linearly independent then* $g_1, \ldots, g_m$ *are* **C**-*linearly independent.*

Naturally, the converses to statements (1) and (2) are not true. However, they are true after taking linear combinations.

Proposition 2.5.7. *Let* $g_1, \ldots, g_m$ *be* **C**-*linearly independent elements of* N. *Then there exists an* $m \times m$-*matrix* $A = (a_{ij})$ *over* **C** *such that, for the new elements* $h_i = \sum_{j=1}^m a_{ij} g_i$, *we have that the initial series* $\mathrm{in}_w(h_1), \ldots, \mathrm{in}_w(h_m)$ *are* **C**-*linearly independent.*

Proof. By Lemma 2.5.6 (1), it suffices to prove that $\text{in}_{\prec_w}(h_1), \ldots, \text{in}_{\prec_w}(h_m)$ are distinct. We show this by induction on m. The case $m = 1$ is trivial. For the inductive step $m - 1 \to m$, we assume that $\text{in}_{\prec_w}(h_1), \ldots, \text{in}_{\prec_w}(h_{m-1})$ are distinct and sorted in increasing order with respect to $\prec_w$. Suppose $\text{in}_{\prec_w}(g_m) = \text{in}_{\prec_w}(h_i)$ for some $i \in \{1, \ldots, m-1\}$. Replace g_m by $g_m - c \cdot h_i$, so that initial monomials cancel and the new initial monomial is higher than $\text{in}_{\prec_w}(h_i)$. Repeating this process at most $m - 1$ times, we arrive at a new series g_m which is non-zero (by the linear independence hypothesis) and satisfies $\text{in}_{\prec_w}(g_m) \neq \text{in}_{\prec_w}(h_i)$ for all $i \in \{1, \ldots, m-1\}$. Now set $h_m = g_m$ and we are done. $\qquad\square$

We next prove the theorem stated in the beginning of Section 2.5.

Proof (of Theorem 2.5.1). Let $r = \text{rank}(I)$. By Corollary 2.4.16, there exists a fundamental set $\{f_1, f_2, \ldots, f_r\}$ of solutions to I on an open ball $\mathcal{U} \subset \mathbf{C}^n$ which can be represented by series in N. The initial series $\text{in}_w(f_1), \ldots, \text{in}_w(f_r)$ represent holomorphic solutions on $\mathcal{U}$ to $\text{in}_{(-w,w)}(I)$, by Theorem 2.5.5. By applying Proposition 2.5.7, we may further assume that $\text{in}_w(f_1), \ldots, \text{in}_w(f_r)$ are $\mathbf{C}$-linearly independent. Therefore the rank of the holonomic D-ideal $\text{in}_{(-w,w)}(I)$ is greater than or equal to r. Theorem 2.2.1 implies that the rank of $\text{in}_{(-w,w)}(I)$ equals $r = \text{rank}(I)$. $\qquad\square$

One surprisingly useful consequence of Theorem 2.5.1 is the following algorithm for computing the holonomic rank of a regular holonomic D-ideal.

Algorithm 2.5.8 (Holonomic rank of a regular holonomic D-ideal I)

1. Take any weight vector w and compute a Gröbner basis G of I with respect to $(-w, w)$ (see Algorithm 1.2.6).
2. Evaluate the rank of $D \cdot \{\text{in}_{(-w,w)}(G)\}$ by, e.g., Algorithm 1.4.17.

We next explain how the complex vectors β_i in the definition of N can be computed from I. Let w be a generic weight vector in $\mathbf{R}^n$ satisfying $u^i \cdot w > 0$ for all i. The commutative polynomial ideal

$$\text{ind}_w(I) \quad := \quad \mathbf{k}(x_1, \ldots, x_n)\langle \partial_1, \ldots, \partial_n \rangle \cdot \text{in}_{(-w,w)}(I) \cap \mathbf{k}[\theta_1, \ldots, \theta_n]$$

is called the *indicial ideal* of I with respect to w. In case of $n = 1$ and $I = D \cdot \{p\}$, the indicial ideal $\text{ind}_w(I)$ with $w = 1$ is a principal ideal and is generated by the indicial polynomial of p at $x = 0$. The indicial ideal is a Frobenius ideal, hence all results in Section 2.3 can be applied to it. The generators of $\text{ind}_w(I)$ are gotten from those of the initial ideal $\text{in}_{(-w,w)}(I)$ by the process of *distraction* as in Theorem 2.3.4. It follows from Theorem 2.3.4 that a holomorphic function is annihilated by $\text{in}_{(-w,w)}(I)$ if and only if it is annihilated by $\text{ind}_w(I)$. Theorem 2.5.1 and Proposition 2.3.6 imply

$$\text{rank}(I) \quad = \quad \text{rank}\big(\text{in}_{(-w,w)}(I)\big) \quad = \quad \text{rank}\big(\text{ind}_w(I)\big).$$

The zeros of the indicial ideal $\text{ind}_w(I)$ in affine n-space $\mathbf{C}^n$ are called the *exponents* of I with respect to w. Thus the rank of the regular holonomic D-ideal I equals the number of exponents counted with multiplicity.

As before let $\prec_w$ be the refinement of the partial term order (2.33) on N by the lexicographic term order. Consider the following finite set of monomials

$$\text{Start}_{\prec_w}(I) := \{\, \text{in}_{\prec_w}(f) : f \in N \text{ is a non-zero solution of } I \,\}.$$

Note that $\text{in}_{\prec_w}(f) = x^a \log(x)^b$ for some $a \in \mathbf{C}^n$ and $b \in \mathbf{N}^n$. We call $\text{Start}_{\prec_w}(I)$ the set of *starting monomials* of I with respect to $\prec_w$.

Lemma 2.5.9. *The number of starting monomials equals* $\text{rank}(I)$.

Proof. By applying Proposition 2.5.7 as in the proof of Theorem 2.5.1, we know that there exists a basis of solutions to the regular holonomic system consisting of elements in N with distinct $\prec_w$-initial monomials. That basis has cardinality $\text{rank}(I)$ and hence so does $\text{Start}_{\prec_w}(I)$. □

Lemma 2.5.10. *The starting monomials are given by the indicial ideal:*

$$\text{Start}_{\prec_w}(I) \;=\; \text{Start}_{\prec_w}\big(\text{in}_{(-w,w)}(I)\big) \;=\; \text{Start}_{\prec_w}\big(\text{ind}_w(I)\big). \quad (2.35)$$

Proof. The first equality in (2.35) follows from (2.34). The second equality holds because $\text{in}_{(-w,w)}(I)$ and $\text{ind}_w(I)$ have the same solutions in N. □

Corollary 2.5.11. *If* $x^a \log(x)^b \in \text{Start}_{\prec_w}(I)$ *then* a *is an exponent of* I *with respect to* w. *For each exponent* a, *the number of starting monomials of the form* $x^a \log(x)^b$ *is the multiplicity of* a *as a root of the indicial ideal* $\text{ind}_w(I)$.

Proof. This follows from (2.35) and the explicit description of solutions to Frobenius ideals given in Theorem 2.3.11. □

We have the following uniqueness result for logarithmic series solutions.

Theorem 2.5.12. *For each* $x^a \log(x)^b \in \text{Start}_{\prec_w}(I)$ *there is a unique element* $f \in N$ *with the following properties:*

(1) f *is annihilated by* I;
(2) $\text{in}_{\prec_w}(f) = x^a \log(x)^b$;
(3) *No starting monomial other than* $x^a \log(x)^b$ *appears in* f *with non-zero coefficient.*

Proof. Let $r = \text{rank}(I)$. The existence of the desired series is proved as follows. Starting with any solution basis for I in N, we first apply the algorithm in the proof of Proposition 2.5.7 to get series $h_1, \ldots, h_r \in N$ such that $\text{in}_{\prec_w}(h_1), \ldots, \text{in}_{\prec_w}(h_r)$ are distinct and increasing in $\prec_w$-order. These r monomials lie in $\text{Start}_{\prec_w}(I)$, and hence they are precisely the elements of

Start$_{\prec_w}(I)$. Let B be the complex $r \times r$ matrix whose (i,j)-entry is the co-efficient of in$_{\prec_w}(h_j)$ in the expansion of h_i. This matrix is upper triangular with 1's on the diagonal and is hence invertible. Transforming the solution basis $(h_1, \ldots, h_r)$ with the inverse of B, we get a solution basis $(f_1, \ldots, f_r)$ with in$_{\prec_w}(f_j) = $ in$_{\prec_w}(h_j)$ and whose r elements satisfy (3).

The uniqueness part of Theorem 2.5.12 is clear: if f and f' are two distinct elements of N satisfying (1), (2), (3), then $f - f'$ is a non-zero solution of I but the $\prec_w$-lowest term of $f - f'$ is not in Start$_{\prec_w}(I)$, a contradiction. $\square$

Example 2.5.13. Consider the following two operators for $n = 2$:

$$P_1 = \underline{\theta_x^2(\theta_x - 2)} - x(\theta_x + \theta_y + a)(\theta_x + b_1)(\theta_x + b_2),$$
$$P_2 = \underline{\theta_y^2(\theta_y - 3)} - y(\theta_x + \theta_y + a)(\theta_y + b'_1)(\theta_y + b'_2)$$

where a, b_i, b'_i are parameters. For any $w \in \mathbf{R}_+^2$, the set Start$_{\prec_w}(I)$ of starting monomials for the regular holonomic system $I = D \cdot \{P_1, P_2\}$ equals

$$\{ 1, \log(x), x^2, \log(y), \log(x)\log(y), x^2\log(y), y^3, y^3\log(x), x^2y^3 \}.$$

This can be seen from the underlined terms since $\{P_1, P_2\}$ is a Gröbner basis. For each of these nine monomials there is a unique series solution to I in $\mathbf{C}[[x, y]][\log(x), \log(y)]$ starting with that monomial and containing none of the other eight.

The series solutions to I described in Theorem 2.5.12 are called the *canonical (series) solutions* for I with respect to $\prec_w$. We close this section by examining the monomials which can appear in the canonical solutions. We shall see that we can take the vectors $\beta_1, \ldots, \beta_p$ in (2.31) to be the exponents, and the vectors $u^1, \ldots, u^n$ can be chosen directly from a Gröbner basis for I.

The *Gröbner cone* of I containing the vector w equals

$$C_w(I) = \{ w' \in \mathbf{R}^n : \text{in}_{(-w,w)}(I) = \text{in}_{(-w',w')}(I) \}.$$

This is a union of open convex polyhedral cones in $\mathbf{R}^n$, since w was assumed to be generic. Let $C_w(I)^*$ be the polar cone of $C_w(I)$. This is a closed convex polyhedral cone in $\mathbf{R}^n$. A vector u lies in $C_w(I)^*$ if and only if in$_{(-w,w)}(I) = $ in$_{(-w',w')}(I)$ implies $u \cdot w' \geq 0$. Since $C_w(I)$ is n-dimensional, the polar cone $C_w(I)^*$ is *strongly convex*, i.e., $C_w(I)^*$ contains no non-zero linear subspace. Consider the monoid consisting of all integer vectors

$$C_w(I)_{\mathbf{Z}}^* := C_w(I)^* \cap \mathbf{Z}^n.$$

We write $\mathbf{C}[[C_w(I)_{\mathbf{Z}}^*]]$ for the ring of formal power series $f = \sum_u c_u x^u$ where $u \in C_w(I)_{\mathbf{Z}}^*$ and $c_u \in \mathbf{C}^*$. Every non-constant term $c_u x^u$ appearing in f satisfies $w \cdot u > 0$. Hence in$_w(f)$ is well-defined for all $f \in \mathbf{C}[[C_w(I)_{\mathbf{Z}}^*]]$.

Theorem 2.5.14. *The canonical solutions to I with respect to $\prec_w$ have the form $x^a \cdot p$ where a is an exponent and p is an element of the ring*

$$\mathbf{C}[[\,C_w(I)^*_{\mathbf{Z}}\,]][\log(x_1),\ldots,\log(x_n)].$$

The degree of the logarithmic series p with respect to $\log(x_i)$ is at most rank$(I) - 1$.

Proof. Let $f \in N$ be any solution to I. Consider the set of all exponent vectors $c \in \mathbf{C}^n$ which appear in the terms $x^c \log(x)^d$ of the expansion of f. From the definition of the ring N in (2.31) it is clear that these exponents c lie in finitely many (say, ν) congruence classes of $\mathbf{C}^n$ modulo its additive subgroup $\mathbf{Z}^n$. Let $f = f_1 + \cdots + f_\nu$ be the corresponding decomposition into subsums, so that the exponent vectors c of any two monomials in f_i differ by an integral vector. Since the operators in D preserve these congruence classes, it follows each of $f_1,\ldots,f_\nu$ is a solution to I. In particular, the original series f contains ν distinct starting monomials in$_{\prec_w}(f_i)$ in its monomial expansion.

Suppose now that f is a canonical solution, and let in$_{\prec_w}(f) = x^a \log(x)^b$. By Corollary 2.5.11, the vector a is an exponent of I with respect to w. By Theorem 2.5.12 (3) and the discussion in the previous paragraph, we have $\nu = 1$. In other words, f has the form $x^a \cdot \sum_u p_u(\log(x_1),\ldots,\log(x_n))\,x^u$, where u runs only over integer vectors and each p_u is a polynomial. The degree of p_u with respect to $\log(x_i)$ is at most rank$(I) - 1$ by Corollary 2.4.14.

Consider any non-zero vector $u \in \mathbf{Z}^n$ which satisfies $p_u \neq 0$. Choose $v \in \mathbf{N}^n$ such that $\log(x)^v$ appears in the expansion of $p_u(\log(x_1),\ldots,\log(x_n))$. Then $x^{a+u}\log(x)^v$ appears in the expansion of f. It is not a starting monomial of I with respect to $\prec_w$. By (2.35) applied to w, it is not a starting monomial of in$_{(-w,w)}(I)$ with respect to the lexicographic term order.

Let w' be any vector in the Gröbner cone $C_w(I)$. Then $x^{a+u}\log(x)^v$ is not a starting monomial of in$_{(-w',w')}(I) = $ in$_{(-w,w)}(I)$ with respect to the lexicographic term order. By applying (2.35) to w' in the reverse direction, we see that $x^{a+u}\log(x)^v$ is not a starting monomial of I with respect to $\prec_{w'}$. Therefore in$_{\prec_{w'}}(f) = $ in$_{\prec_w}(f) = x^a \log(x)^b$, and this implies $w' \cdot u \geq 0$. We conclude that u lies in the polar cone $C_w(I)^*$ to the Gröbner cone. $\square$

For practical purposes one can replace the cone $C_w(I)^*$ by a possibly larger cone, which is still strongly convex, but easier to compute. Let $\mathcal{G}$ be any Gröbner basis of I with respect to $(-w, w)$. Let $C_w(\mathcal{G})$ denote the set of all vectors w' such that in$_{(-w',w')}(g) = $ in$_{(-w,w)}(g)$ for $g \in \mathcal{G}$. Clearly, $C_w(\mathcal{G})$ is an open convex polyhedral cone contained in $C_w(I)$. Therefore its dual $C_w(\mathcal{G})^*$ is a strongly convex closed polyhedral cone containing $C_w(I)^*$. Consider its monoid algebra $\mathbf{C}[[\,C_w(\mathcal{G})^* \cap \mathbf{Z}^n\,]]$. A finite set of algebra generators for this monoid algebra can be calculated directly from the Gröbner basis $\mathcal{G}$.

Definition 2.5.15. Let I be a regular holonomic D-ideal and w a generic weight vector. The ring

$$N_w(I) := \mathbf{C}[[C_w(I)^*_{\mathbf{Z}}]][x^{e^1}, \ldots, x^{e^r}, \log x_1, \ldots, \log x_n] \qquad (2.36)$$

is called the *Nilsson ring* with respect to the weight vector w. Here the set $\{e^1, \ldots, e^r\} \subset \mathbf{C}^n$ is the set of the roots of the indicial ideal $\mathrm{ind}_w(I)$.

Our choice of the name "Nilsson ring" refers to the work of Nilsson [73] in the 1960's. It is customary in complex algebraic geometry to use the term "functions of the Nilsson class" for functions which can be represented by series as above.

The following theorem summarizes our discussion in this section. The convergence is derived from the convergence of the series in Theorem 2.4.12.

Theorem 2.5.16. *Let I be a regular holonomic D-ideal and w a generic weight vector in $\mathbf{R}^n$. Then, there exist $\mathrm{rank}(I)$ many canonical series solutions of I, which lie in the Nilsson ring $N_w(I)$. There exists a point $p \in C_w(I)$ such that the canonical series converge for $x = (x_1, \ldots, x_n) \in \mathbf{C}^n$ satisfying*

$$(-\log|x_1|, \ldots, -\log|x_n|) \in p + C_w(I).$$

Proof. Let $C := C_w(I)$ be the Gröbner cone that contains w. For any given positive number M, there exist a cone $\tilde{C} \subset C$ and a constant p_0 such that

$$p_1\beta_1 + \cdots + p_n\beta_n > |\beta|M \qquad (2.37)$$

holds for all $\beta \in C^*$ and $p \subset \tilde{C}' := \{y \in \tilde{C} \mid |y| \geq p_0\}$. Here $|y|$ abbreviates the Euclidean norm of the real vector y. We shall find a number M such that the series converge on the open set

$$\{(x_1, \ldots, x_n) \in \mathbf{C}^n \mid (-\log|x_1|, \ldots, -\log|x_n|) \in \tilde{C}' + C_w(I)\}.$$

The canonical series solutions are finite sums of monomials in x^{e^i} and $\log x_j$ with coefficients of series of the form $\sum_{\beta \in C^* \cap \mathbf{Z}^n} a_\beta x^\beta$. By Corollary 2.4.16 and the procedure for constructing canonical series solutions, there exist vectors $u^1, \ldots, u^n \in \mathbf{Z}^n$ such that the matrix $(u^1, \ldots, u^n)$ is invertible and $\{x^\beta \mid \beta \in C^* \cap \mathbf{Z}^n\} \subseteq \{(x^{u^1})^{\alpha_1} \cdots (x^{u^n})^{\alpha_n} \mid \alpha \in \mathbf{N}^n\}$. Moreover, the series $\sum_{\beta \in C^* \cap \mathbf{Z}^n} a_\beta x^\beta$ converge when $|x^{u^1}|, \ldots, |x^{u^n}| \ll 1$.

For $\beta \in C^* \cap \mathbf{Z}^n$, there exists the unique $\alpha \in \mathbf{N}^n$ such that $\beta = \alpha_1 u^1 + \cdots + \alpha_n u^n$. Since the matrix $(u^1, \cdots, u^n)$ is invertible and all norms in $\mathbf{R}^n$ is equivalent, there exist positive constants M_1 and M_2 such that $M_1|\alpha| \leq |\beta| \leq M_2|\alpha|$. Hence, there exist positive constants M_3 and $r \geq 1$ such that

$$|a_\beta| \leq M_3 r^{|\beta|},$$

because Cauchy's inequality $|a_{\beta(\alpha)}| \leq M_3' r^{|\alpha|}$ follows from the convergence of the series $\sum a_{\beta(\alpha)} (x^{u^1})^{\alpha_1} \cdots (x^{u^n})^{\alpha_n}$. Here M_3' is a constant. Therefore, the series converge for x satisfying $|x_1|^{\beta_1} \cdots |x_n|^{\beta_n} < \left(\frac{1}{r}\right)^{|\beta|}$ for all $\beta \in C^* \cap \mathbf{Z}^n$. This condition is equivalent to the condition that

$$-\beta_1 \log|x_1| - \cdots - \beta_n \log|x_n| > |\beta| \log r$$

holds for all $\beta \in C^* \cap \mathbf{Z}^n$. The conclusion follows by putting $M := \log r$ and applying (2.37). In fact, $(-\log|x_1|, \ldots, -\log|x_n|) \in p + C$, $p \in \tilde{C}'$ implies $(-\log|x_1| - p_1, \ldots, -\log|x_n| - p_n) \cdot (\beta_1, \ldots, \beta_n) \geq 0$. Then, we have $-\beta_1 \log|x_1| - \cdots - \beta_n \log|x_n| \geq p_1\beta_1 + \cdots + p_n\beta_n > |\beta| \log r$ by (2.37).   □

We close this section with a bibliographic note. For the special case $w = e$ and assuming that the differences of exponents are not integral vectors, Theorem 2.5.1 is a consequence of Kashiwara's result in [61, Theorem 1.1]. Our proof of Theorem 2.5.1 in general is independent of [61] and more elementary.

2.6 Construction of Series Solutions

A system of linear differential equations yields a system of linear difference equations for the coefficients in the series solutions. In order for the difference equations to be compatible, one needs a Gröbner basis for the differential equations. This idea is very old; for instance, it appears in the work of C. Riquier and M. Janet in the 1910's. Their algorithmic methods have been further developed by several authors (see, e.g., [87]) to construct series solutions at a non-singular point for a system of differential equations that is not necessarily linear. We are interested in series solutions to a linear system at certain singular points. That singularity is one reason why we are working in the Weyl algebra D rather than in the ring of differential operators R.

This section gives an algorithm for constructing series solutions to regular holonomic systems. We reconstruct canonical series solutions from solutions to the indicial ideal. Our method is based on the existence and uniqueness of the canonical series solutions, proved in Section 2.5. We determine the coefficients of a canonical series solution with a given starting monomial by solving linear equations over $\mathbf{k}$. These equations are triangularized when the given differential operators form a Gröbner basis with respect to $(-w, w)$.

Theorem 2.6.1. Let $w \in \mathbf{R}^n$ be a generic weight vector and I a regular holonomic D-ideal in $\mathbf{Q}\langle x_1, \ldots, x_n, \partial_1, \ldots, \partial_n \rangle$, given by a Gröbner basis with respect to $(-w, w)$. There exists an algorithm which computes all terms up to a specified w-weight in the canonical series solutions to I with respect to $\prec_w$.

Proof (and algorithm). We first determine the exponents, which are the zeros of the indicial ideal $\mathrm{ind}_w(I)$ in $\mathbf{C}^n$. The coordinates of each exponent A lie in an algebraic extension $\mathbf{k}$ of $\mathbf{Q}$. The extension field $\mathbf{k}$ is a computable field, that is, the field-theoretic operators $+$, $-$, $\times$, $/$ are computable. We will show how to compute the coefficients of the canonical series solution with starting monomial $x^A (\log x)^B$. This is done inductively with respect to the weight w.

For any $p \in \mathbf{Z}^n$, we define the following finite dimensional k-vector space:

$$L_p \quad := \quad x^A \cdot \sum_{0 \le b_i < \text{rank}(I)} \mathbf{k} \cdot x^p (\log x)^b.$$

The action by any torus-fixed operator $f(\theta) \in \mathbf{k}[\theta]$ induces a $\mathbf{k}$-linear map

$$f(\theta) \; : \; L_p \longrightarrow L_p,$$

because $\theta_i \bullet x^{A+p}(\log x)^b = (A_i + p_i)x^{A+p}(\log x)^b + b_i x^{A+p}(\log x)^{b-e_i}$. The monomials in L_p form a $\mathbf{k}$-basis, which is naturally ordered by the chosen term order $\prec_w$ on the Nilsson ring. If we represent the endomorphism $f(\theta)$ of L_p by a square matrix over $\mathbf{k}$ with rows and columns indexed by monomials $x^{A+p}(\log x)^b$ in the order $\prec_w$, then that square matrix is upper triangular.

Let L_p' denote the $\mathbf{k}$-linear subspace of L_p which is spanned by all monomials in L_p which do not lie in the set of starting monomials $\text{Start}_{\prec_w}(I)$. Note that $L_p = L_p'$ for all but at most $\text{rank}(I)$ many integer vectors p.

Let $\{f_1(\theta), \ldots, f_d(\theta)\}$ be any generating set of the indicial ideal $\text{ind}_w(I)$. We restrict the $\mathbf{k}$-linear map $f_i(\theta) : L_p \to L_p$ to the subspace L_p', which amounts to deleting some of the columns in its upper triangular matrix. By vertically concatenating the resulting matrices for $i = 1, 2, \ldots, d$, we obtain the matrix which represents the following $\mathbf{k}$-linear map in the monomial basis:

$$L_p' \to (L_p)^d, \quad v \mapsto (f_1(\theta) \bullet v, \cdots, f_d(\theta) \bullet v). \tag{2.38}$$

It follows from the identity (2.35) that the $\mathbf{k}$-linear map (2.38) is injective.

Let $G = \{g_1, \ldots, g_d\}$ be the given Gröbner basis of I with respect to $(-w, w)$. The Gröbner cone with respect to G is denoted by $C \subset \mathbf{R}^n$. For each element g_i of G, we choose a Laurent monomial $x^{\alpha(i)}$ such that

$$x^{\alpha(i)} g_i = f_i(\theta) - h_i, \quad h_i \in \mathbf{k}\langle x_1^{\pm}, \ldots, x_n^{\pm}, \partial_1, \ldots, \partial_n \rangle \text{ and } \text{ord}_{(-w,w)}(h_i) < 0.$$

Then $f_1, \ldots, f_d$ generate $\text{ind}_w(I)$, and the above maps (2.38) are injective.

We compute the coefficients of the canonical series solution by induction on the w-order k. We start from a canonical solution $x^A (\log x)^B + \cdots$ of the indicial ideal $\text{ind}_w(I)$. Assume that the coefficients c_{pb} of the monomials in L_p' are already known for $0 \le p \cdot w \le k$, $p \in C_{\mathbf{Z}}^*$. The partial solution equals

$$F_k(x) \quad = \quad x^A \cdot \sum_{\substack{0 \le p \cdot w \le k, p \in C_{\mathbf{Z}}^* \\ 0 \le b_j < \text{rank}(I)}} c_{pb} \, x^p (\log x)^b$$

Introduce the infinite-dimensional space $M_k = \sum_{p \cdot w > k, p \in C_{\mathbf{Z}}^*} L_p$. We have

$$(f_i(\theta) - h_i) \bullet F_k(x) = 0, \quad \text{mod } M_k \quad \text{for } i = 1, 2, \ldots, d,$$

and $(f_i(\theta) - h_i) \bullet F_k(x) \mod M_{k+1}$ can be regarded as an element of the finite-dimensional $\mathbf{k}$-vector space $\sum_{p \cdot w = k+1, p \in C_{\mathbf{Z}}^*} L_p$. We will determine the coefficients of monomials in $\sum_{p \cdot w = k+1, p \in C_{\mathbf{Z}}^{\bullet}} L_p'$. Since $\operatorname{ord}_{(-w,w)}(h_i) < 0$, the coefficients are determined by solving the following system of linear equations

$$(f_1(\theta), \cdots, f_d(\theta)) \bullet E(x) = (h_1 - f_1(\theta), \ldots, h_d - f_d(\theta)) \bullet F_k(x) \quad \operatorname{mod} (M_{k+1})^d,$$

where $E(x)$ is an indeterminate element in the space $\sum_{p \cdot w = k+1, p \in C_{\mathbf{Z}}^{\bullet}} L_p'$. Since (2.38) is injective and the existence of the canonical series solution holds (Theorem 2.5.14), there exists a unique solution $E(x)$ in $\sum_{p \cdot w = k+1, p \in C_{\mathbf{Z}}^{\bullet}} L_p'$.
$\square$

Remark 2.6.2. The algorithm in the above proof amounts to solving a k-linear system of equations for each w-degree k. That linear system takes the following form: we are given a vector which lies in the image of the injective linear map (2.38), and our task is to find the unique preimage. Here it is a significant advantage to use the fact that the map (2.38) is a concatenation of upper triangular matrices, one for each generator $f_i(\theta)$ of the indicial ideal. For instance, if $d = 2$ then the matrix representing (2.38) might look like

$$\begin{pmatrix} a & * & * & * \\ 0 & b & * & * \\ 0 & 0 & c & * \\ 0 & 0 & 0 & * \\ 0 & 0 & 0 & d \\ A & * & * & * \\ 0 & * & * & * \\ 0 & B & * & * \\ 0 & 0 & C & * \\ 0 & 0 & 0 & D \end{pmatrix}$$

Here $A \neq 0$ or $a \neq 0$ etc... It is easy to solve such a linear system.

In the remainder of this section we present two examples.

Example 2.6.3. Consider the second order ordinary differential operator

$$\theta(\theta - 3) - x(\theta + a)(\theta + b)$$

Here a, b are parameters and $w = 1$. The starting monomials are $1 = x^0$ and x^3. We first compute the canonical series starting from $F_0(x) = x^3$. We have

$$L_p = L_p' = x^3(kx^p + kx^p \log x) \qquad \text{for all } p \geq 1.$$

By solving the linear equation

$$\theta(\theta-3) \bullet (a_{k+1}x^{k+4} + b_{k+1}x^{k+4} \log x) = x(\theta+a)(\theta+b) \bullet (a_k x^{k+3} + b_k x^{k+3} \log x)$$

inductively, we find $b_k = 0$ and

$$a_k = \frac{(a+3)_k (b+3)_k}{(4)_k (1)_k}.$$

Hence the canonical series solution starting from x^3 equals

$$x^3 \sum_{k=0}^{\infty} \frac{(a+3)_k (b+3)_k}{(4)_k (1)_k} x^k.$$

Next we compute the canonical series starting from $1 = x^0$. Clearly,

$$L_p = \mathbf{k} x^p + \mathbf{k} x^p \log x \quad \text{for all } p \geq 1.$$

Up to the second level everything is as before, and the partial solution equals

$$F_2(x) \quad = \quad 1 + \frac{ab}{1 \cdot (-2)} x + \frac{a(a+1)b(b+1)}{1 \cdot 2 \cdot (-2) \cdot (-1)} x^2.$$

However, since x^3 is a starting monomial, L_3' is spanned by $x^3 \log x$. Thus $L_3' \neq L_3$. The injective $\mathbf{k}$-linear map $L_3' \longrightarrow L_3$ in (2.38) is given by

$$\theta(\theta - 3) \bullet x^3 \log x \quad = \quad 3x^3.$$

The coefficient c of $x^3 \log(x)$ in our canonical series satisfies the equation

$$3c \quad = \quad (a+2)(b+2) \frac{a(a+1)b(b+1)}{1 \cdot 2 \cdot (-2) \cdot (-1)}.$$

We divide by 3 to solve for c. The partial solution at level 3 equals

$$F_3(x) = 1 + \frac{ab}{1 \cdot (-2)} x + \frac{a(a+1)b(b+1)}{1 \cdot 2 \cdot (-2) \cdot (-1)} x^2 + c \cdot x^3 \log x.$$

Can you find the coefficients of the monomials x^4 and $x^4 \log x$ in $F_4(x)$?

Example 2.6.4. We consider the GKZ hypergeometric system associated with

$$A = \begin{pmatrix} 1 & 1 & 1 & 1 & 1 \\ 1 & 1 & 0 & -1 & 0 \\ 0 & 1 & 1 & -1 & 0 \end{pmatrix} \quad \text{and} \quad \beta = \begin{pmatrix} 1 \\ 0 \\ 0 \end{pmatrix}.$$

Let $w = (1, 1, 1, 1, 0)$. A Gröbner basis with respect to $(-w, w)$ consists of

$$(\theta_1 + \theta_2 + \theta_3 + \theta_4 + \theta_5 - 1) \bullet f = 0,$$
$$(\theta_1 + \theta_2 \quad\quad - \theta_4) \bullet f = 0,$$
$$(\quad\quad \theta_2 + \theta_3 - \theta_4) \bullet f = 0,$$
$$(\partial_1 \partial_3 - \partial_2 \partial_5) \bullet f = 0, \quad (\partial_2 \partial_4 - \partial_5^2) \bullet f = 0.$$

We wish to construct the canonical series solutions f to these equations. The indicial ideal for the weights w is generated by five torus-invariant operators:

$$\langle \theta_1 + \theta_2 + \theta_3 + \theta_4 + \theta_5 - 1, \ \theta_1 + \theta_2 - \theta_4, \ \theta_2 + \theta_3 - \theta_4, \ \theta_1 \theta_3, \ \theta_2 \theta_4 \rangle.$$

For this example, this can be checked by applying methods in Sections 4.3 and 4.6 without computing a Gröbner basis in D. Note that generators of I of which initial forms generate the indicial ideal are enough to determine the coefficients of canonical series solutions. Let us find solutions of this indicial ideal. It is primary. The unique root is $(0,0,0,0,1)$. It has multiplicity 4. Algorithm 2.3.14 yields the following solution basis for the indicial ideal:

$$\left\{ x_5, \ x_5 \log(x^q), \ x_5 \log(x^p), \ x_5 \log(x^q)(2\log(x^p) + \log(x^q)) \right\}.$$

Here we abbreviate $p = (1, -1, 1, 0, -1)$ and $q = (0, 1, 0, 1, -2)$. These two vectors come from the operators $\partial_1 \partial_3 - \partial_2 \partial_5$ and $\partial_2 \partial_4 - \partial_5^2$ above. The dual Gröbner cone C^* is the two-dimensional cone in $\mathbf{R}^5$ spanned by p and q.

We wish to extend the four solutions of the indicial ideal to canonical series solutions of the GKZ system. The coefficient of a monomial

$$x_5 x^a (\log x)^b, \quad a \in C^* \cap \mathbf{Z}^5 \tag{2.39}$$

is gotten by solving $\mathbf{Q}$-linear equations, inductively with respect to $a \cdot w$. If we use the general method of Theorem 2.6.1, then (for fixed a) there are $(4-1)^5 = 243$ possible choices for b. The matrix for (2.38) is much too big.

For hypergeometric systems, our general method can be significantly improved by exploiting the homogeneity expressed by the Euler operators, here

$$\theta_1 + \theta_2 + \theta_3 + \theta_4 + \theta_5 - 1, \quad \theta_1 + \theta_2 - \theta_4, \quad \theta_2 + \theta_3 - \theta_4. \tag{2.40}$$

The remaining two operators in the input equations can be replaced by

$$\theta_1 \theta_3 - \frac{x_1 x_3}{x_2 x_5} \theta_2 \theta_5 \ = \ \theta_1 \theta_3 - x^p \theta_2 \theta_5, \tag{2.41}$$

$$\theta_2 \theta_4 - \frac{x_2 x_4}{x_5^2} \theta_5 (\theta_5 - 1) \ = \ \theta_2 \theta_4 - x^q \theta_5 (\theta_5 - 1). \tag{2.42}$$

We show that the solutions can be expressed in terms of these x^p and x^q.

Lemma 2.6.5. *The four canonical series solutions to* $\{$ *(2.40), (2.41), (2.42)*$\}$ *can be written as* $x_5 f(x^p, x^q)$ *where* $f(y_1, y_2)$ *satisfies a regular holonomic D-ideal of rank 4 with the singular locus* $y_1 y_2 = 0$ *near* $y_1 = y_2 = 0$. *Moreover, if we write* C^* *for the cone spanned by* p *and* q, *then the four solutions lie in*

$$x_5 \mathbf{C}[[C_{\mathbf{Z}}^*]][\log x^p, \log x^q]$$

Proof. We consider the following isomorphism of the complex torus $(\mathbf{C}^*)^5$:

$$y_1 = g_1(x) = x^p = \frac{x_1 x_3}{x_2 x_5},$$

$$y_2 = g_2(x) = x^q = \frac{x_2 x_4}{x_5^2},$$

$$y_3 = g_3(x) = x_3, \ y_4 = g_4(x) = x_4, \ y_5 = g_5(x) = x_5.$$

Assume that a function $x_5 f(g_1(x), g_2(x), g_3(x), g_4(x), g_5(x))$ is annihilated by the Euler operators (2.40). Then, the function $f(y_1, \ldots, y_5)$ satisfies

$$\frac{\partial f}{\partial y_3} = \frac{\partial f}{\partial y_4} = \frac{\partial f}{\partial y_5} = 0,$$

which states that f depends only on y_1 and y_2. The singular locus of the given hypergeometric system $\{(2.40), (2.41), (2.42)\}$ is the hypersurface

$$x_1 x_2 x_3 x_4 (27 x_4 x_3^2 x_1^2 + x_5^3 x_3 x_1 - 36 x_5 x_4 x_3 x_2 x_1 - 16 x_4^2 x_2^3 + 8 x_5^2 x_4 x_2^2 - x_5^4 x_2) = 0.$$

We use the inverse transformation

$$x_1 = \frac{y_1 y_2 y_5^3}{y_3 y_4}, \ x_2 = \frac{y_2 y_5^2}{y_4}, \ x_3 = y_3, \ x_4 = y_4, \ x_5 = y_5,$$

to express the singular locus of the system in the new variables on $(\mathbf{C}^*)^5$:

$$27 y_1^2 y_2 - 16 y_2^2 - 36 y_1 y_2 + 8 y_2 + y_1 - 1 = 0$$

Locally at the origin, this hypersurface is contained in $y_1 y_2 = 0$. See e.g. (2.27). Now apply Theorem 2.4.12 or Corollary 2.4.16 to get the desired conclusion. $\square$

Lemma 2.6.5 implies that each canonical series can be written as follows:

$$x_5 \sum_{m=0, n=0}^{\infty} \sum_{0 \le i, j \le 3} c_{mnij} (x^p)^m (x^q)^n (\log x^p)^i (\log x^q)^j.$$

It now makes sense to call $(\log x^p)^i (\log x^q)^j$ a *monomial*. Thus the number of monomials to be considered in (2.39) is reduced from 243 to $4 \times 4 = 16$, a dramatic savings. We can now compute the coefficients c_{mnij} of canonical series solutions f_1, f_2, f_3, f_4 with respect to w by solving linear equations step by step. Here, we assume that the tie-breaking term order $\prec_w$ is given by

$$x_5 (x^p)^m (x^q)^n (\log x^p)^i (\log x^q)^j \succ_w x_5 (x^p)^{m'} (x^q)^{n'} (\log x^p)^{i'} (\log x^q)^{j'}$$
$$\Leftrightarrow (m + n, n, i + j, j) \succ_{\text{lex}} (m' + n', n', i' + j', j').$$

With our new notion of monomial, the starting monomials for $\prec_w$ are

$$\text{Start}_{\prec_w}(f_1) = x_5,$$
$$\text{Start}_{\prec_w}(f_2) = x_5 \log x^q,$$
$$\text{Start}_{\prec_w}(f_3) = x_5 \log x^p,$$
$$\text{Start}_{\prec_w}(f_4) = x_5 (\log x^q)^2.$$

By solving the **Q**-linear equations, we compute the canonical series solutions

$$f_1 = x_5 \tag{2.43}$$

$$f_2 = x_5 \log x^q + x_5 x^p - 2x_5 x^q + \cdots \tag{2.44}$$

$$f_3 = x_5 \log x^p - x_5 x^p - x_5 x^q + \cdots \tag{2.45}$$

$$f_4 = x_5 (\log x^q)^2 + 2x_5 (\log x^p)(\log x^q)$$

$$+2x_5 x^p \log x^q + 2x_5 x^q \log x^q + \cdots \tag{2.46}$$

In the next chapter we shall study GKZ hypergeometric systems and their series solutions for arbitrary matrices A and arbitrary parameters β. The simplification made in Lemma 2.6.5 is available for all GKZ systems. The example studied here is a maximally degenerate hypergeometric system. Such systems are important in toric mirror symmetry ([50], [95]). We shall give an explicit combinatorial construction of their canonical series solutions in Section 3.6. For instance, in the notation of Theorem 3.6.12, the series (2.46) equals $f_4 = 2 \cdot \Psi_{(0,0,0,0,1),\{5\}}$. Its w-initial series (3.47) equals

$$x_5 \log(x^q)(2\log(x^p) + \log(x^q)) = 2x_5 \cdot V(\log x_1 - \log x_5, \ldots, \log x_4 - \log x_5)$$

where V is the volume polynomial for a toric surface as in Proposition 3.6.15:

$$V(w_1, w_2, w_3, w_4) = w_1 w_2 + w_2 w_3 + w_3 w_4 + w_4 w_1 - \frac{1}{2} w_2^2 + \frac{1}{2} w_4^2.$$

3. Hypergeometric Series

When browsing the relevant literature, we find ourselves confronted with numerous competing definitions for the terms "hypergeometric function", "hypergeometric series" and "hypergeometric differential equations". These definitions appear entirely unrelated, yet each of them captures and generalizes some important aspects of the classical hypergeometric functions in one variable, notably of Gauss' function $_2F_1$ which was reviewed in Section 1.3. The general techniques for studying and solving holonomic D-ideals, which were presented in Chapters 1 and 2, are applicable to all of these different formulations of multidimensional hypergeometric differential equations.

In this book we single out one definition of hypergeometric functions, which we find most suitable for problems in computational algebra. These are the *A-hypergeometric functions* which were introduced by Gel'fand, Kapranov and Zelevinsky in the 1980's. The GKZ-hypergeometric system $H_A(\beta)$ is associated with an integer matrix A and a complex parameter vector β. We interpret the matrix A as representing a toric variety. If this toric variety lives in a projective space then $H_A(\beta)$ is regular holonomic, and we shall assume this throughout. The degree of that projective toric variety is denoted by $\mathrm{vol}(A)$, since it coincides with the volume of the polytope spanned by A.

The GKZ system $H_A(\beta)$ is easiest to analyze when the parameters β are generic, and we shall begin with this case in Section 3.1. In Section 3.2 we introduce some essential combinatorial techniques, involving triangulations and monomial ideals, for studying the GKZ system $H_A(\beta)$. The *hypergeometric fan* of a matrix A is the small Gröbner fan of the holonomic D-ideal $H_A(\beta)$ for β generic; this fan is constructed explicitly in Section 3.3. Both Sections 3.2 and 3.3 require familiarity with combinatorial commutative algebra, at the level of Sturmfels' book "Gröbner Bases and Convex Polytopes" [96].

Section 3.4 constructs all convergent hypergeometric series of the form

$$\sum_a c_a \cdot x_1^{a_1} x_2^{a_2} \cdots x_n^{a_n} \tag{3.1}$$

The difficulty here lies in identifying the initial monomial with respect to some weight vector $w \in \mathbf{R}^n$. The coefficients c_a are certain ratios of falling and rising factorials. In Section 3.5 we prove the fundamental inequality

$$\mathrm{rank}\big(H_A(\beta)\big) \quad \geq \quad \mathrm{vol}(A). \tag{3.2}$$

This is done by means of explicit deformations in the space of parameters, which transform the series (3.1) into hypergeometric series with logarithms as in (2.1). In Section 3.6 we apply the same technique to find a solution basis to $H_A(\beta)$ in the "maximally degenerate case" which is of special interest in the interplay of algebraic geometry and mathematical physics. Frequently, the inequality (3.2) will be an equality; Gel'fand, Kapranov and Zelevinsky proved this if β is generic or if the toric variety of A is projectively Cohen-Macaulay. Chapter 4 will be devoted to criteria for equality in (3.2). In this chapter we are primarily concerned with constructing hypergeometric series.

3.1 GKZ-Hypergeometric Ideal for Generic Parameters

In this section we recall the definition of the GKZ-hypergeometric system of differential equations, and we discuss some of its basic properties. Let $A = (a_{ij})$ be an integer $d \times n$-matrix of rank d which satisfies the following *homogeneity assumption*:

$$\text{The } \mathbf{Q}\text{-row span of } A \text{ contains the vector } (1, 1, \ldots, 1). \qquad (3.3)$$

With the matrix A, one associates the following prime ideal of (Krull) dimension d in the commutative polynomial ring $\mathbf{k}[\partial] = \mathbf{k}[\partial_1, \ldots, \partial_n]$:

$$I_A \quad := \quad \langle \, \partial^u - \partial^v \mid Au = Av, \ u, v \in \mathbf{N}^n \, \rangle \quad \subset \quad \mathbf{k}[\partial].$$

We call I_A the *toric ideal* of A. Our hypothesis (3.3) states equivalently that the toric ideal I_A is homogeneous with respect to the usual grading. Thus the zero set of I_A is a toric variety of dimension $d - 1$ in projective space $\mathbf{P}^{n-1}$.

Example 3.1.1. Let $n = 3, d = 2, A = \begin{pmatrix} 2 & 1 & 0 \\ 0 & 1 & 2 \end{pmatrix}$. Then I_A is the principal ideal generated by $\partial_1\partial_3 - \partial_2^2$. Its zero set is the toric curve $(1 : t : t^2)$ in $\mathbf{P}^2$.

Fix a column vector $\beta = (\beta_1, \ldots, \beta_d)^T$ in $\mathbf{k}^d$ and consider the column vector $\theta = (\theta_1, \ldots, \theta_n)^T$, where $\theta_i = x_i\partial_i$. Then $A \cdot \theta - \beta$ is the column vector of length d whose i-th coordinate is $\sum_{j=1}^n a_{ij}\theta_j - \beta_i$. We write $\langle A \cdot \theta - \beta \rangle$ for the ideal in the commutative polynomial ring $\mathbf{k}[\theta] = \mathbf{k}[\theta_1, \ldots, \theta_n]$ generated by these d linear expressions. Let $H_A(\beta)$ denote the left ideal of the Weyl algebra $D = \mathbf{k}\langle x_1, \ldots, x_n, \partial_1, \ldots, \partial_n \rangle$ generated by I_A and $\langle A \cdot \theta - \beta \rangle$. This D-ideal was introduced by Gel'fand, Kapranov and Zelevinsky [38]; it is called the *GKZ-hypergeometric system* with matrix A and parameters β. We also call $H_A(\beta)$ the *A-hypergeometric ideal* with parameters β. A holomorphic function or formal series $\phi(x_1, \ldots, x_n)$ is called *A-hypergeometric function of degree β* if $H_A(\beta) \bullet \phi = 0$.

Example 3.1.2. The toric curve in Example 3.1.1 gives rise to the following GKZ-hypergeometric system of differential equations:

$$H_A(\beta) \quad = \quad D \cdot \{ \partial_1 \partial_3 - \partial_2^2, \, 2x_1 \partial_1 + x_2 \partial_2 - \beta_1, x_2 \partial_2 + 2x_3 \partial_3 - \beta_2 \}.$$

The following expression is an A-hypergeometric function of degree $(-1, 1)$:

$$\frac{-x_2 \pm (x_2^2 - 4x_1 x_3)^{1/2}}{2x_1} \quad = \quad -\frac{x_2}{2x_1} \pm \left(\frac{x_2}{2x_1} - \sum_{m=0}^{\infty} \frac{1}{m+1} \binom{2m}{m} \frac{x_1^m x_3^{m+1}}{x_2^{2m+1}} \right).$$

These are the two roots of a quadratic polynomial $f(t) = x_1 t^2 + x_2 t + x_3$. Indeed, the roots of a polynomial of any degree are A-hypergeometric functions of the coefficients of that polynomial for suitable A, see [97]. An analogous statement is true for systems of n polynomial equations in n variables. Notice the *Catalan numbers* $\frac{1}{m+1}\binom{2m}{m}$ in the right hand series. Coefficients of hypergeometric series are quite interesting combinatorial objects. Their analysis may lead to new algorithms for solving systems of polynomial equations.

Fix the weight vector $w = (1, 0, 0)$. The starting monomials of the two series above are x_2/x_1 and x_3/x_2. They span the solution space of

$$\mathrm{in}_{(-w, w)}(H_A(\beta)) \quad = \quad D \cdot \{ \partial_1 \partial_3, \, 2x_1 \partial_1 + x_2 \partial_2 + 1, x_2 \partial_2 + 2x_3 \partial_3 - 1 \}.$$

Thus the solutions of the initial ideal are the lowest terms in the hypergeometric series. Our aim is to study these objects for arbitrary A, β and w.

Let $A \in \mathbf{Z}^{d \times n}$ as above and $\beta \in \mathbf{k}^d$. For any weight vector $w \in \mathbf{R}^n$ we consider the *initial ideal* $\mathrm{in}_w(I_A) \subset \mathbf{k}[\partial]$. This ideal is a monomial ideal whenever w is sufficiently generic. Recall that $\mathbf{k}$ is an infinite field, so we can talk about "generic" vectors β in $\mathbf{k}^d$. They range over a nonempty Zariski-open subset of $\mathbf{k}^d$. Our first result is a characterization of the initial ideal of the hypergeometric ideal $H_A(\beta)$ under the assumption that β is generic.

Theorem 3.1.3. *Let $w \in \mathbf{R}^n$. For generic parameters β, we have*

$$\mathrm{in}_{(-w, w)}(H_A(\beta)) \quad = \quad D \cdot \mathrm{in}_w(I_A) + D \cdot \langle A \cdot \theta - \beta \rangle. \tag{3.4}$$

Proof. Let $s = (s_1, \ldots, s_d)^T$ be a vector of new indeterminates. Consider the algebra $D[s] = D[s_1, \ldots, s_d]$. This is the $\mathbf{k}$-algebra generated by $\partial_1, \ldots, \partial_n, x_1, \ldots, x_n, s_1, \ldots, s_d$, where the d new variables s_i commute with one another and with the $2n$ generators ∂_i, x_j of the Weyl algebra D. Let $H_A[s]$ be the left ideal in $D[s]$ generated by I_A and the d coordinates of $A \cdot \theta - s$. We define a partial term order $>_w$ on monomials in $D[s]$:

$$s^a x^b \partial^c >_w s^{a'} x^{b'} \partial^{c'} \quad :\Longleftrightarrow \quad -w \cdot b + w \cdot c > -w \cdot b' + w \cdot c', \quad \text{or}$$
$$-w \cdot b + w \cdot c = -w \cdot b' + w \cdot c' \quad \text{and} \quad a >_{lex} a'.$$

Here $>_{lex}$ denotes the lexicographic term order on monomials in $\mathbf{k}[s_1, \ldots, s_d]$. Finally, let $>$ be any term order on monomials in $D[s]$ which refines $>_w$.

Let $\mathcal{G}$ be the reduced Gröbner basis of the toric ideal I_A with respect to the restriction of $>$ to $\mathbf{k}[\partial]$. It consists of binomials $\partial^u - \partial^v$. Let $\mathcal{G}'$ denote the union of $\mathcal{G}$ and the d coordinates of $A \cdot \theta - s$. We claim that $\mathcal{G}'$ is a Gröbner basis for $H_A[s]$ with respect to $>$. To prove this claim we apply Buchberger's algorithm and show that the S-pair formed by any two of the elements in $\mathcal{G}'$ reduces to zero (Theorem 1.1.10). This is automatic for any two binomials $\partial^u - \partial^v$ and $\partial^{u'} - \partial^{v'}$ in $\mathcal{G}$, since $\mathcal{G}$ is a Gröbner basis to begin with. Next we form the S-pair of two elements in $\mathcal{G}'\backslash\mathcal{G}$. Leading terms are underlined:

$$\mathrm{sp}\big(\underline{s_i} - (A\theta)_i, \; \underline{s_j} - (A\theta)_j \big) \quad = \quad s_i(A\theta)_j - s_j(A\theta)_i.$$

It reduces to zero, since Buchberger's First Criterion holds for the commutative polynomial subring $\mathbf{k}[\theta, s]$ of $D[s]$. Note that this criterion does not generally hold in the Weyl algebra. It remains to consider the S-pair of an element in $\mathcal{G}$ and an element in $\mathcal{G}'\backslash\mathcal{G}$:

$$\begin{aligned}
& \mathrm{sp}\big(\underline{\partial^u} - \partial^v, \; \underline{s_i} - (A\theta)_i \big) \\
= \; & s_i(\partial^u - \partial^v) - \partial^u(s_i - (A\theta)_i) \\
= \; & -s_i\partial^v + \partial^u(A\theta)_i \\
= \; & -s_i\partial^v + ((A\theta)_i + (Au)_i)\partial^u \\
= \; & ((A\theta)_i + (Au)_i)(\partial^u - \partial^v) + ((A\theta)_i + (Au)_i)\partial^v - s_i\partial^v \\
= \; & ((A\theta)_i + (Au)_i)(\underline{\partial^u} - \partial^v) - \partial^v(\underline{s_i} - (A\theta)_i) \quad \longrightarrow_{\mathcal{G}'} \quad 0.
\end{aligned}$$

We have proved that $\mathcal{G}'$ is a Gröbner basis of $H_A[s]$ with respect to $>$.

Recall that $>$ refines $>_w$ and that $>_w$ refines the weight $(-w, w, 0)$ on $D[s]$. A variant of Theorem 1.1.6 implies the following identity of $D[s]$-ideals:

$$\mathrm{in}_{(-w,w,0)}\big(H_A[s]\big) \quad = \quad D[s] \cdot \mathrm{in}_w(\mathcal{G}) + D[s] \cdot \langle A \cdot \theta - s \rangle. \tag{3.5}$$

We may now replace $D[s]$ by $D(s) := \mathbf{k}(s)\langle x_1, \dots, x_n, \partial_1, \dots, \partial_n \rangle$, which is the Weyl algebra over the rational function field $\mathbf{k}(s)$. Set $H_A(s) := D(s) \cdot H_A[s]$. The identity (3.5) implies

$$\mathrm{in}_{(-w,w)}\big(H_A(s)\big) \quad = \quad D(s) \cdot \mathrm{in}_w(\mathcal{G}) + D(s) \cdot \langle A \cdot \theta - s \rangle. \tag{3.6}$$

The desired conclusion (3.4) follows directly from (3.6). $\qquad\qquad\square$

For arbitrary parameters β, only the following inclusion holds in (3.4):

$$\mathrm{in}_{(-w,w)}\big(H_A(\beta)\big) \quad \supset \quad D \cdot \mathrm{in}_w(I_A) + D \cdot \langle A \cdot \theta - \beta \rangle. \tag{3.7}$$

The D-ideal on the right hand side is called the *fake initial ideal* of the hypergeometric ideal $H_A(\beta)$ with respect to w. Theorem 3.1.3 states that the (true) initial ideal $\mathrm{in}_{(-w,w)}\big(H_A(\beta)\big)$ coincides with the fake initial ideal when the parameter vector β is sufficiently generic. However, the containment (3.7) may be strict, as the following example shows. The discrepancy between the initial ideal and the fake initial ideal will be a main topic in Chapter 4.

Example 3.1.4. Theorem 3.1.3 does not hold for non-generic β. Take

$$A = \begin{pmatrix} 3 & 2 & 1 & 0 \\ 0 & 1 & 2 & 3 \end{pmatrix} \quad \text{and} \quad w = (1,3,0,0).$$

The initial ideal of the toric ideal I_A equals

$$\text{in}_w\,(I_A) = \langle\, \partial_2^2,\ \partial_2\partial_4,\ \partial_2\partial_3,\ \partial_1\partial_4^2\,\rangle = \langle\partial_1,\partial_2\rangle \cap \langle\partial_2,\partial_4^2\rangle \cap \langle\partial_2^2,\partial_3,\partial_4\rangle. \quad (3.8)$$

Consider the following element of the A-hypergeometric ideal $H_A(\beta)$:

$$\begin{aligned}
p(\beta) &= \theta_2(\theta_2 + 2\theta_3 + 3\theta_4 - \beta_2) \\
&\quad - x_2^2(\partial_2^2 - \partial_1\partial_3) - 2x_2x_3(\partial_2\partial_3 - \partial_1\partial_4) - 3x_2x_4(\partial_2\partial_4 - \partial_3^2) \\
&= -(\beta_2 - 1)\theta_2 + 2x_2x_3\partial_1\partial_4 + x_2^2\partial_1\partial_3 + 3x_2x_4\partial_3^2.
\end{aligned}$$

For any parameter vector $\beta \in \mathbf{k}^2$ with $\beta_2 = 1$, the first term vanishes and

$$\text{in}_{(-w,w)}\,(p(\beta)) \quad = \quad 2x_2x_3\partial_1\partial_4.$$

This monomial does not belong to the fake initial ideal.

Suppose from now on that the weight vector $w \in \mathbf{R}^n$ is generic. This implies that $\text{in}_w(I_A)$ is a monomial ideal in $\mathbf{k}[\partial]$, and, by Theorem 2.3.3 (2), the initial ideal $\text{in}_{(-w,w)}(H_A(\beta))$ is a torus-fixed D-ideal. The *indicial ideal* of $H_A(\beta)$ with respect to w was defined to be the following ideal in the commutative polynomial subring $\mathbf{k}[\theta]$ of the Weyl algebra D:

$$\widetilde{\text{in}}_{(-w,w)}(H_A(\beta)) \quad := \quad R \cdot \text{in}_{(-w,w)}(H_A(\beta)) \cap \mathbf{k}[\theta], \quad (3.9)$$

where $R = \mathbf{k}(x)\langle\partial\rangle$ is the ring of differential operators.

Clearly, the indicial ideal $\widetilde{\text{in}}_{(-w,w)}(H_A(\beta))$ contains the distraction $\widetilde{\text{in}}_w(I_A)$ of the initial ideal $\text{in}_w(I_A)$ of the toric ideal. Recall from Theorem 2.3.4 that the *distraction* of any monomial ideal M in $\mathbf{k}[\partial]$ is the following ideal in $\mathbf{k}[\theta]$:

$$\widetilde{M} \quad := \quad D \cdot M \cap \mathbf{k}[\theta] \quad = \quad \langle\, x^u\partial^u : \partial^u \in M\,\rangle \quad = \quad \langle\, [\theta]_u : \partial^u \in M\,\rangle \subset \mathbf{k}[\theta].$$

If $\partial^a, \partial^b, \ldots, \partial^c$ are the minimal generators of M, then $\{[\theta]_a, [\theta]_b, \ldots, [\theta]_c\}$ is a reduced Gröbner basis for $\widetilde{M}$ with respect to any term order on $\mathbf{k}[\theta]$; see Theorem 3.2.2 below. Combinatorial properties of $\widetilde{M}$ and $\text{in}_w(I_A)$ are essential for studying hypergeometric series. They will be seen in Section 3.2.

Proposition 3.1.5. *The following identity of ideals in $\mathbf{k}[\theta]$ holds for <u>all</u> parameters $\beta \in \mathbf{k}^d$:*

$$\widetilde{\text{in}}_w(I_A) + \langle A \cdot \theta - \beta\rangle \quad = \quad (R \cdot \text{in}_w(I_A) + R \cdot \langle A \cdot \theta - \beta\rangle) \cap \mathbf{k}[\theta]. \quad (3.10)$$

Proof. Lemma 2.3.1 implies that the D-ideal $D \cdot \text{in}_w(I_A) + D \cdot \langle A \cdot \theta - \beta\rangle$ is torus-fixed. Hence the assertion is immediate from Theorem 2.3.4. $\square$

The ideal in (3.10) is denoted by $\widetilde{\mathrm{fin}}_w(H_A(\beta))$ and called the *fake indicial ideal* of $H_A(\beta)$ with respect to w. The roots of the fake indicial ideal in affine n-space are called the *fake exponents* of $H_A(\beta)$ with respect to w.

Corollary 3.1.6. *The indicial ideal* $\widetilde{\mathrm{in}}_{(-w,w)}(H_A(\beta))$ *equals the fake indicial ideal* $\widetilde{\mathrm{fin}}_w(H_A(\beta))$ *for generic parameters* β.

Proof. Immediate from Theorem 3.1.3 and Proposition 3.1.5.   □

Remark 3.1.7. The fake exponents of the hypergeometric ideal $H_A(\beta)$ with respect to any w are defined over the same ground field **k** as the parameter vector β. In particular, if all coordinates of β are rational numbers then the coordinates of the fake exponents are rational numbers as well.

Example 3.1.8. (Continuation of Example 3.1.4)
The distraction of the initial monomial ideal $M = \mathrm{in}_w(I_A)$ in (3.8) equals

$$\widetilde{M} = \langle \theta_2(\theta_2 - 1), \theta_2\theta_4, \theta_2\theta_3, \theta_1\theta_4(\theta_4 - 1) \rangle$$
$$= \langle \theta_1, \theta_2 \rangle \cap \langle \theta_2, \theta_4 \rangle \cap \langle \theta_2, \theta_4 - 1 \rangle \cap \langle \theta_2 - 1, \theta_3, \theta_4 \rangle.$$

For (β_1, β_2) generic, the indicial ideal equals the fake indicial ideal

$$\widetilde{M} + \langle \theta_2 + 2\theta_3 + 3\theta_4 - \beta_2, 3\theta_1 + 2\theta_2 + \theta_3 - \beta_1 \rangle$$
$$= \quad \langle \theta_1, \theta_2, \theta_3 - \beta_1, 3\theta_4 + 2\beta_1 - \beta_2 \rangle \quad \cap$$
$$\langle \theta_2, \theta_4, 2\theta_3 - \beta_2, 6\theta_1 + \beta_2 - 2\beta_1 \rangle \quad \cap$$
$$\langle \theta_2, \theta_4 - 1, 2\theta_3 + 3 - \beta_2, 6\theta_1 + \beta_2 - 2\beta_1 - 3 \rangle.$$

The three roots of this ideal are the exponents. If $\beta_2 = 1$ then the indicial ideal strictly contains the fake indicial ideal. In fact, there are four fake exponents but only three exponents. For $\beta_2 = 1$, the vector $((\beta_1 + 1)/3, 0, -1, 1)$ is a fake exponent, but it is not an exponent (unless $\beta_1 = -1$).

As we have seen in Chapter 1, it sometimes makes more sense to work in the homogenized Weyl algebra $D^{(h)}$ instead of the Weyl algebra D. For given A and β as above, we define the following homogeneous $D^{(h)}$-ideal

$$\overline{H}_A(\beta) \quad := \quad D^{(h)} \cdot I_A + D^{(h)} \cdot \langle A \cdot \theta - h^2\beta \rangle.$$

The dehomogenization of $\overline{H}_A(\beta)$ is the hypergeometric ideal $H_A(\beta)$. The converse is true only partially. The following proposition follows from results to be proved in Section 4.3. We postpone its proof until then.

Proposition 3.1.9. *If the toric ideal I_A is Cohen-Macaulay then the homogenization $H_A(\beta)^{(h)}$ of the hypergeometric ideal $H_A(\beta)$ is equal to $\overline{H}_A(\beta)$.*

To keep things simple, we disregard the issue of Cohen-Macaulayness for now. Throughout Chapter 3 we work with $\overline{H}_A(\beta)$ instead of $H_\beta(A)^{(h)}$. We first note that Theorem 3.1.3 extends to the $D^{(h)}$-ideal $\overline{H}_A(\beta)$.

Theorem 3.1.10. *Let $w \in \mathbf{R}^n$. For generic parameters β, we have*

$$\mathrm{in}_{(0,-w,w)} \left(\overline{H}_A(\beta) \right) \quad = \quad D \cdot \mathrm{in}_w(I_A) + D \cdot \langle A \cdot \theta - h^2 \beta \rangle. \tag{3.11}$$

Proof. The proof is essentially identical to the proof of Theorem 3.1.3, that is, we apply Buchberger's criterion for a suitable term order on $D^{(h)}[s]$. ☐

3.2 Standard Pairs and Triangulations

For our further study of the initial ideal $\mathrm{in}_{(-w,w)} \left(H_A(\beta) \right)$ and of the resulting hypergeometric series, it is necessary to develop more refined combinatorial tools. It is the purpose of this section to provide these tools.

We begin by reviewing some general results about decomposing monomial ideals in a commutative polynomial ring. Let M be any monomial ideal in $\mathbf{k}[\partial] = \mathbf{k}[\partial_1, \ldots, \partial_n]$. We interpret M as a very simple system of partial differential equations, namely, each equation in that system stipulates that a certain mixed partial derivative vanishes. Our goal is to determine the solutions to such a system. As an example, consider the monomial ideal $M = \langle \partial_1^2, \partial_1 \partial_2^2 \rangle$ for $n = 2$. It represents the following differential equations:

$$\frac{\partial^2 f}{\partial x_1^{\,2}} = \frac{\partial^3 f}{\partial x_1 \partial x_2^{\,2}} = 0. \tag{3.12}$$

The general polynomial solution $f(x_1, x_2)$ to these equations looks like this:

$$f(x_1, x_2) \quad = \quad F(x_2) + \alpha x_1 + \beta x_1 x_2,$$

where α, β are scalars and F is any univariate function. Thus there are three types of basic solutions. They correspond to the three standard pairs of M.

A *standard pair* of a monomial ideal M in $\mathbf{k}[\partial]$ is a pair (∂^a, σ), where $a \in \mathbf{N}^n$ and $\sigma \subset \{1, \ldots, n\}$ subject to the following three conditions:

(1) $a_i = 0$ for all $i \in \sigma$;

(2) for all choices of integers $b_j \geq 0$, the monomial $\partial^a \cdot \prod_{j \in \sigma} \partial_j^{b_j}$ is not in M;

(3) For all $l \notin \sigma$, there exist $b_j \geq 0$ such that $\partial^a \cdot \partial_l^{b_l} \cdot \prod_{j \in \sigma} \partial_j^{b_j}$ lies in M.

Let $\mathcal{S}(M)$ denote the set of all standard pairs of M. From this set we can read off a decomposition of M into irreducible (hence primary) monomial ideals. The following identity appears in [98, Equation (3.2)]:

$$M \quad = \quad \bigcap_{(\partial^a, \sigma) \in \mathcal{S}(M)} \langle \partial_i^{a_i+1} : i \notin \sigma \rangle. \tag{3.13}$$

The prime ideal $\langle \partial_i : i \notin \sigma \rangle$ is associated to M if and only if there exists a standard pair of the form $(\,\cdot\,, \sigma)$ in $\mathcal{S}(M)$. We write

$$\text{Ass}(M) \quad := \quad \{\,\sigma \subset \{1, 2, \ldots, n\} \,:\, (\partial^a, \sigma) \in \mathcal{S}(M) \text{ for some } a \in \mathbf{N}^n\,\}$$

by identifying σ and the associated prime $\langle\, \partial_i \,:\, i \notin \sigma\,\rangle$.

In our little example (3.12), the set of standard pairs equals $\mathcal{S}(M) = \{(1, \{2\}), (\partial_1, \emptyset), (\partial_1\partial_2, \emptyset)\}$ and the resulting irreducible decomposition is

$$M \;=\; \langle \partial_1^2, \partial_1\partial_2^2 \rangle \;=\; \langle \partial_1 \rangle \,\cap\, \langle \partial_1^2, \partial_2 \rangle \,\cap\, \langle \partial_1^2, \partial_2^2 \rangle. \qquad (3.14)$$

Note that this intersection is redundant. The middle component can be erased; the minimal irreducible decomposition equals $M = \langle \partial_1 \rangle \,\cap\, \langle \partial_1^2, \partial_2^2 \rangle$. One reason for working with the redundant decomposition (3.13) is that it gives the solutions to the system of differential equations represented by M.

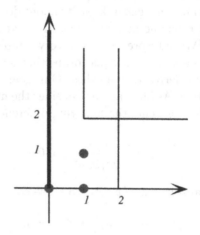

Fig. 3.1. Standard pair decomposition of $M = \langle \partial_1^2, \partial_1\partial_2^2 \rangle$.

Remark 3.2.1. The general polynomial solution to the D-ideal DM equals

$$\sum_{(\partial^a, \sigma) \in \mathcal{S}(M)} c_{a,\sigma} \cdot x^a \cdot F_{a,\sigma}(x_i : i \in \sigma), \qquad \text{where } c_{a,\sigma} \in \mathbf{k}.$$

Let std_M denote the set of all vectors $a \in \mathbf{N}^n$ such that ∂^a is not in M. These are called *standard monomials* modulo M. The previous remark holds because the standard pairs define an irredundant decomposition of the (typically infinite) set std_M into (finitely many) subsets of the form $a + \mathbf{N}^\sigma$.

Figure 3.1 depicts the standard pair decomposition of the monomial ideal (3.14). The vertical thick line starting from the origin represents standard monomials of the form $1 \cdot \partial_2^k$, $k \in \mathbf{N}$, which corresponds to the standard pair $(1, \{2\})$. Two dots represent the standard monomials ∂_1 and $\partial_1\partial_2$ respectively. These correspond to the standard pairs $(\partial_1, \emptyset)$ and $(\partial_1\partial_2, \emptyset)$.

As we saw in the section on Frobenius ideals, it is natural to replace the monomial ideal M by its distraction $\widetilde{M} = D \cdot M \cap \mathbf{k}[\theta]$. The distraction $\widetilde{M}$ has the same solutions as M, when regarded as differential equations, but $\widetilde{M}$ is nicer than M with respect to some algebraic and combinatorial properties.

We next present a theorem which explains these properties. Regarding $\text{std}_M \subset \mathbf{N}^n$ as a subset of the affine space $\mathbf{k}^n$, it makes sense to consider the vanishing ideal of std_M in the polynomial ring $\mathbf{k}[\theta]$. This is the radical ideal in $\mathbf{k}[\theta]$ consisting of all polynomials f which satisfy $f(a) = 0$ for all $a \in \text{std}_M$. Theorem 3.2.2 below is well-known in commutative algebra. It is due to Hartshorne (see Proposition 4.4 on page 294 and Theorem 4.9 on page 297 in [46]). Its variants are frequently used in the study of Hilbert schemes.

Theorem 3.2.2. *The distraction $\widetilde{M}$ of a monomial ideal M equals the vanishing ideal of the set $\text{std}_M \subset \mathbf{N}^n$. In particular, $\widetilde{M}$ is a radical ideal. For any term order, the reduced Gröbner basis of $\widetilde{M}$ consists of the polynomials*

$$[\theta]_a = \prod_{i=1}^{n} \prod_{j=0}^{a_i-1} (\theta_i - j)$$

where $\partial^a = \partial_1^{a_1} \cdots \partial_n^{a_n}$ runs over all minimal generators of M.

Proof. Fix an arbitrary term order $\prec$ on $\mathbf{k}[\theta]$, and let I_M denote the vanishing ideal of std_M. We shall prove that the above polynomials $[\theta]_a$ form a Gröbner basis for I_M with respect to $\prec$. This will imply $I_M = \widetilde{M}$.

We first assume that std_M is finite. Then both $\mathbf{k}[\partial]/M$ and $\mathbf{k}[\theta]/I_M$ are Artinian rings of $\mathbf{k}$-dimension $\#(\text{std}_M)$. Let F be the ideal generated by the polynomials $[\theta]_a$ where ∂^a is a minimal generator of M. Since $[\theta]_a$ vanishes on std_M, we have $F \subset I_M$. This inclusion lifts to initial ideals and we get $\text{in}_\prec(F) \subset \text{in}_\prec(I_M)$. The observation $\text{in}_\prec([\theta]_a) = \theta^a$ implies $\dim_\mathbf{k}(\mathbf{k}[\theta]/\text{in}_\prec(F)) \leq \dim_\mathbf{k}(\mathbf{k}[\partial]/M)$. Consider the following inequalities:

$$\#(\text{std}_M) = \dim_\mathbf{k} \mathbf{k}[\theta]/I_M = \dim_\mathbf{k} \mathbf{k}[\theta]/\text{in}_\prec(I_M) \leq$$
$$\leq \dim_\mathbf{k} \mathbf{k}[\theta]/\text{in}_\prec(F) \leq \dim_\mathbf{k} \mathbf{k}[\partial]/M = \#(\text{std}_M).$$

All inequalities are equalities. Hence $\text{in}_\prec(I_M) = \text{in}_\prec(F) = \langle \theta^a : \partial^a \in M \rangle$, and $F = I_M$. This proves our assertions in the case when std_M is finite.

Next consider the case where std_M is infinite. Suppose, by contradiction, that the polynomials $[\theta]_a$ do not form a Gröbner basis for I_M with respect to $\prec$. Then there exists a non-zero polynomial $f \in I_M$ such that no term of f lies in $\langle \theta^a : \partial^a \in M \rangle$. Let M' be the largest monomial ideal in $\mathbf{k}[\partial]$ such that each term θ^b appearing in f satisfies $b \in \text{std}_{M'}$. Then $\text{std}_{M'}$ is finite and $\text{std}_{M'} \subset \text{std}_M$. We have $f \in I_{M'}$ and no term of f lies in $\langle \theta^a : \partial^a \in M' \rangle$. This is a contradiction to the conclusion of the previous paragraph.

Therefore the polynomials $[\theta]_a$ where ∂^a runs over all minimal generators of M form a Gröbner basis for I_M in both cases. Since θ^a is the only term of

$[\theta]_a$ which lies in $\langle \theta^b : \partial^b \in M \rangle$, we conclude that the given Gröbner basis is actually the reduced Gröbner basis for I_M with respect to $\prec$.   □

Corollary 3.2.3. *The distraction $\widetilde{M}$ equals the intersection of the prime ideals $\langle \theta_i - a_i : i \notin \sigma \rangle$ where (∂^a, σ) runs over $S(M)$. This intersection is irredundant.*

Proof. Note that the ideal $\langle \theta_i - a_i : i \notin \sigma \rangle$ defines the affine space $a + \mathbf{k}^\sigma$. The Zariski closure in $\mathbf{k}^n$ of the set std_M equals the union of the affine spaces $a + \mathbf{k}^\sigma$ where $(\partial^a, \sigma) \in S(M)$. This union is irredundant. Since $\widetilde{M}$ is radical by Theorem 3.2.2, the conclusion follows from Hilbert's Nullstellensatz.   □

For instance, Corollary 3.2.3 tells us that the distraction of the monomial ideal M in (3.14) has the following irredundant prime decomposition

$$\widetilde{M} = \langle \theta_1(\theta_1 - 1), \theta_1\theta_2(\theta_2 - 1) \rangle = \langle \theta_1 \rangle \cap \langle \theta_1 - 1, \theta_2 \rangle \cap \langle \theta_1 - 1, \theta_2 - 1 \rangle.$$

Thus $\widetilde{M}$ is the vanishing radical ideal of a line and two points in the plane.

We next show that distracting a monomial ideal commutes with removing low-dimensional components. Let J be any ideal of dimension d in the commutative polynomial ring $\mathbf{k}[\partial]$ or $\mathbf{k}[\theta]$. Write $\mathrm{top}(J)$ for the intersection of all primary components of J of dimension d. This is well-defined since primary components of dimension d are not embedded and hence unique.

Lemma 3.2.4. *Let M be a monomial ideal of dimension d in $\mathbf{k}[\partial]$, and let $T(M)$ be the set of all standard pairs (∂^a, σ) of M with $\#(\sigma) = d$. Then*

$$\mathrm{top}(M) = \bigcap_{(\partial^a, \sigma) \in T(M)} \langle \partial_i^{a_i+1} : i \notin \sigma \rangle, \qquad and$$

$$\widetilde{\mathrm{top}(M)} = \mathrm{top}(\widetilde{M}) = \bigcap_{(\partial^a, \sigma) \in T(M)} \langle \theta_i - a_i : i \notin \sigma \rangle.$$

Proof. The monomial ideal M itself is the intersection of the ideals $\langle \partial_i^{a_i+1} : i \notin \sigma \rangle$ where (∂^a, σ) runs over all standard pairs. Taking the intersection (3.13) of these irreducible ideals for fixed σ, we get an irredundant primary component of M. Removing all primary components of dimension less than d amounts to deleting all standard pairs (∂^a, σ) with $\#(\sigma) < d$. The same argument applies to the prime decomposition in Corollary 3.2.3.   □

The cardinality of the set $T(M)$ of top-dimensional standard pairs is equal to the degree of M. The cardinality of the set $S(M)$ of standard pairs is called *arithmetic degree* of M; see [51], [98]. Thus, by Corollary 3.2.3, the arithmetic degree of M is the number of affine subspaces which comprise the Zariski closure of the set std_M of (exponent vectors of) standard monomials. For instance, the monomial ideal in (3.14) has arithmetic degree 3.

This raises the question as to how one can compute the arithmetic degree and the standard pairs of an arbitrary monomial ideal M. A combinatorial algorithm for these tasks is given in [51]. We shall describe a more algebraic approach to this computational problem. For $\sigma \subset \{1, 2, \ldots, n\}$ let M_σ denote the monomial ideal in $\mathbf{k}[\partial_i : i \notin \sigma]$ obtained from M by replacing $\partial_j \mapsto 1$ for $j \in \sigma$ in all minimal generators of M. The *saturation* of M_σ is the following ideal quotient, which takes place in $\mathbf{k}[\partial_i : i \notin \sigma]$:

$$\mathrm{sat}(M_\sigma) \quad := \quad \left(M_\sigma : \langle \partial_i : i \notin \sigma \rangle^\infty \right).$$

The three defining conditions for (∂^a, σ) to be a standard pair can be expressed equivalently as follows:

(1) ∂^a is a monomial in $\mathbf{k}[\partial_i : i \notin \sigma]$;
(2) ∂^a does not lie in M_σ;
(3) ∂^a lies in $\mathrm{sat}(M_\sigma)$.

The quotient $\mathrm{sat}(M_\sigma)/M_\sigma$ is a finite-dimensional $\mathbf{k}$-vector space, and our task is to compute the monomial basis of this vector space for each $\sigma \in \mathrm{Ass}(M)$. This can be done, for instance, by computing the ratio of the multi-graded Hilbert series of the two ideals. The arithmetic degree of M is the sum over all σ of the dimensions of the vector spaces $\mathrm{sat}(M_\sigma)/M_\sigma$. This leaves us with the task of enumerating the set $\mathrm{Ass}(M)$ of associated primes of M, which is an interesting combinatorial problem in its own right. In most situations of interest, one can identify a reasonable subset Δ of the power set $2^{\{1,2,\ldots,n\}}$ which contains $\mathrm{Ass}(M)$. Given such Δ, the following algorithm makes sense.

Algorithm 3.2.5 (Computing Standard Pairs)
Input: A monomial ideal $M \subset \mathbf{k}[\partial_1, \ldots, \partial_n]$ and a subset $\Delta \subset 2^{\{1,2,\ldots,n\}}$
    which contains $\mathrm{Ass}(M)$.
Output: The set $\mathcal{S}(M)$ of standard pairs of M.
For each $\sigma \in \Delta$ do
    compute the two monomial ideals M_σ and $\mathrm{sat}(M_\sigma)$ in $\mathbf{k}[\partial_i : i \notin \sigma]$
    for each monomial ∂^a which lies in $\mathrm{sat}(M_\sigma)$ but not in M_σ: output (∂^a, σ).

A family Δ of subsets of $\{1, \ldots, n\}$ is called a *simplicial complex* when the following two conditions are satisfied.

1. if $\sigma \in \Delta$, then any subset of σ belongs to Δ.
2. if $\sigma, \tau \in \Delta$, then $\sigma \cap \tau \in \Delta$.

Each subset is called a *face* of the simplicial complex. The *Stanley-Reisner ideal* of Δ is the ideal in $\mathbf{k}[\partial_1, \ldots, \partial_n]$ generated by monomials $\prod_{i \in \sigma} \partial_i$ where σ runs over all the subsets of $\{1, \ldots, n\}$ such that $\sigma \notin \Delta$. For example, $\Delta = \{\{1,2,3\}, \{2,3,4\}, \{1,2\}, \{1,3\}, \{2,3\}, \{2,4\}, \{3,4\}, \{1\}, \{2\}, \{3\}, \{4\}, \emptyset\}$ is a simplicial complex on $\{1, 2, 3, 4\}$ and its Stanley-Reisner ideal is $\langle \partial_1 \partial_4 \rangle$. We often list only top-dimensional faces (facets) to denote a simplicial complex. For instance, we would write $\Delta = \{\{1, 2, 3\}, \{2, 3, 4\}\}$ for the example above.

Now, let us discuss an obvious choice for the candidate set Δ. Take Δ to be the simplicial complex on $\{1, 2, \ldots, n\}$ whose Stanley-Reisner ideal is the radical of M. The faces of the simplicial complex Δ are the subsets σ with the property that no monomial with support σ lies in M. Computing the simplicial complex Δ amounts to finding the minimal primes of M. Several computer algebra systems have a command for doing this; for instance, in Macaulay 2 [42] the command is decompose(M). Furthermore, we can utilize the following lower bound on the dimension of an associated prime:

$$\text{depth}(M) \leq \#(\sigma) \leq \dim(M) \quad \text{for all} \quad \sigma \in \text{Ass}(M). \qquad (3.15)$$

Recall that the *depth* of a polynomial ideal M is the length of the longest regular sequence modulo M (see, e.g., [47, p.184]). If $\text{depth}(M) = \dim(M)$ then M is *Cohen-Macaulay*. By the Auslander-Buchsbaum formula, the number $n - \text{depth}(M)$ equals the length of the minimal free resolution of M. This length is usually easy to compute; in Macaulay 2 we simply type pdim coker gens M to get $n - \text{depth}(M)$. So, before starting Algorithm 3.2.5, it makes sense to first compute $\text{depth}(M)$ and to delete all σ in the candidate set Δ which satisfy $\#(\sigma) < \text{depth}(M)$.

Example 3.2.6. For $M = \langle \partial_2^2, \partial_2\partial_4, \partial_2\partial_3, \partial_1\partial_4^2 \rangle$ in Example 3.1.4, we have $\text{Ass}(M) = \{\{1, 3\}, \{3, 4\}, \{1\}\}$. The set of the standard pairs is

$$\{(1, \{1, 3\}), (\partial_4, \{1, 3\}), (1, \{3, 4\}), (\partial_2, \{1\})\}.$$

The monomial ideals which appear in the context of GKZ-hypergeometric functions are initial ideals of toric ideals. Such monomial ideals have some special features which are essential for constructing hypergeometric series. These features can be used to speed up Algorithm 3.2.5. We briefly review the basics on initial ideals of toric ideals. Details can be found in [96].

Fix a generic weight vector $w \in \mathbf{R}^n$. Let $M := \text{in}_w(I_A)$ be the initial monomial ideal of the toric ideal. By [96, §8], its radical $\text{rad}(M)$ is the Stanley-Reisner ideal of the regular triangulation Δ_w of A defined by w. Thus Δ_w is a simplicial complex on $\{1, 2, \ldots, n\}$ which defines a triangulation of the polytope $\text{conv}(A)$. The triangulation Δ_w can be computed directly with methods from computational geometry, without the use of computer algebra.

Example 3.2.7. We explain the correspondence above between regular triangulations Δ_w and radical of initial ideals $\text{in}_w(I_A)$ by an example.

$$A \;=\; \begin{array}{c} a \\ b \\ c \end{array} \begin{pmatrix} \begin{array}{ccccc} \partial_1 & \partial_2 & \partial_3 & \partial_4 & \partial_5 \\ 1 & 1 & 1 & 1 & 1 \\ 0 & 2 & 0 & 3 & 1 \\ 0 & 0 & 2 & 2 & 1 \end{array} \end{pmatrix}.$$

By using Algorithm 4.5 in [96], we see that the toric ideal I_A equals

$$\mathbf{k}[\partial_1, \partial_2, \partial_3, \partial_4, \partial_5] \cap \langle \partial_1 - a, \partial_2 - ab^2, \partial_3 - ac^2, \partial_4 - ab^3 c^2, \partial_5 - abc \rangle$$

and is generated by

$$\{ \partial_2 \partial_3 - \partial_5^2, \ \partial_1^3 \partial_4^2 - \partial_2 \partial_5^4 \}. \tag{3.16}$$

The Gröbner basis of I_A with respect to the weight vector $w = (1,1,1,1,0)$ is the set (3.16), and hence

$$\mathrm{rad}(\mathrm{in}_w(I_A)) \quad = \quad \langle \partial_2 \partial_3, \partial_1 \partial_4 \rangle.$$

This tells us that $\overline{23}$ and $\overline{14}$ are not faces (edges) of Δ_w and the regular triangulation is

$$\Delta_w = \{\{1,2,5\}, \{1,3,5\}, \{2,4,5\}, \{3,4,5\}\}.$$

See Figure 3.2 for a picture. It is obtained geometrically by projecting the lower convex hull of the following five points in $\mathbf{R}^3$ into the $y_1 y_2$-plane:

Point number	1	2	3	4	5
w (weight)	1	1	1	1	0
y_1	0	2	0	3	1
y_2	0	0	2	2	1

The standard pairs of the monomial ideal $\langle \partial_2 \partial_3, \partial_1^3 \partial_4^2 \rangle$ are

$$(1, \{3,4,5\}), \ (\partial_1, \{3,4,5\}), \ (\partial_1^2, \{3,4,5\}),$$
$$(1, \{2,4,5\}), \ (\partial_1, \{2,4,5\}), \ (\partial_1^2, \{2,4,5\}),$$
$$(1, \{1,3,5\}), \ (\partial_4, \{1,3,5\}), \ (1, \{1,2,5\}), \ (\partial_4, \{1,2,5\}).$$

In our study of GKZ-hypergeometric functions, we typically use the pair $(M, \Delta) = (\mathrm{in}_w(I_A), \Delta_w)$ as the input for Algorithm 3.2.5. Naturally, we first remove faces of cardinality less than depth(M) from Δ_w. The following general lower bound for depth(M) can be quite useful for that purpose:

$$n - 2^{n-d} + 1 \quad \leq \quad \mathrm{depth}(M) \quad \leq \quad \dim(M) \quad = \quad d = \mathrm{rank}(A).$$

The left inequality follows from [92, Theorem 16.5] and appears in [52, Theorem 3.5]. The investigation of initial ideals of toric ideals is an active area in the algebraic theory of integer programming. A recent advance in this area is the following result on the associated primes of $M = \mathrm{in}_w(I_A)$.

Theorem 3.2.8. (Hoşten-Thomas Chain Theorem [52])
Let M be an initial monomial ideal of a toric ideal I_A. If $\sigma \in \mathrm{Ass}(M)$ with $\#(\sigma) < \dim(M)$ then $\sigma \cup \{i\} \in \mathrm{Ass}(M)$ for some $i \in \{1, 2, \ldots, n\} \backslash \sigma$.

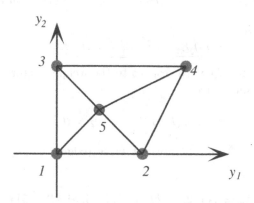

Fig. 3.2. Regular triangulation Δ_w.

This theorem states that the faces of the regular triangulation Δ_w which are associated to $M = \mathrm{in}_w(I_A)$ come in saturated chains. Algorithm 3.2.5 can be arranged to take advantage of this remarkable property. Theorem 3.2.8 is also important for the non-Buchberger algorithm for computing Gröbner bases of I_A presented in [51]. See [96, §12.A and §12.D] for more information.

We shall now return to the (fake) indicial ideals which were discussed in the previous section. We need the following immediate corollary of the above discussion. Recall the definition of $\mathcal{T}(M)$ in Lemma 3.2.4.

Corollary 3.2.9. *Let $M = \mathrm{in}_w(I_A)$ and $\sigma \in \mathrm{Ass}(M)$. Then the columns of A indexed by σ are linearly independent. In particular, if $(\partial^a, \sigma) \in \mathcal{T}(M)$ then the corresponding $d \times d$-submatrix A_σ of A is invertible.*

Proof. The set σ corresponds to a simplex in the regular triangulation $\Delta_w \supset \mathrm{Ass}(M)$, and the set of vertices of a simplex is affinely independent.   □

This corollary implies the following property for each standard pair $(\partial^a, \sigma) \in \mathcal{S}(M)$ and any parameter vector $\beta \in \mathbf{k}^d$: There exists at most one vector $p = (p_1, \ldots, p_n)$ in $\mathbf{k}^n$ satisfying $A \cdot p = \beta$, and $p_i = a_i$ for all $i \notin \sigma$. If this vector p exists then we denote it by $\beta^{(\partial^a, \sigma)}$. In the special case when (∂^a, σ) is a top-dimensional standard pair, that is, when (∂^a, σ) lies in $\mathcal{T}(M)$, then the vector $\beta^{(\partial^a, \sigma)} \in \mathbf{k}^n$ always exists and is unique. Note that the map $\beta \mapsto \beta^{(\partial^a, \sigma)}$ is $\mathbf{k}$-affine; this will become important in Section 3.5.

Theorem 3.2.10. *For generic parameter vectors* $\beta \in \mathbf{k}^d$, *the indicial ideal* $\widetilde{\operatorname{in}}_{(-w,w)}(H_A(\beta))$ *equals the vanishing radical ideal of the finite point set*

$$\{\, \beta^{(\partial^a, \sigma)} \in \mathbf{k}^n \mid (\partial^a, \sigma) \in T(M) \,\}.$$

Proof. By Proposition 3.1.5 and Corollary 3.2.3, the indicial ideal equals

$$\langle A \cdot \theta - \beta \rangle \;+\; \bigcap_{(\partial^a, \sigma) \in \mathcal{S}(M)} \langle \theta_i - a_i : i \notin \sigma \rangle, \qquad (3.17)$$

where the intersection is over <u>all</u> standard pairs. Our claim says that (3.17) equals

$$\bigcap_{(\partial^a, \sigma) \in T(M)} \Big(\langle A \cdot \theta - \beta \rangle + \langle \theta_i - a_i : i \notin \sigma \rangle \Big), \qquad (3.18)$$

since the point $\beta^{(\partial^a, \sigma)}$ corresponds to the maximal ideal in the parentheses in (3.18). The zero set of (3.17) agrees with the zero set of (3.18) since the linear equations $A\theta = \beta$ and $\theta_i = a_i$ for $i \notin \sigma$ have no common solution for generic β when $\#(\sigma) < d$. Moreover, the points $\beta^{(\partial^a, \sigma)}$ are all distinct for $(\partial^a, \sigma) \in T(M)$. Hence the two ideals (3.17) and (3.18) are equal in the local rings at these points. Therefore they are equal in $\mathbf{k}[\theta]$. $\qquad\square$

Here is a description of the initial ideal of the hypergeometric system (for generic parameters) which is more precise than Theorem 3.1.3.

Theorem 3.2.11. *Let w be a generic vector in $\mathbf{R}^n$. For generic parameters β, we have*

$$\operatorname{in}_{(-w,w)}(H_A(\beta)) \;=\; D \cdot \operatorname{top}(\operatorname{in}_w(I_A)) + D \cdot \langle A \cdot \theta - \beta \rangle. \qquad (3.19)$$

Proof. Let $M = \operatorname{in}_w(I_A)$ and $N = \operatorname{top}(M)$. We thus have $M = N \cap Q$, where Q is a monomial ideal of dimension less than d. Lemma 3.2.4 gives an analogous decomposition for the distractions: $\widetilde{M} = \widetilde{N} \cap \widetilde{Q}$. The zero set of $\widetilde{Q}$ is a union of affine subspaces having dimension less than d. The d linear equations $A \cdot \theta = \beta$ have no solution on the zero set of $\widetilde{Q}$, since β is generic. In analogy to the derivation of (3.17) = (3.18), this implies

$$\widetilde{\operatorname{in}}_{(-w,w)}(H_A(\beta)) \;=\; \widetilde{M} + \langle A \cdot \theta - \beta \rangle \;=\; \widetilde{N} + \langle A \cdot \theta - \beta \rangle. \quad (3.20)$$

Since the initial ideal $\operatorname{in}_{(-w,w)}(H_A(\beta))$ contains (3.20), we conclude $\widetilde{N} \subset \operatorname{in}_{(-w,w)}(H_A(\beta))$.

Theorem 3.1.3 implies that the left hand side of (3.19) is contained in the right hand side. The converse follows from a general lemma on distractions of monomial ideals:

Lemma 3.2.12. *Let $M \subset N$ be monomial ideals in $\mathbf{k}[\theta]$, such that* $\operatorname{rad}(M) = \operatorname{rad}(N)$. *Then*

$$D \cdot N = D \cdot M + D \cdot \widetilde{N}.$$

Proof. Clearly, the left hand side contains the right hand side. Conversely, let ∂^a be a monomial in N. For some positive integer r we have $(\partial^a)^r = \partial^{ra} \in M$. It thus suffices to show that ∂^a is a left D-linear combination of ∂^{ra} and $[\theta]_a = x^a \partial^a \in \tilde{N}$. This can be done one variable at a time, by iteratively applying the following lemma for one variable: □

Lemma 3.2.13. *Let $D = \mathbf{k}\langle x, \partial \rangle$ be the scalar Weyl algebra and let $0 < s < r$ be integers. Then ∂^s is a left D-linear combination of ∂^r and $[\theta]_s = x^s \partial^s$.*

Proof. Direct calculation. This completes the proof of Theorem 3.2.11.   □

Corollary 3.2.14. *Suppose β is generic in $\mathbf{k}^d$ and $w, w' \in \mathbf{R}^n$. Then*

$$\mathrm{in}_{(-w,w)}(H_A(\beta)) = \mathrm{in}_{(-w',w')}(H_A(\beta)) \quad \text{holds in } D$$

if and only if

$$\mathrm{top}(\mathrm{in}_w(I_A)) = \mathrm{top}(\mathrm{in}_{w'}(I_A)) \quad \text{holds in } \mathbf{k}[\partial].$$

Proof. It suffices to show this for generic weight vectors w, w', since the small Gröbner fan is determined by its maximal faces. The if-direction follows from Theorem 3.2.11. For the only-if direction we show how to reconstruct $\mathrm{top}(\mathrm{in}_w(I_A))$ from $\mathrm{in}_{(-w,w)}(H_A(\beta))$. First compute the indicial ideal by intersecting with $\mathbf{k}[\theta]$. Next list its roots. They are the vectors $\beta^{(\partial^a,\sigma)}$ in Theorem 3.2.10. Since β is generic, we can recover each standard pair (∂^a, σ) from its corresponding root $\beta^{(\partial^a,\sigma)}$; namely, a is the integer part of $\beta^{(\partial^a,\sigma)}$ and σ is the support of $\beta^{(\partial^a,\sigma)} - a$. We have thus reconstructed the set $\mathcal{T}(\mathrm{in}_w(I_A))$ from $\mathrm{in}_{(-w,w)}(H_A(\beta))$. Lemma 3.2.4 now yields $\mathrm{top}(\mathrm{in}_w(I_A))$.   □

We close this section with a proof of Theorem 2.4.11. We assume $A \in \mathbf{Z}^{d \times n}$ is a rank d matrix, but do not assume that I_A is homogeneous.

Lemma 3.2.15. *For generic β, we have*

$$\mathrm{rank}\big(\mathrm{in}_{(-e,e)}(H_A(\beta))\big) \quad = \quad \mathrm{degree}(\mathrm{in}_e(I_A))$$

and $\quad \mathrm{rank}\big(\mathrm{in}_{(e,-e)}(H_A(\beta))\big) = \begin{cases} \mathrm{degree}(\mathrm{in}_{-e}(I_A)) & \text{if } \dim \mathrm{in}_{-e}(I_A) = d \\ 0 & \text{otherwise.} \end{cases}$

Proof. The proof of Theorem 3.1.3 is valid for inhomogeneous A as well. Hence we have for generic β

$$\mathrm{in}_{(e,-e)}(H_A(\beta)) \quad = \quad D \cdot \mathrm{in}_{-e}(I_A) + D \cdot \langle A\theta - \beta \rangle.$$

We first show that the D-module $D/(D \cdot \mathrm{in}_{-e}(I_A) + D \cdot \langle A\theta - \beta \rangle)$ is monodromic. Let M denote the ideal of $\mathbf{k}[\partial]$ generated by all the monomials in $\mathrm{in}_{-e}(I_A)$. Then we see that if (∂^u, σ) is a standard pair of M, then σ is contained in a facet of conv(A). Each minimal prime of the ideal $\widetilde{M}+\langle A\theta-\beta\rangle$ is of the form $\langle\theta_i - u_i : i \notin \sigma\rangle + \langle A\theta - \beta\rangle$ where (∂^u, σ) is a standard pair of M. Since σ is contained in a facet, the above prime ideal contains an element $\sum_{i\in\sigma} \theta_i - c$ for some $c \in \mathbf{k}$. Therefore that D-module is monodromic.

We conclude that the D-ideal $D \cdot \mathrm{in}_{-e}(I_A) + D \cdot \langle A\theta - \beta\rangle$ is regular holonomic in the same way as the proof of Theorem 2.4.9. Take a generic weight vector $w \in \mathbf{R}^n$. Theorem 2.5.1 implies

$$\mathrm{rank}(D\cdot\mathrm{in}_{-e}(I_A)+D\cdot\langle A\theta-\beta\rangle) = \mathrm{rank}(\mathrm{in}_{(-w,w)}(D\cdot\mathrm{in}_{-e}(I_A)+D\cdot\langle A\theta-\beta\rangle)).$$

Arguments similar to the proof of Theorem 3.1.3 show that, for generic β,

$$\mathrm{in}_{(-w,w)}(D \cdot \mathrm{in}_{-e}(I_A) + D \cdot \langle A\theta - \beta\rangle) = D \cdot \mathrm{in}_w(\mathrm{in}_{-e}(I_A)) + D \cdot \langle A\theta - \beta\rangle.$$

Note that $\mathrm{in}_w(\mathrm{in}_{-e}(I_A)) = \mathrm{in}_{-e+\varepsilon w}(I_A)$ for sufficiently small $\varepsilon > 0$ by Lemma 2.1.6, and it is a monomial ideal whose degree equals degree($\mathrm{in}_{-e}(I_A)$). Since β is generic, we have as in Theorem 3.2.10

$$\mathrm{rank}(\widetilde{\mathrm{in}}_{-e+\varepsilon w}(I_A) + \langle A\theta - \beta\rangle) = \begin{cases} \text{degree}(\mathrm{in}_{-e}(I_A)) & \text{if dim } \mathrm{in}_{-e}(I_A) = d \\ 0 & \text{otherwise.} \end{cases}$$

Theorem 2.3.9 combined with Proposition 3.1.5 implies

$$\mathrm{rank}(D \cdot \mathrm{in}_{-e+\varepsilon w}(I_A) + D \cdot \langle A\theta - \beta\rangle) = \mathrm{rank}(\widetilde{\mathrm{in}}_{-e+\varepsilon w}(I_A) + \langle A\theta - \beta\rangle).$$

This completes the proof of the second equality.

For the first equality, the proof goes in the same way. By Lemma 2.2.2 we have dim $\mathrm{in}_e(I_A) = d$ in this case. $\qquad\square$

Proof (of Theorem 2.4.11). Suppose that $A = (a_1, \ldots, a_n)$ has rank d, and assume that I_A is not homogeneous. Let β be a generic parameter. We introduce the following matrices of sizes $(d + 1) \times n$ and $(d + 1) \times (n + 1)$ respectively:

$$\mathrm{aff}(A) = \begin{pmatrix} 1 & 1 & \cdots & 1 \\ a_1 & a_2 & \cdots & a_n \end{pmatrix} \text{ and } \mathrm{aff}(0, A) = \begin{pmatrix} 1 & 1 & 1 & \cdots & 1 \\ 0 & a_1 & a_2 & \cdots & a_n \end{pmatrix}.$$

The ideal $I_{\mathrm{aff}(A)}$ is the largest homogeneous ideal contained in I_A, and $I_{\mathrm{aff}(0,A)} \subset \mathbf{k}[\partial_0, \partial_1, \ldots, \partial_n]$ is the homogenization of I_A. In what follows, all ideals in $\mathbf{k}[\partial_1, \ldots, \partial_n]$ will be identified with their extension in $\mathbf{k}[\partial_0, \partial_1, \ldots, \partial_n]$. We examine the initial ideals of $I_{\mathrm{aff}(0,A)}$ with respect to $(0, e)$ and $(0, -e)$ respectively. They have the same degree, and their dehomogenizations by setting $x_0 = 1$ are $\mathrm{in}_e(I_A)$ and $\mathrm{in}_{-e}(I_A)$ respectively. In fact, the first initial ideal has no x_0 appearing in its minimal generators, and hence

$$\mathrm{in}_{(0,e)}(I_{\mathrm{aff}(0,A)}) = \mathrm{in}_e(I_A).$$

The second initial ideal has the form

$$\mathrm{in}_{(0,-e)}(I_{\mathrm{aff}(0,A)}) = I_{\mathrm{aff}(A)} + M$$

where M is a monomial ideal each of whose generators contains x_0. This implies

$$\mathrm{in}_{(0,-e)}(I_{\mathrm{aff}(0,A)}) \subset \left(I_{\mathrm{aff}(A)} + \langle x_0 \rangle\right) \cap \mathrm{in}_{-e}(I_A).$$

Since both the matrices $\mathrm{aff}(A)$ and $\mathrm{aff}(0, A)$ are of rank $d+1$, the dimensions of $\mathrm{in}_{(0,-e)}(I_{\mathrm{aff}(0,A)})$ and $I_{\mathrm{aff}(A)} + \langle x_0 \rangle$ equal $d+1$. The dimension of $\mathrm{in}_{-e}(I_A)$ in $\mathbf{k}[x_0, x_1, \ldots, x_n]$ is less than or equal to $d+1$ by Lemma 2.2.2. We write $\mathrm{degree}(\mathrm{in}_{-e}(I_A)) = 0$ if the dimension of $\mathrm{in}_{-e}(I_A)$ in $\mathbf{k}[\partial_0, \partial_1, \ldots, \partial_n]$ is less than $d+1$. Then by the additivity of the degree, we have

$$\begin{aligned}
\mathrm{degree}(\mathrm{in}_{-e}(I_A)) &\leq \mathrm{degree}(\mathrm{in}_{(0,-e)}(I_{\mathrm{aff}(0,A)})) - \mathrm{degree}(I_{\mathrm{aff}(A)}) \\
&< \mathrm{degree}(\mathrm{in}_{(0,-e)}(I_{\mathrm{aff}(0,A)})) \\
&= \mathrm{degree}(\mathrm{in}_{(0,e)}(I_{\mathrm{aff}(0,A)})) \\
&= \mathrm{degree}(\mathrm{in}_e(I_A)).
\end{aligned}$$

Lemma 3.2.15 implies $\mathrm{rank}(\mathrm{in}_{(-e,e)}(H_A(\beta))) > \mathrm{rank}(\mathrm{in}_{(e,-e)}(H_A(\beta)))$. From Theorem 2.5.1, we conclude that $H_A(\beta)$ is not regular holonomic.   □

3.3 The Hypergeometric Fan

Fix a $d \times n$-matrix A satisfying the condition (3.3) above. Let β be a generic parameter vector. We define the *hypergeometric fan* of A to be the small Gröbner fan of the homogeneous $D^{(h)}$-ideal $\overline{H}_A(\beta)$. This is well-defined since initial ideals of $\overline{H}_A(\beta)$ are independent of β, for β ranging in some non-empty Zariski-open subset of $\mathbf{k}^d$. The hypergeometric fan is a complete fan in $\mathbf{R}^n$. By our results in the previous chapter, this fan carries important asymptotic information about the solution space of $H_A(\beta)$. It is the aim of this section to study the combinatorial structure of this fan. The material to be presented requires some experience with combinatorics of polyhedra. Readers mainly interested in D-modules and hypergeometric functions may prefer to skip this section and move on to Section 3.4 first.

Proposition 3.3.1. *Two vectors $w, w' \in \mathbf{R}^n$ lie in the same cone of the hypergeometric fan if and only if*

$$\mathrm{top}(\mathrm{in}_w(I_A)) = \mathrm{top}(\mathrm{in}_{w'}(I_A)) \quad \text{holds in } \mathbf{k}[\partial].$$

Proof. Immediate from Theorem 3.1.10 and Corollary 3.2.14.   □

In the combinatorial literature one can find two closely related fans associated with the matrix A, namely, the Gröbner fan and the secondary fan. We recall their definitions. Two vectors $w, w' \in \mathbf{R}^n$ lie in the same cone of the *Gröbner fan* of A if $\mathrm{in}_w(I_A) = \mathrm{in}_{w'}(I_A)$, and, they lie in the same cone of the *secondary fan* if any of the following three equivalent conditions holds:

- the regular subdivisions Δ_w and $\Delta_{w'}$ coincide;
- $\mathrm{rad}(\mathrm{in}_w(I_A)) = \mathrm{rad}(\mathrm{in}_{w'}(I_A))$;
- $\mathrm{rad}(\mathrm{top}(\mathrm{in}_w(I_A))) = \mathrm{rad}(\mathrm{top}(\mathrm{in}_{w'}(I_A)))$.

Thus the Gröbner fan always refines the secondary fan; see [96, Proposition 8.15]. We obtain the following stronger result from Proposition 3.3.1:

Corollary 3.3.2. *For any $d \times n$-matrix A, the Gröbner fan refines the hypergeometric fan, and the hypergeometric fan refines the secondary fan.*

A common feature of all three fans is that they live in $\mathbf{R}^n$ but their natural dimension is $n - d$, since each of the above equivalence classes contains the row space of the matrix A. For a better understanding of these fans, one must divide out by the row space of A and work in $\mathbf{R}^{n-d}$ instead of $\mathbf{R}^n$. This can be done by the technique of *Gale duality*. We shall explain this technique and then recast the actors in our hypergeometric drama in their Gale dual guise. This will lead to a new combinatorial construction of the hypergeometric fan.

Abbreviate $m = n - d$ and consider an arbitrary $n \times m$-integer matrix B of rank m. Let $b_1, \ldots, b_n$ denote the row vectors of B. They span $\mathbf{R}^m$. We define the following *lattice ideal* in $\mathbf{k}[\partial] = \mathbf{k}[\partial_1, \ldots, \partial_n]$.

$$J_B := \langle \partial^{(Bu)+} - \partial^{(Bu)-} : u \in \mathbf{Z}^m \rangle.$$

Here, ∂^{v+} means $\prod_{v_i > 0} \partial_i^{v_i}$ and ∂^{v-} means $\prod_{v_i < 0} \partial_i^{-v_i}$ for $v \in \mathbf{Z}^n$. It is not difficult to show that the lattice ideal J_B is prime if and only if $\mathbf{Z}\{b_1, \ldots, b_n\} = \mathbf{Z}^m$. This is equivalent to the existence of a $d \times n$-integer matrix A which makes the following sequence of free abelian groups exact:

$$0 \longrightarrow \mathbf{Z}^{n-d} \xrightarrow{B} \mathbf{Z}^n \xrightarrow{A} \mathbf{Z}^d. \qquad (3.21)$$

If (3.21) is exact, then J_B equals our familiar toric ideal I_A, and we call B a *Gale transform* of A. An m-subset τ of $\{1, \ldots, n\}$ is called a *row basis* of B if the set $B_\tau := \{b_i : i \in \tau\}$ is a basis of $\mathbf{R}^m$. We write $\mathrm{pos}(B)$ for the open convex polyhedral cone spanned by the vectors $b_1, \ldots, b_n$ in $\mathbf{R}^m$, and similarly we write $\mathrm{pos}(B_\tau)$ for the open cone spanned by the subset B_τ. Note that the homogeneity assumption (3.3) on A is equivalent to

$$b_1 + b_2 + \cdots + b_n = 0.$$

This condition implies $\mathrm{pos}(B) = \mathbf{R}^m$, but in what follows we shall consider arbitrary matrices B whose cone $\mathrm{pos}(B)$ may be a proper subset of $\mathbf{R}^m$.

Let Σ_B denote the fan in $\mathbf{R}^m$ which is the common refinement of the simplicial cones $\mathrm{pos}(B_\tau)$ where τ runs over all bases of B. The fan Σ_B is a polyhedral subdivision of the cone $\mathrm{pos}(B)$. We call Σ_B the *secondary fan* of the matrix B. Using the exact sequence (3.21), it was shown by Billera, Filliman and Sturmfels [16] that the preimage of Σ_B under the transpose matrix B^T coincides with the secondary fan of A as defined above.

Likewise the Gröbner fan of the lattice ideal J_B can be represented in $\mathbf{R}^m$ instead of $\mathbf{R}^n$. We state this precisely in the following lemma.

Lemma 3.3.3. *Every initial ideal of J_B with respect to a partial term order on $\mathbf{k}[\partial]$ comes from a vector w lying in the cone $\mathrm{pos}(B) \subset \mathbf{R}^m$ as follows:*

$$
\mathrm{in}_w(J_B) \quad = \quad \langle \partial^{(By)+} : y \in \mathbf{Z}^m, \, w \cdot y > 0 \rangle
$$
$$
+ \; \langle \partial^{(By)+} - \partial^{(By)-} : y \in \mathbf{Z}^m, \, w \cdot y = 0 \rangle.
$$

Proof. Every partial term order for J_B can be represented by a non-negative weight vector $v \in \mathbf{R}^n_+$. The m-vector $w := B^T \cdot v$ lies in $\mathrm{pos}(B)$. Using the identity $v^T \cdot (By) = (Bv)^T \cdot y = w^T \cdot y$, we see that $\mathrm{in}_w(J_B)$ consists precisely of the initial forms with respect to v of the binomials in J_B. Conversely, for every $w \in \mathrm{pos}(B)$ we can find $v \in \mathbf{R}^n_+$ with the above properties. $\quad\square$

Two vectors $w, w' \in \mathbf{R}^m$ are called *B-equivalent* if $\mathrm{in}_w(J_B) = \mathrm{in}_{w'}(J_B)$. The equivalence classes of $\mathbf{R}^m$ are the cones in the *Gröbner fan* $\mathcal{G}_B$ of the matrix B. To recover the familiar Gröbner fan of J_B in $\mathbf{R}^n$ one again takes the preimage of $\mathcal{G}_B$ under the transpose matrix B^T.

Example 3.3.4. (2 × 2-matrices)
We illustrate the concepts introduced above by giving an explicit geometric construction of the Gröbner fan $\mathcal{G}_B$ in the special case $m = n = 2$. We write

$$
B \;\; = \;\; \begin{pmatrix} b_{11} & b_{12} \\ b_{21} & b_{22} \end{pmatrix} \qquad \text{where} \; \det(B) = b_{11}b_{22} - b_{12}b_{21} > 0.
$$

The cone $\mathrm{pos}(B)$ consists of all non-negative linear combinations of the rows of B. Let H be the minimal generating set ("Hilbert basis") of the semigroup $\mathrm{pos}(B) \cap \mathbf{Z}^2$. The Hilbert basis H is the set of lattice points on the bounded edges of the convex hull of $\mathrm{pos}(B) \cap (\mathbf{Z}^2 \backslash \{0\})$; see e.g. [84, Proposition 1.21]. Draw the ray $\mathbf{R}_+ \cdot h$ into $\mathrm{pos}(B)$ for each $h \in H$. We shall prove that the resulting decomposition of $\mathrm{pos}(B)$ equals the Gröbner fan $\mathcal{G}_B$.

Let $w = (w_1, w_2) \in \mathbf{R}^2$ be a generic vector in $\mathrm{pos}(B)$. Let $h = (h_1, h_2)$ resp. $h' = (h'_1, h'_2)$ be the unique elements of H immediately adjacent to the left resp. right of w. All 2×2-minors of the following matrix are non-negative:

$$
\begin{pmatrix} b_{11} & h_1 & w_1 & h'_1 & b_{21} \\ b_{12} & h_2 & w_2 & h'_2 & b_{22} \end{pmatrix}.
$$

The reduced Gröbner basis of J_B with respect to w equals

$$
\{ \underline{x_1^{b_{11}h'_2 - b_{12}h'_1}} - x_2^{b_{22}h'_1 - b_{21}h'_2}, \; \underline{x_2^{b_{22}h_1 - b_{21}h_2}} - x_1^{b_{11}h_2 - b_{12}h_1}, \; x_1^{a_1}x_2^{a_2} - 1 \},
$$

where (a_1, a_2) is the w-smallest non-zero non-negative integer vector in the lattice spanned by the columns of B. By construction, the four exponents in the first two binomials are non-negative, and the underlined terms are the w-leading terms. If one of the coordinates of (a_1, a_2) is zero, then $x_1^{a_1}x_2^{a_2} - 1$ coincides with one of the other two binomials; otherwise the above Gröbner basis has cardinality 3. The fact that the reduced Gröbner basis has at most three elements in this situation was observed in [99, Remark 4.6]. The correctness of the exact description above follows from the properties of the Hilbert basis H. More precisely, it can be checked that the S-pair of $x_1^{a_1} x_2^{a_2} - 1$ with either of the other binomials equals the remaining binomial. The above reduced Gröbner bases shows that the cone of the Gröbner fan $\mathcal{G}_B$ containing w is spanned by h and h'. This proves our claim.

Returning to our general discussion, we shall next present an interpretation of the Gröbner fan $\mathcal{G}_B$ and of the secondary fan Σ_B in terms of linear and integer programming. This will provide a combinatorial explanation for the fact that $\mathcal{G}_B$ is always a refinement of the secondary fan Σ_B. For any $u \in \mathbf{Z}^n$ we define the following two convex polyhedra in $\mathbf{R}^m$:

$$P_u := \{ y \in \mathbf{R}^m : B \cdot y \geq u \} \text{ and } Q_u := \text{conv}\{ y \in \mathbf{Z}^m : B \cdot y \geq u \}. \quad (3.22)$$

Thus Q_u is the convex hull of all lattice points in P_u. For w in the interior of $\text{pos}(B)$ let $\text{face}_w(P_u)$ denote the set of all points in P_u at which the linear functional $y \mapsto w \cdot y$ attains its minimum. Then $\text{face}_w(P_u)$ is a bounded face of P_u, and all bounded faces of P_u have this form, for some $w \in \text{pos}(B)$. The same holds for bounded faces of Q_u: they have the form $\text{face}_w(Q_u)$, for some $w \in \text{pos}(B)$. The following result appeared in [99, Theorem 5.4].

Proposition 3.3.5. *Two vectors $w, w' \in \text{pos}(B) \subset \mathbf{R}^m$ lie in the same cone of the secondary fan Σ_B if and only if $\text{face}_w(P_u) = \text{face}_{w'}(P_u)$ for all $u \in \mathbf{Z}^n$, and they lie in the same cone of the Gröbner fan $\mathcal{G}_B$ if and only if $\text{face}_w(Q_u) = \text{face}_{w'}(Q_u)$ for all $u \in \mathbf{Z}^n$.*

Proof. We may assume that w and w' are generic vectors, i.e., they lie in open cones of the relevant fans. This means in particular that the faces of P_u and Q_u defined by these vectors are vertices. For the first claim we note that the w-optimal vertex of P_u is given by some row basis τ of B as follows:

$$\text{face}_w(P_u) \quad = \quad \{ y \in P_u : b_i \cdot y = u_i \text{ for } i \in \tau \}. \quad (3.23)$$

This implies that $w \in \text{pos}(B_\tau)$. Reason: if $w \notin \text{pos}(B_\tau)$ then there exists $\tilde{y} \in \mathbf{R}^m$ with $w \cdot \tilde{y} < 0$ and $b_i \cdot \tilde{y} > 0$ for $i \in \tau$, so that, for $y \in \text{face}_w(P_u)$, the vector $y + \tilde{y}$ also lies in P_u and is w-better than y. Conversely, for every row basis τ there exists a vector $u \in \mathbf{R}^m$ such that the identity (3.23) holds; for instance, set $u_i := 0$ for $i \in \tau$ and set $u_j := -M$ for $j \notin \tau$ where $M \gg 0$. Thus from the representation (3.23) we conclude that $\text{face}_w(P_u) = \text{face}_{w'}(P_u)$ holds for all u if and only if the set $\{ \tau : w \in \text{pos}(B_\tau) \}$ equals $\{ \tau : w' \in \text{pos}(B_\tau) \}$. This proves the first assertion.

For the second assertion, we recall how to solve integer programs using Gröbner bases. Without loss of generality we may assume that the polytope Q_u is non-empty and contains the origin. This means that the right hand side vector u has only non-positive coordinates. Then the optimal vertex

$\text{face}_w(Q_u) = \{y\} \subset \mathbf{Z}^m$ is characterized by the condition that the monomial ∂^{By-u} equals the w-normal form of the monomial ∂^{-u} modulo J_B. Knowing these normal forms for all monomials ∂^{-u} is equivalent to knowing the initial monomial ideal $\text{in}_w(J_B)$. Therefore $\text{in}_w(J_B) = \text{in}_{w'}(J_B)$ holds if and only if $\text{face}_w(Q_u) = \text{face}_{w'}(Q_u)$ for all u. This is precisely our second assertion. $\square$

Proposition 3.3.5 gives a combinatorial proof for the fact that the Gröbner fan refines the secondary fan as follows: For $u \in \mathbf{Z}^n$, there exists a positive integer N such that all the vertices of $P_{Nu} = NP_u$ are lattice points, because the set of vertices of P_u is finite. We have $P_{Nu} = Q_{Nu}$ for such N. Let $w, w' \in \mathbf{R}^m$ lie in the same cone of the Gröbner fan $\mathcal{G}_B$. Then we have $\text{face}_w(Q_{Nu}) = \text{face}_{w'}(Q_{Nu})$ by Proposition 3.3.5. Hence $\text{face}_w(P_{Nu}) = \text{face}_{w'}(P_{Nu})$ and thus $\text{face}_w(P_u) = \text{face}_{w'}(P_u)$. We see w and w' lie in the same cone of the secondary fan Σ_B, again by Proposition 3.3.5.

We are now prepared for our main definition: The *hypergeometric fan* $\mathcal{H}_B$ of the matrix B is the common refinement of the Gröbner fans $\mathcal{G}_{B_\tau}$ where τ runs over all bases of B. The following theorem, which is the main result in this section, implies that this choice of name is justified, since the fan $\mathcal{H}_B$ is characterized by the same geometric condition as in Proposition 3.3.1.

Theorem 3.3.6. *Two vectors $w, w' \in \text{pos}(B) \subset \mathbf{R}^m$ lie in the same cone of the hypergeometric fan $\mathcal{H}_B$ if and only if $\text{top}(\text{in}_w(J_B)) = \text{top}(\text{in}_{w'}(J_B))$.*

Proof. This theorem is proved using the technique of localization in integer programming, proposed in [96, §12.D] and [99, §6], and developed in full detail in [51] and [52]. This theory tells us that, for generic $w \in \mathbf{R}^m$ and a row basis τ of B, the localization of $\text{in}_w(J_B)$ by the monomial prime ideal $\langle \partial_i : i \in \tau \rangle$ can be identified with $\text{in}_w(J_{B_\tau})$. Hence for generic w, the top-part of the initial monomial ideal has the following primary decomposition:

$$\text{top}(\text{in}_w(J_B)) \quad = \quad \bigcap_{\substack{\tau \,:\, \text{basis s.t.} \\ w \in \text{pos}(B_\tau)}} \text{in}_w(J_{B_\tau}). \qquad (3.24)$$

This identity can also be seen directly by applying [52, Theorem 4.2].

Clearly, the hypergeometric fan $\mathcal{H}_B$ refines the secondary fan Σ_B. Suppose that $w, w' \in \text{pos}(B)$ are generic vectors in the same maximal cone of Σ_B. Then they lie in the same cone of $\mathcal{H}_B$ if and only if $\text{in}_w(J_{B_\tau}) = \text{in}_{w'}(J_{B_\tau})$ for all τ such that $w, w' \in \text{pos}(B_\tau)$. In view of (3.24), this holds if and only if $\text{top}(\text{in}_w(J_B)) = \text{top}(\text{in}_{w'}(J_B))$, because $\text{in}_w(J_{B_\tau})$ can be recovered from the intersection (3.24) by localizing at the prime ideal $\langle \partial_i : i \in \tau \rangle$. $\square$

We next present an example of a homogeneous toric ideal $I_A = J_B$ for which the hypergeometric fan $\mathcal{H}_B$ differs from both the Gröbner fan $\mathcal{G}_B$ and the secondary fan Σ_B.

Example 3.3.7. Let $d = 2$, $n = 5$ and $A = \begin{pmatrix} 0 & 1 & 2 & 3 & 4 \\ 4 & 3 & 2 & 1 & 0 \end{pmatrix}$. The following matrix B is a Gale transform of A:

$$B = \begin{pmatrix} 1 & 0 & 0 \\ -2 & 1 & 0 \\ 1 & -2 & 1 \\ 0 & 1 & -2 \\ 0 & 0 & 1 \end{pmatrix}.$$

The toric ideal defines the rational quartic curve in projective 4-space:

$$I_A = J_B = \langle \underline{\partial_1\partial_3 - \partial_2^2}, \underline{\partial_1\partial_4 - \partial_2\partial_3}, \underline{\partial_1\partial_5 - \partial_3^2}, \underline{\partial_2\partial_4 - \partial_3^2}, \underline{\partial_2\partial_5 - \partial_3\partial_4}, \underline{\partial_3\partial_5 - \partial_4^2} \rangle.$$

The secondary polytope of the configuration A is combinatorially equivalent to the 3-dimensional cube. (Note that $m = 3$). Hence the secondary fan Σ_B has eight maximal cones, corresponding to the eight triangulations of the configuration A. The hypergeometric fan $\mathcal{H}_B$ has $28 = 1+2+2+3+5+5+3+7$ maximal cones, and the Gröbner fan $\mathcal{G}_B$ has $42 = 1+2+2+5+8+8+3+13$ maximal cones. They are distributed as follows:

(1) The finest triangulation $\{\{1,2\},\{2,3\},\{3,4\},\{4,5\}\}$ is unimodular and hence supports a unique initial ideal, which is square-free. Its six monomial generators are underlined above.

(2a) The triangulation $\{\{1,3\},\{3,4\},\{4,5\}\}$ supports 2 initial ideals; their tops are distinct.

(2b) The triangulation $\{\{1,2\},\{2,3\},\{3,5\}\}$ supports 2 initial ideals; their tops are distinct.

(3) The triangulation $\{\{1,2\},\{2,4\},\{4,5\}\}$ supports 5 initial ideals; they give 3 distinct tops.

(4a) The triangulation $\{\{1,4\},\{4,5\}\}$ supports 8 initial ideals; they give 5 distinct tops.

(4b) The triangulation $\{\{1,2\},\{2,5\}\}$ supports 8 initial ideals; they give 5 distinct tops.

(5) The triangulation $\{\{1,3\},\{3,5\}\}$ supports 3 initial ideals; their tops are distinct.

(6) The triangulation $\{\{1,5\}\}$ supports 13 initial ideals giving 7 distinct tops.

We explain case (3) in detail. The given triangulation corresponds to the radical ideal $M = \langle \partial_3, \partial_1\partial_4, \partial_1\partial_5, \partial_2\partial_5 \rangle$. The corresponding cone of the Gröbner fan consists of all vectors w in $\mathbf{R}^3$ such that $\mathrm{rad}(\mathrm{in}_w(J_B)) = M$. This cone splits into five cones in the Gröbner cone $\mathcal{G}_B$, corresponding to distinct initial ideals $\mathrm{in}_w(J_B)$. The first two are

$$I_1 = \langle \partial_3\partial_5, \partial_3\partial_4, \partial_1\partial_5, \partial_1\partial_4, \partial_1\partial_3, \partial_2\partial_5^2, \partial_3^2 \rangle \quad \text{and}$$
$$I_2 = \langle \partial_3\partial_5, \partial_3\partial_4, \partial_1\partial_5, \partial_2\partial_3, \partial_1\partial_3, \partial_2\partial_5^2, \partial_1\partial_4^2, \partial_1^2\partial_4, \partial_3^2 \rangle.$$

These two ideals have the same top-part:

$$\text{top}(I_1) = \text{top}(I_2) = \langle \partial_3, \partial_1\partial_4, \partial_1\partial_5, \partial_2\partial_5^2 \rangle.$$

Next we get two more initial ideals I_3 and I_4 by applying the permutation $(15)(24)$ to the indices in I_1 and I_2. The fifth initial ideal is symmetric and has no embedded components:

$$I_5 = \text{top}(I_5) = \langle \partial_3\partial_5, \partial_2\partial_5, \partial_1\partial_5, \partial_1\partial_4, \partial_1\partial_3, \partial_3^2 \rangle.$$

If β is generic then we obtain three distinct fake indicial ideals $\widetilde{I}_j + \langle A\cdot\theta - \beta \rangle$.

Our next result implies that Example 3.3.7 has minimal size among homogeneous toric ideals whose Gröbner fan strictly refines the hypergeometric fan. This result was first conjectured by Hoşten and Thomas [51].

Theorem 3.3.8. *If $m = 2$ then the Gröbner fan $\mathcal{G}_B$ equals the hypergeometric fan $\mathcal{H}_B$.*

Proof. Both fans are in the plane $\mathbf{R}^2$. We give explicit descriptions which show that they are equal. Consider any row basis $\tau = \{i, j\}$. Then $\text{pos}(B_{\{i,j\}})$ is the closed cone spanned by b_i and b_j. Let $H^{i,j}$ be the Hilbert basis of the semigroup $\text{pos}(B_{\{i,j\}}) \cap \mathbf{Z}^2$. Draw the ray $\mathbf{R}_+ \cdot h$ into the plane for each $h \in H^{i,j}$. We know from Example 3.3.4 that the resulting decomposition of $\text{pos}(B_{\{i,j\}})$ coincides with the Gröbner fan $\mathcal{G}_{B_{\{i,j\}}}$. Suppose now that the vectors $b_1, b_2, \ldots, b_n$ are listed in cyclic order in $\mathbf{Z}^2$. If $\text{pos}(B) = \mathbf{R}^2$ then we set $b_{n+1} := b_1$ and we also replace n by $n+1$. Let $i < j$. The fan $\mathcal{G}_{B_{\{i,j\}}}$ is clearly refined by the concatenation of the fans $\mathcal{G}_{B_{\{l,l+1\}}}$ for $i \le l < j$. The common refinement of all the fans $\mathcal{G}_{B_{\{i,j\}}}$ is the hypergeometric fan $\mathcal{H}_B$, by definition. We conclude that $\mathcal{H}_B$ equals the concatenation of the adjacent fans $\mathcal{G}_{B_{\{l,l+1\}}}$ for $l = 1, 2, \ldots, n-1$.

Next we prove each edge of a polygon Q_u in (3.22) is perpendicular to one of the Hilbert basis vectors $h \in H^{i,i+1}$, for some i. Let E be an edge of Q_u with consecutive lattice points L, L'. Suppose E is not defined by $b_i \cdot y = u_i$ for any i; otherwise there is nothing to prove. Let E_l (E_r, respectively) be the nearest edge among the edges on the left (right, respectively) of E defined by $b_i \cdot y = u_i$ for some i. Let $b_j \cdot y = u_j$ ($b_k \cdot y = u_k$, respectively) be the defining equation of E_l (E_r, respectively). Draw four lines through L or L' parallel to E_l or E_r. Then the only lattice points of the obtained parallelogram are L and L'. Rotate the picture by $90°$ around L. Then the vector from L to L' is rotated to one of the Hilbert basis vectors $h \in H^{j,k}$. By the explicit description of $H^{j,k}$ above, we see $h \in H^{i,i+1}$ when $h \in \text{pos}(B_{\{i,i+1\}})$.

Proposition 3.3.5 states that the Gröbner fan $\mathcal{G}_B$ is the common refinement of the normal fans of the polygons Q_u. Therefore each ray in the Gröbner fan $\mathcal{G}_B$ is generated by a vector $h \in H^{i,i+1}$. The construction in the first paragraph of this proof shows that $\mathcal{H}_B$ refines $\mathcal{G}_B$. □

3.4 Logarithm-free Hypergeometric Series

We shall explicitly describe all logarithm-free hypergeometric series. By this we mean series solutions to the hypergeometric system $H_A(\beta)$ which lie in a Nilsson ring as in Section 2.5 but do not contain $\log(x_i)$ for any i.

Let $v = (v_1, \ldots, v_n)$ be a vector in $\mathbf{k}^n$ and $u = (u_1, \ldots, u_n)$ a vector in $\mathbf{Z}^n$. We decompose u into positive and negative part, $u = u_+ - u_-$, where u_+ and u_- are non-negative vectors with disjoint support. Consider the following two scalars in $\mathbf{k}$, which are conveniently abbreviated using falling factorials:

$$[v]_{u_-} = \prod_{i:u_i<0} \prod_{j=1}^{-u_i} (v_i - j + 1)$$

$$[u+v]_{u_+} = \prod_{i:u_i>0} \prod_{j=1}^{u_i} (v_i + j)$$

The following equivalent formulation will also be useful:

$$\partial^{u_-} \bullet x^v = [v]_{u_-} \cdot x^{v-u_-} \tag{3.25}$$

$$\partial^{u_+} \bullet x^{v+u} = [u+v]_{u_+} \cdot x^{v-u_-} \tag{3.26}$$

In what follows we shall first assume that no coordinate v_i is a negative integer; in symbols, $v \in (\mathbf{k}\backslash\mathbf{Z}_-)^n$. This assumption guarantees that

$$[u+v]_{u_+} \neq 0 \qquad \text{for all } u \in \mathbf{Z}^n. \tag{3.27}$$

We set $L := \ker_{\mathbf{Z}}(A)$. This is a rank $n - d$ sublattice of $\mathbf{Z}^n$. Finding a basis of L amounts to computing a Gale transform B as in the previous section.

Proposition 3.4.1. *Let* $v \in (\mathbf{k}\backslash\mathbf{Z}_-)^n$ *and* $\beta = Av$. *Then the formal series*

$$\phi_v := \sum_{u \in L} \frac{[v]_{u_-}}{[v+u]_{u_+}} \cdot x^{v+u} \tag{3.28}$$

is well-defined and is annihilated by the hypergeometric D-ideal $H_A(\beta)$.

Proof. The series ϕ_v is well-defined by (3.27). It is annihilated by the Euler operators $A \cdot \theta - \beta$ because $A \cdot (v+u) = \beta$ for all occurring monomials x^{v+u}. We must show that ϕ_v is annihilated by $\partial^{\tilde{u}_+} - \partial^{\tilde{u}_-}$ for all $\tilde{u} = \tilde{u}_+ - \tilde{u}_- \in L$. In view of (3.25) and (3.26), this claim is equivalent to the identity

$$\frac{(\partial^{u_-} \bullet x^v)}{(\partial^{u_+} \bullet x^{v+u})} \cdot (\partial^{\tilde{u}_-} \bullet x^{v+u}) = \frac{(\partial^{(u+\tilde{u})_-} \bullet x^v)}{(\partial^{(u+\tilde{u})_+} \bullet x^{v+u+\tilde{u}})} \cdot (\partial^{\tilde{u}_+} \bullet x^{v+u+\tilde{u}}). \tag{3.29}$$

We may verify (3.29) separately for each coordinate x_i. There are six cases:
Case 1: $u_i \geq 0$ and $\tilde{u}_i \geq 0$;
Case 2: $u_i \geq 0$, $\tilde{u}_i \leq 0$ and $u_i + \tilde{u}_i \geq 0$;
Case 3: $u_i \geq 0$, $\tilde{u}_i \leq 0$ and $u_i + \tilde{u}_i \leq 0$;
Case 4: $u_i \leq 0$, $\tilde{u}_i \geq 0$ and $u_i + \tilde{u}_i \geq 0$;

Case 5: $u_i \leq 0$, $\tilde{u}_i \geq 0$ and $u_i + \tilde{u}_i \leq 0$;

Case 6: $u_i \leq 0$ and $\tilde{u}_i \leq 0$.

In case 1, the contribution of the i-th coordinate to (3.29) is the easily checked identity

$$\frac{x_i^{v_i}}{(\partial_i^{u_i} \bullet x_i^{v_i+u_i})} \cdot x_i^{v_i+u_i} = \frac{x_i^{v_i}}{(\partial_i^{u_i+\tilde{u}_i} \bullet x_i^{v_i+u_i+\tilde{u}_i})} \cdot (\partial_i^{\tilde{u}_i} \bullet x_i^{v_i+u_i+\tilde{u}_i}).$$

In case 2, the contribution of the i-th coordinate to (3.29) is the identity

$$\frac{x_i^{v_i}}{(\partial_i^{u_i} \bullet x_i^{v_i+u_i})} \cdot (\partial_i^{-\tilde{u}_i} \bullet x_i^{v_i+u_i}) = \frac{x_i^{v_i}}{(\partial_i^{u_i+\tilde{u}_i} \bullet x_i^{v_i+u_i+\tilde{u}_i})} \cdot x_i^{v_i+u_i+\tilde{u}_i}.$$

The other four cases 3,4,5 and 6 are analogous. □

For general vectors $v \in \mathbf{k}^n$ all the numerators $[v]_{u_-}$ in the sum defining ϕ_v will be non-zero. Such series are totally uninteresting for us, since they do not lie in any Nilsson ring and hence do not represent a holomorphic function anywhere. To turn the formal series ϕ_v into something that is analytically meaningful, we need to identify those vectors v for which the support of ϕ_v lies in a pointed cone. This is accomplished in the next theorem. We fix an arbitrary parameter vector $\beta \in \mathbf{k}^d$ and a generic weight vector $w \in \mathbf{R}^n$.

Theorem 3.4.2. *Let $v \in (\mathbf{k}\backslash\mathbf{Z}_-)^n$ be a fake exponent of $H_A(\beta)$ with respect to a generic $w \in \mathbf{R}^n$. Then ϕ_v lies in the corresponding Nilsson ring, and the initial series of ϕ_v with respect to w is equal to the monomial $\mathrm{in}_w(\phi_v) = x^v$.*

Proof. The support of our hypergeometric series ϕ_v equals

$$\mathrm{supp}(\phi_v) = \{u \in L : [v]_{u_-} \neq 0\}.$$

We shall prove the following claim

$$u \in \mathrm{supp}(\phi_v) \quad \text{and} \quad w \cdot u \leq 0 \quad \text{implies} \quad u = 0. \qquad (3.30)$$

This assertion states that x^v is the unique w-smallest monomial appearing in ϕ_v, which is sufficient to prove both parts of Theorem 3.4.2.

Set $M = \mathrm{in}_w(I_A)$. By Lemma 4.1.3, there exists a standard pair $(\partial^a, \sigma) \in \mathcal{S}(M)$ such that $v_i = a_i \in \mathbf{N}$ for $i \notin \sigma$. If $u \in \mathrm{supp}(\phi_v)$ then $u_i + a_i \geq 0$ for all $i \notin \sigma$. (Otherwise $u_i \leq u_i + a_i < 0$ implies $\partial^{-u_i} \bullet x_i^{v_i} = \partial^{-u_i} \bullet x_i^{a_i} = 0$ and hence $[v]_{u_-} = \partial^{u_-} \bullet x^v = 0$.) Therefore

$$\mathrm{supp}(\phi_v) \subset \{u \in L : u_i + a_i \geq 0 \text{ for all } i \notin \sigma\} \qquad (3.31)$$

Let u be a vector in the right hand side of (3.31) which satisfies $w \cdot u \leq 0$. Let u_σ denote the vector in $\mathbf{N}^n$ whose i-th coordinate equals $-\min\{u_i, 0\}$ if $i \in \sigma$ and equals 0 if $i \notin \sigma$. Then both $a+u+u_\sigma$ and $a+u_\sigma$ are non-negative integer vectors, and they satisfy

$$w \cdot (a + u_\sigma) \geq w \cdot (a + u + u_\sigma) \quad \text{and} \quad \partial^{a+u_\sigma} - \partial^{a+u+u_\sigma} \in I_A. \quad (3.32)$$

Since (∂^a, σ) is a standard pair and $\text{supp}(u_\sigma) \subset \sigma$, the monomial ∂^{a+u_σ} does not lie in $M = \text{in}_w(I_A)$. This is a contradiction to (3.32), unless $u = 0$. □

The hypotheses of Theorem 3.4.2 imply in particular that the vector v is automatically an exponent. This has the following remarkable consequence:

Corollary 3.4.3. *A fake exponent which is not an exponent possesses at least one negative integer coordinate.*

Let us assume now that β is generic and $M = \text{in}_w(I_A)$ a monomial ideal as before. Then all fake exponents are exponents, and the exponents are precisely the vectors $v = \beta^{(\partial^a, \sigma)}$, where $(\partial^a, \sigma) \in T(M)$ runs over all top-dimensional standard pairs. Moreover, from Theorem 3.2.10 we see that

$$\text{rank}(H_A(\beta)) \quad = \quad \# T(M) \quad = \quad \text{degree}(I_A) \quad = \quad \text{vol}(A).$$

If β is generic, then no fake exponent v has a negative integer coordinate. This implies that the series constructed above span the solution space of $H_A(\beta)$.

Proposition 3.4.4. *If β is generic then the set of canonical series solutions of $H_A(\beta)$ with respect to w equals $\{ \phi_v : v = \beta^{(\partial^a, \sigma)} \text{ and } (\partial^a, \sigma) \in T(M) \}$.*

Retaining the above notation and genericity assumption on β and w, we abbreviate $\phi_a^\sigma := \phi_v$ for $v = \beta^{(\partial^a, \sigma)}$ with $(\partial^a, \sigma) \in T(M)$. The series ϕ_a^σ are called the *generic A-hypergeometric series* with respect to w.

Example 3.4.5. Let A and w be as in Example 3.1.4. Then $T(M) = \{(1, \{3, 4\}), (1, \{1, 3\}), (\partial_4, \{1, 3\})\}$, and we find

$$\phi_{(0000)}^{\{3,4\}} \quad = \quad x_3^{\beta_1} x_4^{(\beta_2 - 2\beta_1)/3} + w\text{-higher terms}$$

$$\phi_{(0000)}^{\{1,3\}} \quad = \quad x_1^{(2\beta_1 - \beta_2)/6} x_3^{\beta_2/2} + w\text{-higher terms}$$

$$\phi_{(0001)}^{\{1,3\}} \quad = \quad x_1^{(2\beta_1 - \beta_2 + 3)/6} x_3^{(\beta_2 - 3)/2} x_4 + w\text{-higher terms}$$

We shall now determine the supports of the generic A-hypergeometric series and see how they are related to the hypergeometric fan of A.

Lemma 3.4.6. *If β is generic and $(\partial^a, \sigma) \in T(M)$ then*

$$\text{supp}(\phi_a^\sigma) \quad = \quad \{ u \in L : u_i + a_i \geq 0 \text{ for all } i \notin \sigma \}.$$

Proof. The vector $v = \beta^{(\partial^a, \sigma)}$ has the property that v_i is not an integer for $i \in \sigma$ and $v_i = a_i$ if $i \notin \sigma$. If u lies in the right hand side of (3.31) then $v_i + u_i \geq 0$ for all $i \notin \sigma$ and $v_i + u_i$ is not an integer for $i \in \sigma$. This implies $[v]_{u_-} \neq 0$ and hence $u \in \text{supp}(\phi_a^\sigma)$. Hence equality holds in (3.31). □

Fix β generic. The convex hull of $\operatorname{supp}(\phi_a^\sigma)$ in $\mathbf{R}^n$ is a closed convex polyhedron of dimension $\leq n - d$, called the *Newton polyhedron* of ϕ_a^σ. In the next theorem we vary the weight vector $w \in \mathbf{R}^n$, so that we get all possible initial monomial ideals $\operatorname{in}_w(I_A)$ of the toric ideal I_A. This gives us all possible *generic A-hypergeometric series*.

Theorem 3.4.7. *The hypergeometric fan of A is the common refinement of the normal fans of the Newton polyhedra of all generic A-hypergeometric series.*

Proof. To prove this result we use the Gale diagram technique introduced in the previous section. Construct an exact sequence as in (3.21) and identify L with $\mathbf{Z}^{n-d} = \mathbf{Z}^m$ via the $n \times m$-matrix B. Under this identification, the Newton polyhedron of ϕ_a^σ is precisely

$$Q_a^\sigma \;=\; \operatorname{conv}\big\{ y \in \mathbf{Z}^m \;:\; (B \cdot y)_i + a_i \geq 0 \ \text{ for all } \ i \notin \sigma \big\}. \qquad (3.33)$$

This is a closed convex polyhedron in $\mathbf{R}^m$. Its normal fan is the subdivision of $\operatorname{pos}(B)$ consisting of the cones $\{ w \in \operatorname{pos}(B) : \operatorname{face}_w(Q_a^\sigma) = F \}$ where F runs over all faces of Q_a^σ.

The hypergeometric fan $\mathcal{H}_B$ is the common refinement of the Gröbner fans $\mathcal{G}_{B_\tau}$ where τ runs over all row bases of B. By Proposition 3.3.5, the Gröbner fan $\mathcal{G}_{B_\tau}$ is the common refinement of the normal fans of the polyhedra $Q_u = \operatorname{conv}\{ y \in \mathbf{Z}^m : B_\tau \cdot y \geq u \}$ where u runs over $\mathbf{Z}^m$. For all $y \in \mathbf{Z}^m$, the polyhedron Q_{u+By} is a translate of Q_u and hence has the same normal fan. Therefore the right hand side vector u needs to run only over the finite set of residue classes of $\mathbf{Z}^m$ modulo the finite index sublattice $\operatorname{image}_\mathbf{Z}(B_\tau)$. We conclude that $\mathcal{H}_B$ is the common refinement of the normal fans of the polyhedra Q_u where τ runs over all row bases of B and u runs over the finite abelian group $\mathbf{Z}^m / \operatorname{image}_\mathbf{Z}(B_\tau)$.

Note that a d-subset σ of $\{1, \ldots, n\}$ is a column basis of A if and only if its complement $\tau = \{1, 2, \ldots, n\} \backslash \sigma$ is a row basis of B. Thus $\mathcal{H}_B$ is the common refinement of the polyhedra (3.33) where σ runs over column bases of A and $(a_i : i \notin \sigma)$ runs over the group $\mathbf{Z}^m / \operatorname{image}_\mathbf{Z}(B_\tau)$. Now identify I_A with the ideal J_B in Section 3.3 and consider the primary decomposition (3.24) of one of its initial ideals. If $w \in \operatorname{pos}(B_\tau)$ then σ is a facet in the regular triangulation Δ_w and the corresponding standard pairs are (∂^a, σ) where ∂^a runs over the finite set of monomials in $\mathbf{k}[\partial_i \;:\; i \in \tau] \backslash \operatorname{in}_w(J_{B_\tau})$. The exponent vectors a of these monomials form a system of representatives for the group $\mathbf{Z}^m / \operatorname{image}_\mathbf{Z}(B_\tau)$; see e.g. [99, Corollary 3.5]. Therefore $\mathcal{H}_B$ is the common refinement of the normal fans of the polyhedra (3.33) where σ runs over column bases of A and $a = (a_i : i \notin \sigma)$ runs over the vectors indexing the generic A-hypergeometric series ϕ_a^σ. $\qquad \square$

Example 3.4.8. As before let $A = \begin{pmatrix} 3 & 2 & 1 & 0 \\ 0 & 1 & 2 & 3 \end{pmatrix}$. The toric ideal $I_A = \langle \partial_1\partial_3 - \partial_2^2, \ \partial_1\partial_4 - \partial_2\partial_3, \ \partial_2\partial_4 - \partial_3^2 \rangle$ has eight distinct initial monomial ideals:

- $M_1 = \langle \partial_3^2, \partial_2^3, \partial_1 \partial_3, \partial_2 \partial_3 \rangle,\ \mathcal{T}(M_1) = \{(1, \{1, 4\}), (\partial_2, \{1, 4\}), (\partial_2^2, \{1, 4\})\}$
- $M_2 = \langle \partial_2^2, \partial_3^2, \partial_2 \partial_3 \rangle,\ \ \mathcal{T}(M_2) = \{(1, \{1, 4\}), (\partial_2, \{1, 4\}), (\partial_3, \{1, 4\})\}$
- $M_3 = \langle \partial_2^2, \partial_3^3, \partial_2 \partial_3, \partial_2 \partial_4 \rangle,\ \mathcal{T}(M_3) = \{(1, \{1, 4\}), (\partial_3, \{1, 4\}), (\partial_3^2, \{1, 4\})\}$
- $M_4 = \langle \partial_2^2, \partial_2 \partial_3, \partial_2 \partial_4, \partial_1 \partial_4^2 \rangle,\ \mathcal{T}(M_4) = \{(1, \{1, 3\}), (\partial_4, \{1, 3\}), (1, \{3, 4\})\}$
- $M_5 = \langle \partial_2^2, \partial_1 \partial_4, \partial_2 \partial_4 \rangle,\ \ \mathcal{T}(M_5) = \{(1, \{1, 3\}), (\partial_2, \{1, 3\}), (1, \{3, 4\})\}$
- $M_6 = \langle \partial_3^2, \partial_1 \partial_3, \partial_2 \partial_3, \partial_1^2 \partial_4 \rangle,\ \mathcal{T}(M_6) = \{(1, \{1, 2\}), (1, \{2, 4\}), (\partial_1, \{2, 4\})\}$
- $M_7 = \langle \partial_3^2, \partial_1 \partial_3, \partial_1 \partial_4 \rangle,\ \ \mathcal{T}(M_7) = \{(1, \{1, 2\}), (1, \{2, 4\}), (\partial_3, \{2, 4\})\}$
- $M_8 = \langle \partial_1 \partial_3, \partial_1 \partial_4, \partial_2 \partial_4 \rangle,\ \ \mathcal{T}(M_8) = \{(1, \{1, 2\}), (1, \{2, 3\}), (1, \{3, 4\})\}$

The set of all generic A-hypergeometric series is indexed by $\bigcup_{i=1} \mathcal{T}(M_i)$. This set has cardinality 14. Hence there are precisely 14 generic A-hypergeometric series. The common refinement of the supports of their normal fan is a planar fan with eight maximal cones. It equals the Gröbner fan by Theorem 3.3.8.

From now on we consider non-generic parameter vectors β. One interesting special case when the hypothesis of Theorem 3.4.2 holds is when the fake exponent v is a non-negative integer vector, in symbols, $v \in \mathbf{N}^n$. Consider the *integer programming problem*

$$\text{minimize } u \cdot w \text{ subject to } A \cdot u = \beta,\ u \in \mathbf{N}^n \qquad (3.34)$$

This program is *feasible* if and only if the following set is non-empty:

$$F_{A,\beta} \ := \ \{u \in \mathbf{N}^n : Au = \beta\}.$$

The next lemma states that solving the integer programming problem (3.34) is equivalent to finding the unique non-negative integer exponent for $H_A(\beta)$.

Lemma 3.4.9. *If $F_{A,\beta} = \emptyset$ then no exponent lies in $\mathbf{N}^n$. Otherwise there is a unique exponent v in $\mathbf{N}^n$ which is the solution to the integer program (3.34).*

Proof. The first statement is obvious since $F_{A,\beta}$ corresponds to all possible monomials of A-degree β. If there are no such monomials then there is no exponent in $\mathbf{N}^n$. A vector $v \in F_{A,\beta}$ is an exponent if and only if ∂^v is a standard monomial of $\mathrm{in}_w(I_A)$. But ∂^v is a standard monomial of $\mathrm{in}_w(I_A)$ if and only if v is the optimal solution to (3.34); see [96, Algorithm 5.6]. $\square$

For $u = (u_1, \ldots, u_n) \in \mathbf{N}^n$ we introduce the multinomial coefficient

$$\binom{|u|}{u} := \frac{(u_1 + u_2 + \cdots + u_n)!}{u_1!\, u_2! \cdots u_n!}.$$

Lemma 3.4.10. *If $v \in \mathbf{N}^n$ is an exponent then the series ϕ_v is a polynomial, namely,*

$$\phi_v \ = \ \frac{1}{\binom{|v|}{v}} \cdot \sum_{u \in F_{A,\beta}} \binom{|u|}{u} \cdot x^u.$$

Proof. The sum defining ϕ_v is only over the finite set $F_{A,\beta} = \{p + v \in \mathbf{N}^n :$ $p \in L\}$ because $[v]_{p_-} = 0$ if $v + p$ has a negative integer coordinate. It remains to show the following identity for all $p \in L$ with $v + p \in \mathbf{N}^n$:

$$\frac{[v]_{p_-}}{[v + p]_{p_+}} = \frac{\binom{|v+p|}{v+p}}{\binom{|v|}{v}}. \tag{3.35}$$

Since p has coordinate sum zero by our hypothesis (3.3), we have $|v| = |v+p|$, and so the right hand side of (3.35) equals

$$\frac{v_1!}{(v_1 + p_1)!} \frac{v_2!}{(v_2 + p_2)!} \cdots \frac{v_n!}{(v_n + p_n)!}.$$

By grouping this product into two parts, according to $p_i > 0$ or $p_i < 0$, we see that it equals the left hand side of (3.35). $\qquad\square$

The previous two lemmas immediately imply the following result.

Proposition 3.4.11. *The vector space of polynomial solutions to the A-hypergeometric system $H_A(\beta)$ is at most one-dimensional. It is spanned by the polynomial ϕ_v where v is the optimal solution to the problem (3.34).*

Our next goal is to describe all possible logarithm-free hypergeometric series. For this purpose we have to consider series ϕ_v where v has some negative integer coordinates. We have to carefully redefine what this means. For any vector $v = (v_1, \ldots, v_n) \in \mathbf{k}^n$ we define the *negative support* as follows:

$$\text{nsupp}(v) \quad := \quad \{i \in \{1, 2, \ldots, n\} : v_i \text{ is a negative integer}\}.$$

We also introduce the following subset of $L = \ker_{\mathbf{Z}}(A)$:

$$N_v \quad := \quad \{u \in L : \text{nsupp}(v + u) = \text{nsupp}(v)\}.$$

If v lies in $\mathbf{N}^n$ then N_v is finite and in bijection with the set $F_{A,\beta}$ above, but, in general, the set N_v will be infinite. We now redefine our series:

$$\phi_v \quad := \quad \sum_{u \in N_v} \frac{[v]_{u_-}}{[v + u]_{u_+}} \cdot x^{v+u}. \tag{3.36}$$

The next lemma shows that this definition coincides with the old definition (3.28) in the case $\text{nsupp}(v) = \emptyset$, which was our assumption in Theorem 3.4.2.

Lemma 3.4.12. *Let $v \in \mathbf{k}^n$ and $u \in L$. Then $[v]_{u_-} \neq 0$ if and only if $\text{nsupp}(v + u)$ is a subset of $\text{nsupp}(v)$.*

Proof. If $\text{nsupp}(v+u)$ is not a subset of $\text{nsupp}(v)$, then there exists an index i such that $v_i + u_i \in \mathbf{Z}_-$ and $v_i \in \mathbf{N}$. This implies $\partial^{-u_i} \bullet x^{v_i} = 0$ and hence $[v]_{u_-} = 0$ by (3.25). Conversely, $[v]_{u_-} = 0$ implies $v_i + u_i \in \mathbf{Z}_-$ and $v_i \in \mathbf{N}$ for some index i, and hence $\text{nsupp}(v + u)$ is not a subset of $\text{nsupp}(v)$. $\qquad\square$

A vector $v \in \mathbf{k}^n$ is said to have *minimal negative support* if there is no element $u \in L$ such that $\operatorname{nsupp}(v + u)$ is a proper subset of $\operatorname{nsupp}(v)$.

Proposition 3.4.13. *The series ϕ_v defined in (3.36) is annihilated by $H_A(\beta)$ for $\beta = Av$ if and only if the vector $v \in \mathbf{k}^n$ has minimal negative support.*

Proof. We apply Lemma 3.4.12 for both directions of the proof. The if-direction follows with the same argument as in the proof of Proposition 3.4.1. For the only-if direction suppose that ϕ_v is A-hypergeometric but $\operatorname{nsupp}(v+u)$ is a proper subset of $\operatorname{nsupp}(v)$ for some $u \in L$. Then ∂^{u-} does not annihilate x^v. Since $\partial^{u+} - \partial^{u-}$ annihilates ϕ_v, the monomial x^{v+u} must appear with non-zero coefficient in ϕ_v. This is a contradiction since $v + u \notin N_v$. $\square$

We now fix a generic weight vector $w \in \mathbf{R}^n$ and consider the initial ideal $M = \operatorname{in}_w(I_A)$ of the toric ideal as before. Proposition 3.4.13 implies the following improvement of Theorem 3.4.2.

Theorem 3.4.14. *Let $v \in \mathbf{k}^n$ be a fake exponent of $H_A(\beta)$ which has minimal negative support. Then v is an exponent, and the series ϕ_v defined in (3.36) is a canonical solution to the A-hypergeometric system $H_A(\beta)$.*

Proof. The only thing left to prove is that ϕ_v is a canonical solution, which means that x^v is the only starting monomial appearing with non-zero coefficient in ϕ_v. Suppose that $v+u$ is a fake exponent for some $u \in N_v$. We assume that $u \cdot w > 0$; otherwise interchange the roles of v and $v + u$. Since $v + u$ is a fake exponent, $\partial^{u+} \bullet x^{v+u} = 0$. Hence there exists an index i such that $u_i > 0$, $v_i + u_i \in \mathbf{N}$, and v_i is a negative integer. This contradicts $u \in N_v$. $\square$

Fix A, β and w. Let $\operatorname{Minex}_{A,\beta,w} \subset \mathbf{k}^n$ denote the set of fake exponents of $H_A(\beta)$ which have minimal negative support. The converse to Theorem 3.4.14, which generalizes Proposition 3.4.4 is also true:

Corollary 3.4.15. *The set of logarithm-free canonical series solutions of $H_A(\beta)$ with respect to $w \in \mathbf{R}^n$ equals $\big\{ \phi_v : v \in \operatorname{Minex}_{A,\beta,w} \big\}$.*

Proof. Theorem 3.4.14 states that each element in $\big\{ \phi_v : v \in \operatorname{Minex}_{A,\beta,w} \big\}$ is a canonical series solution. For the converse let Ψ be any canonical series solution with respect to a term order on the Nilsson ring N which refines w. Then $\operatorname{in}_w(\Psi) = x^v$ for some exponent $v \in \mathbf{k}^n$. By the same argument as in the proof of Proposition 3.4.13 we see that v must have minimal negative support, i.e., $\operatorname{nsupp}(v + u)$ is not a subset of $\operatorname{nsupp}(v)$ for any $u \in L$. Now, both Ψ and ϕ_v are canonical series solutions having the same starting term. This implies that $\Psi = \phi_v$, and we are done. $\square$

The next proposition will be used in Section 3.6.

Proposition 3.4.16. *Let $v \in \mathbf{k}^n$ be a fake exponent of $H_A(\beta)$ with respect to a generic weight vector $w \in \mathbf{R}^n$. Then the intersection of $v + L$ and $\mathrm{Minex}_{A,\beta,w}$ is not empty. In particular, if $H_A(\beta)$ has only one fake exponent v with respect to w, then $\{v\} = \mathrm{Minex}_{A,\beta,w}$ and ϕ_v is the unique logarithm-free canonical series solution of $H_A(\beta)$ with respect to w.*

Proof. If v has minimal negative support, then there is nothing to prove. Suppose v does not have minimal negative support. Let $v' = v - u' \in v + L$ have minimal negative support $\mathrm{nsupp}(v')$, which is strictly contained in $\mathrm{nsupp}(v)$. Then Lemma 3.4.12 implies $[v]_{u'_+} \neq 0$. Since v is a fake exponent, this means $w \cdot u' < 0$. Hence we have $w \cdot v < w \cdot v'$.

Choose $v' \in v + L$ above so that the value of $w \cdot v'$ is as small as possible. We claim that v' is a fake exponent. To prove this, it is sufficient to show that $u \in L$ and $w \cdot u > 0$ imply that the coefficient $[v']_{u_+}$ in $\partial^{u_+} \bullet x^{v'}$ is zero. Suppose $[v']_{u_+} \neq 0$ with $u \in L$ and $w \cdot u > 0$. Then we see $\mathrm{nsupp}(v' - u) \subseteq \mathrm{nsupp}(v')$ from Lemma 3.4.12, and thus $-u \in N_{v'}$. The inequality $w \cdot (v' - u) < w \cdot v'$ contradicts the choice of v'.   □

3.5 Lower Bound for the Holonomic Rank

In this section we prove the following basic inequality.

Theorem 3.5.1. *Let A be an integer $d \times n$-matrix of rank d which has $(1, 1, \ldots, 1)$ in its row space. For any parameter vector $\beta \in \mathbf{k}^d$ we have*

$$\mathrm{rank}\big(H_A(\beta)\big) \quad \geq \quad \mathrm{vol}(A).$$

Proof. We shall construct an explicit set of $\mathrm{vol}(A)$ many linearly independent logarithmic series solutions to $H_A(\beta)$. These will be elements of the Nilsson ring N introduced in Chapter 2, for some generic weight vector $w \in \mathbf{R}^n$.

We divide the proof into two parts. In the first part of the proof we consider only the special case where none of the exponents v of $H_A(\beta)$ has a negative integer coordinate.

Choose a generic vector β' in $\mathbf{k}^d$ and consider the hypergeometric ideal $H_A(\beta + \varepsilon\beta')$ for small $\varepsilon > 0$. Its fake indicial ideal $\widetilde{\mathrm{fin}}_w(H_A(\beta + \varepsilon\beta'))$ is radical and has precisely $\mathrm{vol}(A)$ many distinct roots. Each of these roots has the form $v + \varepsilon v'$ where v is a root of $\widetilde{\mathrm{fin}}_w(H_A(\beta))$. Since none of the integer coordinates of v is negative, by hypothesis, the same holds true for $v + \varepsilon v'$.

We now fix an exponent v of $H_A(\beta)$ and we write $v + \varepsilon v_1, \ldots, v + \varepsilon v_r$ for the corresponding distinct exponents of $H_A(\beta + \varepsilon\beta')$. Here $v_i = (v_{i1}, \ldots, v_{in}) \in \mathbf{k}^n$. From Theorem 3.4.2 we get r linearly independent hypergeometric series

$$\phi_{v + \varepsilon v_i} \quad = \quad \sum_{u \in L} \frac{[v + \varepsilon v_i]_{u_-}}{[v + u + \varepsilon v_i]_{u_+}} \cdot x^{v + u + \varepsilon v_i}.$$

These r series are solutions to $H_A(\beta + \varepsilon\beta')$, and their initial terms are distinct:

$$\text{in}_w(\phi_{v+\varepsilon v_i}) \quad = \quad x^v \cdot (x^{v_i})^\varepsilon \quad = \quad x^v \cdot \exp(\varepsilon \cdot \log(x^{v_i})).$$

Every appearing coefficient is a rational function in ε,

$$q_{i,u}(\varepsilon) \quad := \quad \frac{[v + \varepsilon v_i]_{u_-}}{[v + u + \varepsilon v_i]_{u_+}} \quad \in \quad \mathbf{k}(\varepsilon),$$

which does not have a pole at $\varepsilon = 0$. We can therefore expand the series

$$\phi_{v+\varepsilon v_i} \quad = \quad x^v \cdot \exp(\varepsilon \cdot \log(x^{v_i})) \cdot \sum_{u \in L} q_{i,u}(\varepsilon) \cdot x^u$$

in the ring $N[[\varepsilon]]$, using the familiar series for the exponential function:

$$\exp(\varepsilon \cdot \log(x^{v_i})) \quad = \quad \sum_{\ell=0}^{\infty} \frac{\varepsilon^\ell}{\ell!} \left(v_{i1} \log(x_1) + \cdots + v_{in} \log(x_n) \right)^\ell \text{ in } N[[\varepsilon]].$$

We introduce a term order on the ring $N[[\varepsilon]]$ as follows: First order terms by their ε-degree, then refine this using the w-order in Definition 2.5.4, and finally break ties on monomials using a lexicographic order.

Using this term order, we now apply Proposition 2.5.7 to the $\mathbf{k}$-linearly independent elements $\phi_{v+\varepsilon v_1}, \ldots, \phi_{v+\varepsilon v_r}$ of the extended Nilsson ring $N[[\varepsilon]]$. There exists an invertible $r \times r$-matrix C over $\mathbf{k}$ such that the coordinates of the vector $(\psi_1(\varepsilon, x), \ldots, \psi_r(\varepsilon, x)) = (\phi_{v+\varepsilon v_1}, \ldots, \phi_{v+\varepsilon v_r}) \cdot C$ are series in $N[[\varepsilon]]$ having distinct initial terms. Each of the new series begins as follows:

$$\psi_i(\varepsilon, x) = \varepsilon^{d_i} \cdot x^v \cdot p_i(\log(x_1), \ldots, \log(x_n)) \quad + \quad \text{higher terms in } \varepsilon,$$

where p_i is a homogeneous polynomial of degree d_i, and each of the higher terms in ε is also higher than x^v with respect to w. By construction, the polynomials $p_1, \ldots, p_r$ are $\mathbf{k}$-linearly independent. For $i = 1, \ldots, r$ we set

$$\widetilde{\psi}_i(x) \quad := \quad \lim_{\varepsilon \to 0} \varepsilon^{-d_i} \cdot \psi_i(\varepsilon, x).$$

This is an element of the Nilsson ring N which begins as follows:

$$\widetilde{\psi}_i(x) = x^v \cdot p_i(\log(x_1), \ldots, \log(x_n)) + \text{higher terms w.r.t. } w .$$

The series $\widetilde{\psi}_1(x), \ldots, \widetilde{\psi}_r(x)$ are $\mathbf{k}$-linearly independent by construction. We claim that they are annihilated by the A-hypergeometric ideal $H_A(\beta)$.

To show this we consider the generators $\partial^u - \partial^v$ and $(A\theta)_j - \beta_j - \varepsilon\beta'_j$ of the D-ideal $H_A(\beta + \varepsilon\beta')$. They annihilate $\psi_i(\varepsilon, x)$ by construction. We regard the series $\psi_i(\varepsilon, x)$ as a power series in ε whose coefficients are elements of N. These coefficients are annihilated by $\partial^u - \partial^v$ since that binomial operator does not contain ε. In particular, the lowest coefficient $\widetilde{\psi}_i(x)$ is annihilated by

$\partial^u - \partial^v$. The operator $(A\theta)_j - \beta_j$ maps $\psi_i(\varepsilon, x)$ to its multiple $\varepsilon\beta'_j \cdot \psi_i(\varepsilon, x)$. In particular, the lowest coefficient $\widetilde{\psi}_i(x)$ is mapped by $(A\theta)_j - \beta_j$ to a multiple of ε, which is necessarily zero. We conclude that $H_A(\beta)$ annihilates $\widetilde{\psi}_i(x)$.

If we carry out the above construction for all the distinct exponents v of $H_A(\beta)$, then we get vol(A) many linearly independent logarithmic series solutions to $H_A(\beta)$. This completes the proof of Theorem 3.5.1 in the special case when no exponent has a negative integer coordinate.

Example 3.5.2. The deformation technique in the above proof for constructing logarithmic series will be demonstrated for the following explicit example. Let $n = 5, d = 3$ and $A = \begin{pmatrix} 1 & 1 & 1 & 1 & 1 \\ -1 & 1 & 1 & -1 & 0 \\ -1 & -1 & 1 & 1 & 0 \end{pmatrix}$ and $\beta = \begin{pmatrix} 1 \\ 0 \\ 0 \end{pmatrix}$. Let $w = (1,1,1,1,0)$, so that I_A has the reduced Gröbner basis $\{\underline{\partial_1\partial_3} - \partial_5^2, \underline{\partial_2\partial_4} - \partial_5^2\}$. This is a maximally degenerate case: there exists only one exponent $v = (0,0,0,0,1)$, whose multiplicity in the fake indicial ideal equals 4. Set $\beta' = \begin{pmatrix} 0 \\ 2 \\ 0 \end{pmatrix}$. For the generic parameter vector $\beta + \varepsilon\beta'$ we get the four distinct exponents $v + \varepsilon v_1, v + \varepsilon v_2, v + \varepsilon v_3$ and $v + \varepsilon v_4$, where

$$v_1 := (-1,1,0,0,0), \quad v_2 := (0,1,1,0,-2),$$

$$v_3 := (0,0,1,-1,0), \quad v_4 := (-1,0,0,-1,2).$$

Two of the four corresponding hypergeometric series are just monomials:

$$
\begin{aligned}
\phi_{v+\varepsilon v_1} &= & x_1^{-\varepsilon} x_2^\varepsilon x_5^1, \\
\phi_{v+\varepsilon v_2} &= x_2^\varepsilon x_3^\varepsilon x_5^{1-2\varepsilon} + \tfrac{2\varepsilon(2\varepsilon-1)}{\varepsilon+1} x_1 x_2^\varepsilon x_3^{1+\varepsilon} x_5^{-1-2\varepsilon} + \cdots \\
\phi_{v+\varepsilon v_3} &= & x_3^\varepsilon x_4^{-\varepsilon} x_5^1, \\
\phi_{v+\varepsilon v_4} &= x_1^{-\varepsilon} x_4^{-\varepsilon} x_5^{1+2\varepsilon} + \tfrac{2\varepsilon(2\varepsilon+1)}{1-\varepsilon} x_1^{1-\varepsilon} x_3 x_4^{-\varepsilon} x_5^{-1+2\varepsilon} + \cdots.
\end{aligned}
$$

We expand these four series into logarithmic series with respect to ε, e.g.,

$$\phi_{v+\varepsilon v_1} = x_5 + (\log(x_2)-\log(x_1))\cdot x_5 \cdot \varepsilon + \frac{(\log(x_2) - \log(x_1))^2}{2} \cdot x_5 \cdot \varepsilon^2 + \cdots.$$

The ε-expansions of the other three hypergeometric series $\phi_{v+\varepsilon v_i}$ have the same initial term x_5, but by taking suitable linear combinations we get series with distinct initial terms:

$$\psi_2(\varepsilon, x) := \phi_{v+\varepsilon v_1} - \phi_{v+\varepsilon v_2} = \left(\log(x_5^2/x_1 x_3) \cdot x_5 + \cdots\right) \cdot \varepsilon + O(\varepsilon^2)$$

$$\psi_3(\varepsilon, x) := \phi_{v+\varepsilon v_2} - \phi_{v+\varepsilon v_3} = \left(\log(x_2 x_4/x_5^2) \cdot x_5 + \cdots\right) \cdot \varepsilon + O(\varepsilon^2)$$
$$\psi_4(\varepsilon, x) :=$$
$$\phi_{v+\varepsilon v_1} - \phi_{v+\varepsilon v_2} + \phi_{v+\varepsilon v_3} - \phi_{v+\varepsilon v_4}$$
$$= \left(\log(x_5^2/x_1 x_3) \cdot \log(x_2 x_4/x_5^2) \cdot x_5 + \cdots\right) \cdot \varepsilon^2 + O(\varepsilon^3)$$

For each series we consider the lowest coefficient with respect to ε:

$$\lim_{\varepsilon \to 0} \frac{1}{\varepsilon} \cdot \psi_2(\varepsilon, x) = \log(x_5^2/x_1 x_3) \cdot x_5 + 2x_2 x_4 x_5^{-1} + 2x_1 x_3 x_5^{-1} + \cdots$$

$$\lim_{\varepsilon \to 0} \frac{1}{\varepsilon} \cdot \psi_3(\varepsilon, x) = \log(x_2 x_4/x_5^2) \cdot x_5 - 2x_2 x_4 x_5^{-1} - 2x_1 x_3 x_5^{-1} + \cdots$$

$$\lim_{\varepsilon \to 0} \frac{1}{\varepsilon^2} \cdot \psi_4(\varepsilon, x) = \log(x_5^2/x_1 x_3) \cdot \log(x_2 x_4/x_5^2) \cdot x_5 - x_1^2 x_3^2 x_5^{-1} + \cdots$$

These series plus $\phi_v = x_5$ are four linearly independent solutions to $H_A(\beta)$.

Proof (continued). We now prove Theorem 3.5.1 in the general case. Let μ denote the maximum number of negative integer coordinates of any fake exponent v of $H_A(\beta)$ with respect to w. The case $\mu = 0$ has been dealt with, so we may assume $\mu > 0$. As before, we introduce the perturbation $\beta \to \beta + \varepsilon \beta'$ and we consider the resulting vol(A) many hypergeometric series

$$\phi_{v+\varepsilon v'} = \sum_{u \in L} q_u(\varepsilon) \cdot x^{v+u+\varepsilon v'}.$$

Each coefficient

$$q_u(\varepsilon) = \frac{[v + \varepsilon v']_{u_-}}{[v + u + \varepsilon v']_{u_+}}$$

is a rational function in ε which may have a pole of order at most μ at $\varepsilon = 0$. Thus $\varepsilon^\mu \cdot \phi_{v+\varepsilon v'}$ can be expanded as an element of the extended Nilsson ring $N[[\varepsilon]]$. We use the same term order on $N[[\varepsilon]]$ as before: first by ε-degree, then by w-weight, etc... Consider the initial term under this term order:

$$\text{in}(\varepsilon^\mu \cdot \phi_{v+\varepsilon v'}) = c \cdot x^a \cdot \log(x)^b \cdot \varepsilon^i \qquad \text{where} \quad c \in \mathbf{k}^*.$$

In the previous case $\mu = 0$ we always had $a = v$ but this need not be true now. The vector a is still an exponent of $H_A(\beta)$, but x^a might be strictly higher than x^v in the w-order. That is the reason why we must consider all exponents simultaneously now.

Let $\nu = \text{vol}(A)$. We apply Proposition 2.5.7 to the ν **k**-linearly independent series $\varepsilon^\mu \cdot \phi_{v+\varepsilon v'}$ in $N[[\varepsilon]]$. Let $\Psi_1(\varepsilon, x), \ldots, \Psi_\nu(\varepsilon, x)$ be the transformed series. They are solutions of $H_A(\beta + \varepsilon \beta')$. Their initial monomials $\text{in}(\Psi_i(\varepsilon, x))$ are distinct, but, if we are unlucky then it can happen that

$$\text{in}(\Psi_i(\varepsilon, x)) = x^a \cdot \log(x)^b \cdot \varepsilon^{c_i} \quad \text{and} \quad \text{in}(\Psi_j(\varepsilon, x)) = x^a \cdot \log(x)^b \cdot \varepsilon^{c_j}.$$

In this case we remove extraneous ε-factors, so that $\Psi_i(0, x) \neq 0$ for all i, and then we apply Proposition 2.5.7 again. This process terminates after finitely many iterations because the ν original series $\phi_{v+\varepsilon v'}$ are in fact linearly independent over the field of fractions of $\mathbf{k}[[\varepsilon]]$.

From this process we obtain ν series $\Psi_1(\varepsilon, x), \ldots, \Psi_\nu(\varepsilon, x)$ in $N[[\varepsilon]]$, which are annihilated by $H_A(\beta + \varepsilon\beta')$, and with the property that their initial monomials $\mathrm{in}(\Psi_i(\varepsilon, x)) = x^a \cdot \log(x)^b$ are distinct and do not contain ε. This shows that $\Psi_1(0, x), \ldots, \Psi_\nu(0, x)$ are linearly independent elements of N which are solutions to the original hypergeometric system $H_A(\beta)$.   □

Example 3.5.3. The following example demonstrates the interplay between different exponents which may occur for $\mu > 0$ and which lead to interesting cancellations in the series. Let $d = 2, n = 3$, $A = \begin{pmatrix} 1 & 1 & 1 \\ 0 & 1 & 2 \end{pmatrix}$, $\beta = \begin{pmatrix} 10 \\ 8 \end{pmatrix}$ and $w = (1, 0, 0)$. The toric ideal I_A is principal with generator $\partial_1\partial_3 - \partial_2^2$. Hence $\mathrm{rank}(H_A(\beta)) = 2$. The two exponents are $(2, 8, 0)$ and $(0, 12, -2)$. From the first (nonnegative) exponent we get a polynomial solution to $H_A(\beta)$:

$$\phi_{(2,8,0)} = x_1^2 x_2^8 + \frac{56}{3} x_1^3 x_3 x_2^6 + 70 x_1^4 x_3^2 x_2^4 + 56 x_1^5 x_3^3 x_2^2 + \frac{14}{3} x_1^6 x_3^4.$$

To construct a second solution, we perturb β in the direction of $\beta' = \begin{pmatrix} 0 \\ 1 \end{pmatrix}$. For the perturbed parameter vector $\beta + \varepsilon\beta' = \begin{pmatrix} 10 \\ 8+\varepsilon \end{pmatrix}$ we get the two hypergeometric series $\phi_{(2-\varepsilon,8+\varepsilon,0)}(x) = \phi_{(2,8,0)}(x) + O(\varepsilon)$ and $\phi_{(0,12-\varepsilon,\varepsilon-2)}(x) =$

$$x_2^{12-\varepsilon} x_3^{\varepsilon-2} + \frac{(\varepsilon-12)(\varepsilon-11)}{\varepsilon-1} x_1 x_2^{10-\varepsilon} x_3^{\varepsilon-1} + \frac{(\varepsilon-12)\cdots(\varepsilon-9)}{2\varepsilon(\varepsilon-1)} x_1^2 x_2^{8-\varepsilon} x_3^{\varepsilon} + \cdots$$

$$= -5940 \cdot \phi_{(2,8,0)}(x) \cdot \varepsilon^{-1} + \left(x_2^{12} x_3^{-2} - 132 x_1 x_2^{10} x_3^{-1} + \cdots \right) + O(\varepsilon).$$

The initial monomial of $\varepsilon \cdot \phi_{(0,12-\varepsilon,\varepsilon-2)}(x)$ in our term order on $N[[\varepsilon]]$ equals $x_1^2 x_2^8$ and hence coincides with the initial monomial of $\phi_{(2-\varepsilon,8+\varepsilon,0)}(x)$. We now cancel initial monomials by taking the linear combination

$$\psi(\varepsilon, x) := \phi_{(0,12-\varepsilon,\varepsilon-2)}(x) + 5940 \cdot \varepsilon^{-1} \cdot \phi_{(2-\varepsilon,8+\varepsilon,0)}(x).$$

This is an element of $N[[\varepsilon]]$. Its lowest coefficient with respect to ε equals

$$\psi(0, x) = x_2^{12} x_3^{-2} - 132 x_1 x_2^{10} x_3^{-1} - 5940 \log(x_2 x_3) x_1^2 x_2^8 + 81180 x_1^3 x_2^6 x_3 + \cdots.$$

This logarithmic series is the desired second solution to the system $H_A(\beta)$.

3.6 Unimodular Triangulations

In this section we study the special case of matrices A and generic weights $w \in \mathbf{R}^n$ such that the initial ideal $\mathrm{in}_w(I_A)$ of the toric ideal $I_A \subset \mathbf{k}[\partial]$ is radical. In other words we now assume that $\text{in}_w(I_A)$ is an ideal generated by square-free monomials in the indeterminates $\partial_1, \partial_2, \ldots, \partial_n$. This hypothesis is equivalent, by [96, Corollary 8.9], to saying that the regular triangulation Δ_w of the configuration A is *unimodular*, which means that all facets σ of Δ_w have normalized volume equal to 1. Following [96, Corollary 8.4], we can express the radical ideal $\text{in}_w(I_A)$ directly in terms of the simplicial complex Δ_w in the following two complementary ways:

$$
\begin{aligned}
\text{in}_w(I_A) &= \langle \partial_{i_1}\partial_{i_2}\cdots\partial_{i_r} : \{i_1, i_2, \ldots, i_r\} \text{ is not a face of } \Delta_w \rangle \\
&= \bigcap_{\sigma \in \Delta_w} \langle \partial_j : j \notin \sigma \rangle
\end{aligned}
$$

In the first formula above it suffices to take only minimal non-faces $\{i_1, \ldots, i_r\}$ of Δ_w, and in the second formula it suffices to take the maximal faces (facets) σ of Δ_w. The zero set of the radical ideal $\text{in}_w(I_A)$ can be identified with Δ_w. A natural example for the above formula is the square-free initial ideal appearing in (1.42) and (1.43) in Section 1.5. The triangulation Δ_w defined by (1.43) is the "staircase triangulation" of the product of simplices $\Delta_1 \times \Delta_{m-1}$.

We shall prove that, under the above hypothesis, the hypergeometric system $H_A(\beta)$ has rank equal to $\text{vol}(A)$ for all $\beta \in \mathbf{k}^d$, and we explicitly determine the indicial ideal $\widetilde{\text{in}}_{(-w,w)}(H_A(\beta))$ and its solutions. Furthermore, assuming that the exponents have no negative integer coordinates, we shall construct the canonical solution basis of $H_A(\beta)$ in the Nilsson ring N corresponding to w. These results apply in particular to the case when A represents a reflexive polytope with midpoint β, and hence we recover results of Hosono, Lian, and Yau [50] and Stienstra [95] related to toric mirror symmetry. In the context of this book, the reflexive case is rather special, but because of importance we discuss it in some detail in the end of this section.

Many important families of hypergeometric systems arise from configurations A which possess unimodular triangulations. We list some examples. The first one, and perhaps most important, is the *Aomoto-Gel'fand system*, i.e., the A-hypergeometric system whose configuration A is the vertex set of a product of two simplices. The toric ideal I_A is the ideal of 2×2-minors of a matrix of indeterminates, and the configuration A is *unimodular*, which means that every initial monomial ideal $\text{in}_w(I_A)$ is square-free [96, Remark 8.10]. The projective variety defined by I_A is the Segre embedding of the product of two projective spaces. See [96, Example 5.1] for a combinatorial discussion of this example. A special case of the Aomoto-Gel'fand system was discussed in Section 1.5.

Many classical varieties of algebraic geometry are toric: for instance, rational normal scrolls, Veronese varieties, and any embedding of a product of (more than two) projective spaces. The defining ideals of all of these varieties possess square-free initial ideals and hence the results of this section apply to the corresponding hypergeometric functions. See [96, §14.A] for an explicit square-free quadratic initial ideal of all varieties of Veronese type. These also include the toric ideals defined by hypersimplices, which appeared in [38].

We now return to our general discussion. For the rest of this section we assume that $\mathrm{in}_w(I_A)$ is square-free and hence Δ_w is unimodular. Each facet σ of Δ_w is the index set of an invertible $d \times d$-submatrix A_σ of A. Its inverse A_σ^{-1} is a well-defined $d \times d$-matrix with entries in the rational numbers, and hence $A_\sigma^{-1}\beta$ is a vector in $\mathbf{k}^d$. We identify $A_\sigma^{-1}\beta$ with the vector in $\mathbf{k}^n$ whose σ-coordinates agree with $A_\sigma^{-1}\beta$ and whose other coordinates are zero. In what follows, the symbol $A_\sigma^{-1}\beta$ always denotes that vector in $\mathbf{k}^n$.

Lemma 3.6.1. *The fake exponents of $H_A(\beta)$ with respect to w are precisely the vectors $A_\sigma^{-1}\beta$, where σ runs over all facets of Δ_w.*

Proof. The distraction of any square-free monomial ideal is equal to that very same ideal. Hence the distraction of $\mathrm{in}_w(I_A)$ equals

$$\widetilde{\mathrm{in}}_w(I_A) \;=\; \bigcap_{\sigma \in \Delta_w} \langle \theta_j : j \notin \sigma \rangle.$$

Therefore the fake indicial ideal equals

$$\widetilde{\mathrm{fin}}_w(H_A(\beta)) \;=\; \left(\bigcap_{\sigma \in \Delta_w} \langle \theta_j : j \notin \sigma \rangle \right) + \langle A\theta - \beta \rangle. \qquad (3.37)$$

The radical of this ideal equals

$$\mathrm{rad}\left(\widetilde{\mathrm{fin}}_w(H_A(\beta)) \right) \;=\; \bigcap_{\sigma \in \Delta_w} \left(\langle \theta_j : j \notin \sigma \rangle + \langle A\theta - \beta \rangle \right).$$

The zeros of this radical ideal are precisely the vectors $A_\sigma^{-1}\beta \in \mathbf{k}^n$ where σ runs over all facets of Δ_w. □

Let $p \in \mathbf{k}^n$ be one of the fake exponents characterized in the previous lemma. We define

$$\mathcal{U}_p \;=\; \{ \sigma \in \Delta_w : A_\sigma^{-1}\beta = p \}.$$

This is a subset of the facets of Δ_w. We regard $\mathcal{U}_p$ as a simplicial complex on $\{1, 2, \ldots, n\}$, just like Δ_w itself. Note that $\mathrm{supp}(p) \subset \{1, 2, \ldots, n\}$ is a face of $\mathcal{U}_p$ and hence of Δ_w. More precisely, we can observe the following fact.

Remark 3.6.2. The simplicial complex $\mathcal{U}_p$ equals the star of $\mathrm{supp}(p)$ in Δ_w, i.e., the facets of $\mathcal{U}_p$ are those facets of Δ_w which contain $\mathrm{supp}(p)$.

We introduce the Stanley-Reisner ideal of $\mathcal{U}_p = \mathrm{star}_{\mathrm{supp}(p)}(\Delta_w)$:

$$I_p \;=\; \bigcap_{\sigma \in \mathcal{U}_p} \langle \theta_j : j \notin \sigma \rangle.$$

Remark 3.6.2 implies that the square-free monomial ideal I_p is simply the image of the given ideal $\widetilde{\mathrm{in}}_w(I_A)$ under replacing $\theta_i \mapsto 1$ for all $i \in \mathrm{supp}(p)$.

Example 3.6.3. Let $d = 3, n = 9$ and $A = \begin{pmatrix} 1 & 1 & 1 & 1 & 1 & 1 & 1 & 1 & 1 \\ 0 & 1 & 2 & 0 & 1 & 2 & 0 & 1 & 2 \\ 0 & 0 & 0 & 1 & 1 & 1 & 2 & 2 & 2 \end{pmatrix}$.

The reduced Gröbner basis of I_A with respect to $w = (2, 0, 0, 0, -1, 0, 0, 0, 2)$ consists of the following twenty quadrics. Initial terms are underlined:

$$\underline{\partial_1 \partial_5} - \partial_2 \partial_4, \; \underline{\partial_1 \partial_8} - \partial_4 \partial_5, \; \underline{\partial_2 \partial_7} - \partial_4 \partial_5, \; \underline{\partial_4 \partial_8} - \partial_5 \partial_7, \; \underline{\partial_1 \partial_3} - \partial_2^2,$$

$$\underline{\partial_1 \partial_6} - \partial_2 \partial_5, \; \underline{\partial_3 \partial_4} - \partial_2 \partial_5, \; \underline{\partial_1 \partial_9} - \partial_5^2, \; \underline{\partial_2 \partial_8} - \partial_5^2, \; \underline{\partial_3 \partial_7} - \partial_5^2,$$

$$\underline{\partial_4 \partial_6} - \partial_5^2, \; \underline{\partial_4 \partial_9} - \partial_5 \partial_8, \; \underline{\partial_6 \partial_7} - \partial_5 \partial_8, \; \underline{\partial_7 \partial_9} - \partial_8^2, \; \underline{\partial_2 \partial_6} - \partial_3 \partial_5,$$

$$\underline{\partial_2 \partial_9} - \partial_5 \partial_6, \; \underline{\partial_3 \partial_8} - \partial_5 \partial_6, \; \underline{\partial_5 \partial_9} - \partial_6 \partial_8, \; \underline{\partial_3 \partial_9} - \partial_6^2, \; \underline{\partial_1 \partial_7} - \partial_4^2.$$

The projective toric variety cut out by these equations is the embedding of $\mathbf{P}^1 \times \mathbf{P}^1$ into $\mathbf{P}^8$ given by the line bundle $\mathcal{O}(2, 2)$. The underlined initial terms define a unimodular triangulation of the square with side length two:

$$\Delta_w \;\; = \;\; \{\, 124, \; 235, \; 245, \; 356, \; 457, \; 568, \; 578, \; 689 \,\}.$$

Let $\beta = (1, 2, 1)^T$. This parameter vector has precisely five fake exponents:

- $p = (-2, 2, 0, 1, 0, 0, 0, 0, 0)$ with $\mathcal{U}_p = \{124\}$ and $I_p = \langle \theta_3, \theta_5, \theta_6, \theta_7, \theta_8, \theta_9 \rangle$.
- $p = (0, -1, 1, 0, 1, 0, 0, 0, 0)$ with $\mathcal{U}_p = \{235\}$ and $I_p = \langle \theta_1, \theta_4, \theta_6, \theta_7, \theta_8, \theta_9 \rangle$.
- $p = (0, 0, 0, 0, 1, 0, -1, 1, 0)$ with $\mathcal{U}_p = \{578\}$ and $I_p = \langle \theta_1, \theta_2, \theta_3, \theta_4, \theta_6, \theta_9 \rangle$.
- $p = (0, 0, 0, -1, 2, 0, 0, 0, 0)$, $\mathcal{U}_p = \{245, 457\}$, $I_p = \langle \theta_2 \theta_7, \theta_1, \theta_3, \theta_6, \theta_8, \theta_9 \rangle$.
- $p = (0, 0, 0, 0, 0, 1, 0, 0, 0)$ with $\mathcal{U}_p = \{356, 568, 689\}$ and
  $I_p = \langle \theta_3 \theta_8, \theta_3 \theta_9, \theta_5 \theta_9, \theta_1, \theta_2, \theta_4, \theta_7 \rangle$.

We invite the reader to prepare a similar chart for the three fake exponents arising from the parameter vector $\beta = (1, 1, 1)^T$.

Proposition 3.6.4. *The primary component supported at the point p of the fake indicial ideal $\widetilde{\mathrm{fin}}_w(H_A(\beta))$ is equal to $I_p + \langle A \cdot \theta - \beta \rangle$.*

Proof. Localizing the fake indicial ideal at the point p amounts to erasing those components $\langle \theta_j : j \notin \sigma \rangle$ of the intersection in (3.37) which satisfy $p_j \neq 0$ for some $j \notin \sigma$. The remaining σ are precisely the facets of $\mathcal{U}_p$, by Remark 3.6.2. We therefore obtain the desired primary decomposition

$$\widetilde{\mathrm{fin}}_w(H_A(\beta)) \;\; = \;\; \bigcap_{\substack{p \, : \, \text{fake} \\ \text{exponent}}} \Big(I_p + \langle A \cdot \theta - \beta \rangle \Big). \tag{3.38}$$

To further analyze the primary component supported at p, it is convenient to translate p to the origin. This translation replaces $I_p + \langle A \cdot \theta - \beta \rangle$ by the homogeneous Artinian ideal $I_p + \langle A \cdot \theta \rangle$.

The ideal I_p is unmixed of dimension d, and its degree is equal to the number $\#\mathcal{U}_p$ of facets of the corresponding simplicial complex $\mathcal{U}_p$. Since the triangulation Δ_w is unimodular, we have the following combinatorial identity

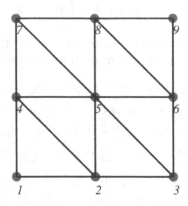

Fig. 3.3. Unimodular triangulation

$$\text{vol}(A) \quad = \quad \#\Delta_w \quad = \sum_{\substack{p\,:\,\text{fake}\\ \text{exponent}}} \#\mathcal{U}_p. \qquad (3.39)$$

Lemma 3.6.5. *The ideal I_p is Cohen-Macaulay, and hence*

$$\text{rank}\big(I_p + \langle A \cdot \theta \rangle\big) \quad = \quad \text{degree}(I_p) \quad = \quad \#\mathcal{U}_p. \qquad (3.40)$$

Proof. The regular triangulation Δ_w is a shellable simplicial complex. Moreover, the star of any face in a shellable complex is again shellable. Therefore $\mathcal{U}_p$ is a shellable simplicial complex. By [94, Theorem III.2.5] this implies that $\mathcal{U}_p$ is a Cohen-Macaulay complex, which means that its Stanley-Reisner ideal I_p is Cohen-Macaulay. The identity (3.40) follows also from [94, Theorem III.2.5] since the cardinality of the k-basis given there equals $\#\mathcal{U}_p$.   □

We are now prepared to state the first main result in this section.

Theorem 3.6.6. *Suppose that $\text{in}_w(I_A)$ is square-free. Then the indicial ideal coincides with the fake indicial ideal and it has the primary decomposition*

$$\widetilde{\text{in}}_{(-w,w)}(H_A(\beta)) \quad = \bigcap_{p\,:\,\text{exponent}} \Big(I_p + \langle A \cdot \theta - \beta \rangle\Big). \qquad (3.41)$$

Moreover, the rank of this Frobenius ideal is equal to $\text{vol}(A)$.

Proof. From (3.38) we see that the rank of the fake indicial ideal equals the sum of the numbers (3.40) as p ranges over all fake exponents. In view of (3.39), this means that the fake indicial ideal $\widetilde{\text{fin}}_w(H_A(\beta))$ has rank equal to $\text{vol}(A)$. The indicial ideal $\widetilde{\text{in}}_{(-w,w)}(H_A(\beta))$ contains $\widetilde{\text{fin}}_w(H_A(\beta))$

and hence has rank at most vol(A). On the other hand, by the result of the previous section and Theorem 2.5.1, the indicial ideal has rank at least vol(A). This implies that the indicial ideal has rank equal to vol(A), and $\widetilde{\text{in}}_{(-w,w)}(H_A(\beta)) = \widetilde{\text{fin}}_w(H_A(\beta))$. Thus each fake exponent p is actually an exponent, and the formula (3.41) follows from (3.38).   □

The above theorem allows us to write down a solution basis to the differential equations represented by $\widetilde{\text{in}}_{(-w,w)}(H_A(\beta))$. We abbreviate $Q_p := I_p + \langle A \cdot \theta \rangle$, and we write $Q_p^\perp$ for its orthogonal complement as in Section 2.3. Using Theorem 2.3.11 we get the following corollary to Theorem 3.6.6.

Corollary 3.6.7. *The space of solutions to* $\widetilde{\text{in}}_{(-w,w)}(H_A(\beta))$ *is spanned by the functions* $g\big(\log(x_1), \ldots, \log(x_n)\big) x_1^{p_1} x_2^{p_2} \cdots x_n^{p_n}$ *where* $p = (p_1, \ldots, p_n)$ *runs over the set* $\{A_\sigma^{-1}\beta \in \mathbf{k}^n : \sigma \in \Delta_w\}$ *of exponents and* g *runs over* $Q_p^\perp$.

We next show how to pick a certain canonical basis for the vector space $Q_p^\perp$. Recall that $\mathcal{U}_p$ is a shellable complex. This means that there exists an ordering $\sigma_1, \ldots, \sigma_r$ of the facets of $\mathcal{U}_p$ such that, for each i, the simplex σ_i has a unique minimal face η_i which is not contained in the simplicial complex generated by $\{\sigma_1, \ldots, \sigma_{i-1}\}$. The faces $\eta_1, \eta_2, \ldots, \eta_r$ play a special role in the decomposition of the complex $\mathcal{U}_p$. In particular, the following result is well-known in combinatorial commutative algebra [94, Theorem III.2.5].

Proposition 3.6.8. *The square-free monomials* $\theta^{\eta_1}, \ldots, \theta^{\eta_r}$ *form a* $\mathbf{k}$-*vector space basis for the graded Artinian ring* $\mathbf{k}[\theta]/Q_p = \mathbf{k}[\theta]/(I_p + \langle A\theta \rangle)$.

Note that $\eta_1 = \emptyset$ and hence $x^{\eta_1} = 1$. We call $\theta^{\eta_1}, \ldots, \theta^{\eta_r}$ the *shelling monomials*. Any other monomial θ^a has a unique $\mathbf{k}$-linear expansion in terms of the shelling monomials modulo Q_p:

$$\theta^a - \sum_{i=1}^r c_{a,i} \cdot \theta^{\eta_i} \in Q_p. \tag{3.42}$$

The rational numbers $c_{a,i}$ are uniquely determined from this property, and they can be computed by Gröbner bases. Clearly, since Q_p is a homogeneous ideal, we have $c_{a,i} = 0$ unless the degree of θ^a equals the degree of θ^{η_i}, which is the cardinality of the set η_i. For $i \in \{1, 2, \ldots, r\}$ define the polynomial

$$g_i(x) := x^{\eta_i} + \sum_a \frac{c_{a,i}}{a!} \cdot x^a, \tag{3.43}$$

where the sum is over all non-shelling monomials of degree equal to $\#\eta_i$.

Proposition 3.6.9. *The polynomials* $g_1, \ldots, g_r$ *form a* $\mathbf{k}$-*basis for* $Q_p^\perp$.

Proof. The proof of this is the same as the one in the end of Section 2.3, where we used Gröbner duality to construct a basis of $I^\perp$. Here we simply use the shelling basis of Proposition 3.6.8 instead of the $\mathbf{k}$-basis consisting of the standard monomials with respect to some term order.   □

We call the homogeneous polynomials $g_1, \ldots, g_r \in \mathbf{k}[x]$ the basis of $Q_p^\perp$ induced by the shelling $\sigma_1, \ldots, \sigma_r$, or just the *shelling basis* for short. Note, however, that different shelling orders will give rise to different shelling bases.

We fix a shelling for each of the complexes $\mathcal{U}_p$. By combining the bases of Proposition 3.6.9, we obtain the following result from Corollary 3.6.7.

Theorem 3.6.10. *The functions*

$$g_i(\log(x_1), \ldots, \log(x_n)) \cdot x_1^{p_1} \cdots x_n^{p_n},$$

for all exponents p and all indices i form a solution basis of $\widetilde{\mathrm{in}}_{(-w,w)}(H_A(\beta))$.

The basis constructed in this theorem is called the *shelling solution basis*. It is a solution basis for the Gröbner deformation $\mathrm{in}_{(-w,w)}(H_A(\beta))$ of our hypergeometric system $H_A(\beta)$, and hence it can be used as the lowest order terms in a solution basis to $H_A(\beta)$. In particular, if we fix a term order on the Nilsson ring N as in Section 2.5, then we can read off the starting monomials for the canonical series solution to $H_A(\beta)$. To illustrate the above results, let us now construct the shelling basis in the case of our earlier example.

Example 3.6.11. Let A and w be as in Example 3.6.3 and set $\beta = (1,1,1)^T$. We compute the shelling solution basis to the indicial ideal $\widetilde{\mathrm{in}}_{(-w,w)}(H_A(\beta))$. The exponent $p = (0,0,0,0,1,0,0,0,0)$ gives the simplicial complex $\mathcal{U}_p = \{\, 235, 356, 568, 578, 457, 245 \,\}$ which has the Stanley-Reisner ideal

$$I_p \;=\; \langle\, \theta_1, \theta_9, \theta_2\theta_6, \theta_2\theta_7, \theta_2\theta_8, \theta_3\theta_4, \theta_3\theta_7, \theta_3\theta_8, \theta_4\theta_6, \theta_4\theta_8, \theta_6\theta_7 \rangle.$$

The given ordering of the six triangles is a shelling of $\mathcal{U}_p$. The resulting shelling basis for the algebra $\mathbf{k}[\theta]/Q_p$ is given in the second row of the following table:

g_1	g_2	g_3	g_4	g_5	g_6
235	356	568	578	457	245
1	θ_6	θ_8	θ_7	θ_4	$\theta_2\theta_4$

Using formulas (3.42) and (3.43) we find the following shelling basis of $Q_p^\perp$:

$$g_1 = 1, \quad g_2 = \underline{x_6} - x_3 - x_5 + x_2, \quad g_3 = \underline{x_8} - 2x_5 + x_2,$$

$$g_4 = \underline{x_7} + x_3 - 2x_5, \quad g_5 = \underline{x_4} + x_3 - x_5 - x_2,$$

$$g_6 = x_2x_4 + x_2x_3 + x_3x_6 + x_4x_7 + x_6x_8 + x_7x_8 - x_2x_5 - x_3x_5 - x_4x_5$$
$$- x_5x_6 - x_5x_7 - x_5x_8 + 3x_5^2 - \tfrac{1}{2}(x_2^2 + x_3^2 + x_4^2 + x_6^2 + x_7^2 + x_8^2)$$

In view of Theorem 3.6.10, we conclude that the six functions

$$g_i(\log(x_1), \ldots, \log(x_6)) \cdot x_5, \quad i = 1, \ldots, 6$$

are linearly independent solutions to the indicial ideal $\widetilde{\mathrm{in}}_{(-w,w)}(H_A(\beta))$. There are two more Laurent monomial solutions coming from the two triangles of Δ_w which do not contain the point 5. The triangle 124 contributes the solution x_2x_4/x_1 and the triangle 689 contributes the solution x_6x_8/x_9.

We now state the main result in this section. We assume that the columns of A span $\mathbf{Z}^d$. This means that $\det(A_\sigma) = \pm 1$ for all facets σ of Δ_w. Recall that a face not in the boundary is called an interior face. Let τ be any interior face of the unimodular triangulation Δ_w. Then its link,

$$\mathrm{link}_\tau(\Delta_w) \quad = \quad \{\sigma \in \Delta_w \ : \ \sigma \cap \tau = \emptyset \text{ and } \sigma \cup \tau \in \Delta_w\}$$

is a simplicial sphere of dimension $d - \#(\tau) - 1$. In what follows we list the facets σ of $\mathrm{link}_\tau(\Delta_w)$ as ordered lists in their natural order so that they form a homology cycle in Δ_w. For an indeterminate parameter vector $\beta = (\beta_1, \ldots, \beta_d)^T$ we consider the n-vector $A^{-1}_{\tau,\sigma}\beta$. The d coordinates of $A^{-1}_{\tau,\sigma}\beta$ indexed by $\tau \cup \sigma$ contain non-zero linear forms with integer coefficients in $\beta_1, \ldots, \beta_d$, while the remaining $n - d$ coordinates are zero. We use the symbol

$$\prod_\sigma A^{-1}_{\tau,\sigma}\beta \qquad (3.44)$$

to denote the product of the coordinates indexed by σ of the vector $A^{-1}_{\tau,\sigma}\beta$. Thus (3.44) is a homogeneous polynomial of degree $\#(\sigma)$ in $\mathbf{Z}[\beta_1, \ldots, \beta_d]$.

Suppose that p is an exponent of the hypergeometric system $H_A(Ap)$. We shall approximate the given parameter vector Ap by the generic parameter vector $\Lambda p + \beta$ and then analyze what happens for $\beta \to 0$. Recall that $\mathrm{supp}(p)$ is a face of Δ_w. Fix any interior face τ of Δ_w which contains $\mathrm{supp}(p)$. For each facet σ of $\mathrm{link}_\tau(\Delta_w)$, the vector $p + A^{-1}_{\tau,\sigma}\beta$ is an exponent of the hypergeometric system $H_A(Ap + \beta)$. It gives rise to a logarithm-free hypergeometric series $\phi_{p+A^{-1}_{\tau,\sigma}\beta}$ as defined in Section 3.4. For fixed generic β, the scalar multiple

$$\frac{1}{\prod_\sigma A^{-1}_{\tau,\sigma}\beta} \cdot \phi_{p+A^{-1}_{\tau,\sigma}\beta} \qquad (3.45)$$

is a series solution to $H_A(Ap + \beta)$. If we sum the expressions (3.45) over all $\sigma \in \mathrm{link}_\tau(\Delta_w)$ then we clearly get another solution to $H_A(Ap + \beta)$. It turns out that this new solution is well-behaved with respect to our deformation, provided that none of the coordinates of the exponent p is a negative integer.

Theorem 3.6.12. *Let* $p \in (\mathbf{k}\backslash\mathbf{Z}_-)^n$ *be an exponent of* $H_A(Ap)$ *with respect to* w *and let* τ *be an interior face of* Δ_w *which contains* $\mathrm{supp}(p)$. *Then the limit*

$$\Psi_{p,\tau} \quad = \quad \lim_{\beta \to 0} \sum_{\sigma \in \mathrm{link}_\tau(\Delta_w)} \frac{1}{\prod_\sigma A^{-1}_{\tau,\sigma}\beta} \cdot \phi_{p+A^{-1}_{\tau,\sigma}\beta} \qquad (3.46)$$

is well-defined and represents a solution to the hypergeometric ideal $H_A(Ap)$.

If τ is a full-dimensional simplex of Δ_w then $\Psi_{p,\tau}$ is equal to the (logarithm-free) series ϕ_p, which is well-defined since $p \in (\mathbf{k}\backslash\mathbf{Z}_-)^n$. Otherwise the limit in Theorem 3.6.12 produces a logarithmic series in the Nilsson ring for $H_A(Ap)$ and w. We demonstrate how this works in a simple example.

Example 3.6.13. Let $d = 2, n = 3, A = \begin{pmatrix} 1 & 1 & 1 \\ 0 & 1 & 2 \end{pmatrix}$, $w = (1,0,0)$. The triangulation Δ_w consists of two unit segments $\{1,2\}$ and $\{2,3\}$. It corresponds to the underlined initial ideal of $I_A = \langle \partial_1 \partial_3 - \partial_2^2 \rangle$. We choose $p = (0,1,0)$ and $\tau = \text{supp}(p) = \{2\}$, which is an interior vertex of Δ_w. Since $\text{link}_\tau(\Delta_w) = \{\{1\}, \{3\}\}$, the two submatrices $A_{\tau,\sigma}$ to be considered are $A_{21} = \begin{pmatrix} 1 & 1 \\ 1 & 0 \end{pmatrix}$ and $A_{23} = \begin{pmatrix} 1 & 1 \\ 1 & 2 \end{pmatrix}$. They spawn the hypergeometric series

$$\phi_{p+A_{21}^{-1}\beta} = x_1^{\beta_1-\beta_2} x_2^{\beta_2+1} + \frac{\beta_2(\beta_2+1)}{\beta_1-\beta_2+1} x_1^{\beta_1-\beta_2+1} x_2^{\beta_2-1} x_3$$
$$+ \frac{(\beta_2+1)\beta_2(\beta_2-1)(\beta_2-2)}{2(\beta_1-\beta_2+1)(\beta_1-\beta_2+2)} x_1^{\beta_1-\beta_2+2} x_2^{\beta_2-3} x_3^2 + \cdots$$

$$\phi_{p+A_{23}^{-1}\beta} = x_2^{2\beta_1-\beta_2+1} x_3^{\beta_2-\beta_1} + \frac{(2\beta_1-\beta_2)(2\beta_1-\beta_2+1)}{\beta_1-\beta_2+1} x_1 x_2^{2\beta_1-\beta_2-1} x_3^{\beta_2-\beta_1+1}$$
$$+ \frac{(2\beta_1-\beta_2+1)\cdots\cdots(2\beta_1-\beta_2-2)}{2(\beta_1-\beta_2+1)(\beta_1-\beta_2+2)} x_1^2 x_2^{2\beta_1-\beta_2-3} x_3^{\beta_2-\beta_1+2} + \cdots$$

The denominator polynomials (3.44) appearing in (3.45) are just linear forms:

$$\prod_{\{1\}} A_{21}^{-1} \cdot \beta = \beta_1 - \beta_2 \quad \text{and} \quad \prod_{\{3\}} A_{23}^{-1} \cdot \beta = \beta_2 - \beta_1.$$

The sum $\frac{1}{\beta_1-\beta_2}\left(\phi_{p+A_{21}^{-1}\beta} - \phi_{p+A_{23}^{-1}\beta}\right)$ in (3.46) is best computed by w-degree:

$$\frac{1}{\beta_1-\beta_2}\left(x_1^{\beta_1-\beta_2} x_2^{\beta_2+1} - x_2^{2\beta_1-\beta_2+1} x_3^{\beta_2-\beta_1}\right) = \log\left(\frac{x_1 x_3}{x_2^2}\right) \cdot x_2 + O(\beta_1, \beta_2).$$

Combining the w-second terms from the series $\phi_{p+A_{21}^{-1}\beta}$ and $\phi_{p+A_{23}^{-1}\beta}$ gives

$$\frac{\beta_2(\beta_2+1)}{(\beta_1-\beta_2)(\beta_1-\beta_2+1)} x_1^{\beta_1-\beta_2+1} x_2^{\beta_2-1} x_3 - \frac{(2\beta_1-\beta_2)(2\beta_1-\beta_2+1)}{(\beta_1-\beta_2)(\beta_1-\beta_2+1)} x_1 x_2^{2\beta_1-\beta_2-1} x_3^{\beta_2-\beta_1+1}$$
$$= -2 \cdot x_1 x_2^{-1} x_3 + O(\beta_1, \beta_2).$$

Similarly, the w-third terms give $-x_1^2 x_2^{-3} x_3^2 + O(\beta_1, \beta_2)$. We conclude

$$\Psi_{p,\tau} = \log\left(\frac{x_1 x_3}{x_2^2}\right) \cdot x_2 - 2 \cdot x_1 x_2^{-1} x_3 - x_1^2 x_2^{-3} x_3^2 + \cdots.$$

The proof of Theorem 3.6.12 will be based on a careful analysis of what happens to the w-initial terms of the summands in (3.46) under the limit $\beta \to 0$. The main point is to construct the polynomial V_τ with the property

$$\text{in}_w(\Psi_{p,\tau}) = V_\tau(\log(x_1), \ldots, \log(x_n)) \cdot x^p. \tag{3.47}$$

For instance, in Example 3.6.13 we saw that $V_\tau(x_1, x_2, x_3) = x_1 - 2x_2 + x_3$. For the general case we need some further combinatorial preparations. Fix $w = (w_1, \ldots, w_n)$ in the interior of the secondary cone of Δ_w. This means that the Δ_w coincides with the normal fan of the unbounded simple polyhedron

$$P_{A,w} := \{u \in \mathbf{R}^d : u \cdot A \le w\}.$$

The bounded i-dimensional faces of $P_{A,w}$ are in bijection with the interior $(d - i - 1)$-dimensional faces τ of Δ_w. The face corresponding to τ equals:

$$\text{face}_\tau(P_{A,w}) = \{u \in \mathbf{R}^d : u \cdot a_i = w_i \text{ if } i \in \tau \text{ and } u \cdot a_j \le w_j \text{ if } j \notin \tau\}.$$

Note that $\text{face}_\tau(P_{A,w})$ is a simple lattice polytope whose normal fan equals $\text{link}_\tau(\Delta_w)$. We consider the *normalized i-dimensional volume* of this polytope:

$$V_\tau(w_1, \ldots, w_n) := \text{volume}(\text{face}_\tau(P_{A,w})).$$

Here "normalized" means that a simplex whose vertices affinely span the lattice $\{u \in \mathbf{Z}^d : u \cdot a_i = w_i \text{ if } i \in \tau\}$ has volume 1; cf. [96, Theorem 4.16]. Note that V_τ is a polynomial function as $w = (w_1, \ldots, w_n)$ ranges over the secondary cone of Δ_w. We call $V_\tau(w_1, \ldots, w_n)$ the *volume polynomial* of τ.

Remark 3.6.14. The volume polynomial V_τ is homogeneous of degree $d - \#(\tau)$.

A non-trivial example of a volume polynomial is the quadric g_6 in Example 3.6.11. If $\tau = \{5\}$ then $\text{face}_\tau(P_{A,w})$ is a centrally symmetric hexagon whose normalized area equals $g_6(w_1, \ldots, w_9) = V_\tau(w_1, \ldots, w_9)$. In Lemma 3.6.20 below we shall see that the vector spaces $Q_p^\perp$ are always spanned by volume polynomials. Our first goal is to express V_τ in terms of exponential sums. Such formulas are related to toric geometry, and they have been studied by many authors, including Barvinok, Brion-Vergne, Khovanski-Puhlikov, Lawrence and Morelli. Our discussion will follow the work of Barvinok in [9].

Proposition 3.6.15. (cf. [9, Section 3]) *The volume polynomial for τ equals*

$$V_\tau(w_1, \ldots, w_n) = \lim_{\beta \to 0} \sum_{\sigma \in \text{link}_\tau(\Delta_w)} \frac{1}{\prod_\sigma A_{\tau,\sigma}^{-1} \beta} \cdot \exp(w \cdot A_{\tau,\sigma}^{-1} \beta) \quad (3.48)$$

Proof. Consider the following integral where du denotes the normalized volume measure on the affine space $\{u \in \mathbf{R}^d : u \cdot a_i = w_i \text{ if } i \in \tau\}$:

$$\int_{u \in \text{face}_\tau(P_{A,w})} \exp(u \cdot \beta) \, du. \quad (3.49)$$

If we set $\beta \to 0$ in (3.49) then we clearly get the left hand side of (3.48). Therefore it remains to be shown that the integral (3.49) is equal to the sum on the right hand side of (3.48). This is a consequence of [9, Theorem 2.6]. $\square$

Example 3.6.16. Let A, Δ_w and $\tau = \{5\}$ as in Example 3.6.11. We display the formula (3.48) for the quadratic volume polynomial $g_6 = V_\tau$. Note that

$$\text{link}_\tau(\Delta) = \{ \{2, 3\}, \{3, 6\}, \{6, 8\}, \{8, 7\}, \{7, 4\}, \{4, 2\} \}.$$

If we list the six facets σ of $\text{link}_\tau(\Delta)$ in the order given above, then the six summands on the right hand side of (3.48) are found to be

$$\frac{\exp\Big((2\beta_1-\beta_2-\beta_3)w_2 + (-\beta_1+\beta_2)w_3 + \beta_3 w_5\Big)}{(2\beta_1-\beta_2-\beta_3)(-\beta_1+\beta_2)}$$

$$+ \frac{\exp\Big((\beta_1-\beta_3)w_3 + (-2\beta_1+\beta_2+\beta_3)w_6 + (2\beta_1-\beta_2)w_5\Big)}{(\beta_1-\beta_3)(-2\beta_1+\beta_2+\beta_3)}$$

$$+ \frac{\exp\Big((-\beta_1+\beta_2)w_6 + (-\beta_1+\beta_3)w_8 + (3\beta_1-\beta_2-\beta_3)w_5\Big)}{(-\beta_1+\beta_2)(-\beta_1+\beta_3)}$$

$$+ \frac{\exp\Big((-2\beta_1+\beta_2+\beta_3)w_8 + (\beta_1-\beta_2)w_7 + (2\beta_1-\beta_3)w_5\Big)}{(-2\beta_1+\beta_2+\beta_3)(\beta_1-\beta_2)}$$

$$+ \frac{\exp\Big((-\beta_1+\beta_3)w_7 + (2\beta_1-\beta_2-\beta_3)w_4 + \beta_2 w_5\Big)}{(-\beta_1+\beta_3)(2\beta_1-\beta_2-\beta_3)}$$

$$+ \frac{\exp\Big((\beta_1-\beta_2)w_4 + (\beta_1-\beta_3)w_2 + (-\beta_1+\beta_2+\beta_3)w_5\Big)}{(\beta_1-\beta_2)(\beta_1-\beta_3)}.$$

If we expand each of the six occurrences of the exponential function into the usual series then we see that the poles in $\beta_1, \beta_2, \beta_3$ miraculously disappear, and the above sum equals $V_\tau(w_1, \ldots, w_9) + O(\beta_1, \beta_2, \beta_3)$.

We shall now relate the volume polynomials to the hypergeometric system.

Corollary 3.6.17. *If* $\text{supp}(p) \subset \tau$ *then* $V_\tau\big(\log(x_1), \ldots, \log(x_n)\big)$ *is annihilated by* $Q_p = I_p + \langle A\theta \rangle$.

Proof. For any row vector $v = (v_1, \ldots, v_d)$, the volume polynomial satisfies

$$V_\tau(w + v \cdot A) \quad = \quad V_\tau(w).$$

This means that $V_\tau(w)$ is annihilated by the d coordinates of $A \cdot \partial_w$, and therefore $V_\tau\big(\log(x_1), \ldots, \log(x_n)\big)$ is annihilated by the d coordinates of $A\theta$. To prove that $V_\tau\big(\log(x_1), \ldots, \log(x_n)\big)$ is annihilated by I_p, we will show that $V_\tau(w)$ is annihilated by $\frac{\partial^s}{\partial w_{i_1} \cdots \partial w_{i_s}}$ whenever $\{i_1, \ldots, i_s\}$ is not a face of

$$\text{star}_\tau(\Delta_w) \quad = \quad \tau \star \text{link}_\tau(\Delta_w) \subset \mathcal{U}_p.$$

Equivalently, the support of each monomial occurring in the expansion of $V_\tau(w)$ lies in $\text{star}_\tau(\Delta_w)$. This follows from Proposition 3.6.15. □

We are now ready to prove our main result.

Proof (of Theorem 3.6.12). It follows from Corollary 3.6.17 that

$$V_\tau\big(\log(x_1), \ldots, \log(x_n)\big) \cdot x^p$$

is annihilated by the initial ideal $\text{in}_{(-w,w)}(H_A(Ap))$. Since the rank of $H_A(Ap)$ coincides with the rank of $H_A(Ap + \beta)$, it follows from the limit construction of logarithmic solutions in the previous section that there exist meromorphic functions $c_\sigma(\beta) = c_\sigma(\beta_1, \ldots, \beta_d)$ with the properties

$$\sum_{\sigma \in \mathrm{link}_\tau(\Delta_w)} c_\sigma(\beta) \cdot \mathrm{in}_w(\phi_{p+A_{\tau,\sigma}^{-1}\beta}) = V_\tau\big(\log(x_1), \ldots, \log(x_n)\big) \cdot x^p + O(\beta) \quad (3.50)$$

and

$$\sum_{\sigma \in \mathrm{link}_\tau(\Delta_w)} c_\sigma(\beta) \cdot \phi_{p+A_{\tau,\sigma}^{-1}\beta} \quad \text{is a Nilsson series solution to } H_A(Ap+\beta). \quad (3.51)$$

Moreover, the identity (3.50) uniquely characterizes the multipliers $c_\sigma(\beta)$ modulo $O(\beta)$. Now, if we substitute $w_i = \log(x_i)$ into (3.48) then we see that (3.50) holds for the $c_\sigma(\beta) = 1/\prod_\sigma A_{\tau,\sigma}^{-1}\beta$. We are done in view of (3.51). □

In the notation of Theorem 3.6.12 we can record the following fact:

Corollary 3.6.18. *The Nilsson series $\Psi_{p,\tau}$ is specified uniquely by (3.47).*

We shall next prove that the series $\Psi_{p,\tau}$ span the space of all hypergeometric series associated with the exponent $p \in (\mathbf{k}\backslash\mathbf{Z}_-)^n$ and we select a subset which forms a basis. As before, let $\sigma_1, \sigma_2, \ldots, \sigma_r$ be a shelling of $\mathcal{U}_p$ and let Σ_i denote the simplicial complex spanned by $\{\sigma_1, \ldots, \sigma_i\}$ for all i. Then σ_i has a unique smallest face η_i which is not in Σ_{i-1}. We write $\tau_i := \sigma_i\backslash\eta_i$ for the complementary face of σ_i. It is easy to check that $\mathrm{supp}(p) \subset \tau_i$. Note also that $\tau_1 = \sigma_1$.

Lemma 3.6.19. *The face τ_i is an interior face of Σ_i and hence also of $\mathcal{U}_p$ and of the triangulation Δ_w. Moreover, $\mathrm{star}_{\tau_i}(\Delta_w)$ is a subcomplex of Σ_i.*

Proof. We may assume $i > 1$. Every facet of σ_i which contains τ_i is also a face of Σ_{i-1}. The set of these facets equals the star of τ_i in the boundary of the complex Σ_{i-1}. Therefore, after attaching σ_i to form Σ_i, the face τ_i has become an interior face. The second assertion follows from the fact that Δ_w is a d-ball containing the d-ball Σ_i. Namely, $\mathrm{link}_{\tau_i}(\Sigma_i) \subset \mathrm{link}_{\tau_i}(\Delta_w)$ are simplicial spheres of the same dimension, and hence they must be equal. □

The previous lemma ensures that the volume polynomial V_{τ_i} is well-defined and that $V_{\tau_i}(\log(x_1), \ldots, \log(x_n))$ is a solution of $Q_p = I_p + \langle A\theta \rangle$.

Lemma 3.6.20. *The volume polynomials $V_{\tau_1}, \ldots, V_{\tau_r}$ are a $\mathbf{k}$-basis of $Q_p^\perp$.*

Proof. First note that $V_{\tau_i}(x_1, \ldots, x_n)$ is a homogeneous polynomial of degree $d - \#(\tau_i) = \#(\eta_i)$. By Lemma 3.6.19, the support of every monomial appearing in the expansion of V_{τ_i} lies in Σ_i. Therefore V_{τ_i} is a $\mathbf{k}$-linear combination of the first i polynomials $g_1, \ldots, g_i$ in the shelling basis of Proposition 3.6.9. Moreover, it follows from Proposition 3.6.15 that the monomial x^{η_i} appears with coefficient 1 in the expansion of V_{τ_i}. We conclude that the $r \times r$-matrix which expresses the volume polynomials $V_{\tau_1}, \ldots V_{\tau_r}$ in the shelling basis $g_1, \ldots, g_r$ of $Q_p^\perp$ is upper-triangular with 1's on the main diagonal. □

Theorem 3.6.21. *Let $p \in (k \backslash \mathbf{Z}_-)^n$ be an exponent of $H_A(Ap)$ and $\tau_1, \ldots, \tau_r$ as above. The hypergeometric series $\Psi_{p,\tau_1}, \ldots, \Psi_{p,\tau_r}$ constructed in Theorem 3.6.12 form a $\mathbf{k}$-basis for the subspace of solutions to $H_A(Ap)$.*

Proof. This follows from Corollary 3.6.18 and Lemma 3.6.20.   □

Corollary 3.6.22. *Suppose that $\mathrm{in}_w(I_A)$ is square-free and none of the exponents p of $H_A(\beta)$ has a negative integer coordinate. Then the hypergeometric series Ψ_{p,τ_i} gotten by shelling of each $\mathcal{U}_p$ form a solution basis for $H_A(\beta)$.*

This corollary applies to the case of *reflexive polytopes* which play an important role in mirror symmetry. The parameters of interest there are called *maximally degenerate*. We refer to Hosono, Lian, and Yau [49], [50] and Stienstra [95] for details on this theory and many references. In our notation the maximally degenerate case is described combinatorially as follows:

The *core* of the unimodular triangulation Δ_w is defined as the intersection of all facets of Δ_w; see [95, equation (16)]. Equivalently, core(Δ_w) consists of all indices i such that no minimal generator of $\mathrm{in}_w(I_A)$ contains x_i. Clearly,

$$\mathrm{star}_{\mathrm{core}(\Delta_w)}(\Delta_w) \;=\; \Delta_w.$$

We say that a parameter vector $\beta \in \mathbf{k}^d$ lies in the core of Δ_w if β is a $\mathbf{k}$-linear combination of the column vectors a_i, $i \in \mathrm{core}(\Delta_w)$. It is easy to see that if β lies in the core of Δ_w then the hypergeometric system $H_A(\beta)$ has only one fake exponent $p = p(\beta)$ in direction w. Note that the zero vector lies in the core of any unimodular triangulation Δ_w, so the following theorem applies to the parameter vector $\beta = (0, 0, \ldots, 0)^T$ whenever $\mathrm{in}_w(I_A)$ is square-free.

Theorem 3.6.23. *Fix a shelling of the unimodular triangulation Δ_w with special faces $\tau_1, \ldots, \tau_r$ as above and suppose that $\beta \in \mathbf{k}^d$ is in the core of Δ_w. Then the hypergeometric series $\Psi_{p,\tau_1}, \ldots, \Psi_{p,\tau_r}$ are a solution basis to $H_A(\beta)$.*

Proof. Since there is only one fake exponent $p = p(\beta)$, we no longer need to require that p have no negative integer coordinates by Proposition 3.4.16, and Corollary 3.6.22 does apply in this situation. Note that $\mathrm{supp}(p) \subset \mathrm{core}(\Delta_w) \subset \tau_i$ for all i. Hence the Nilsson series Ψ_{p,τ_i} are well-defined and have the required properties.   □

4. Rank versus Volume

Let A be a $d \times n$-integer matrix of rank d and $I_A \subset \mathbf{k}[\partial]$ its toric ideal as before. Throughout this chapter we assume the homogeneity condition (3.3). This means that I_A is a homogeneous ideal, i.e., I_A is generated by binomials $\partial_1^{a_1} \partial_2^{a_2} \cdots \partial_n^{a_n} - \partial_1^{b_1} \partial_2^{b_2} \cdots \partial_n^{b_n}$, where $a_1 + a_2 + \cdots + a_n = b_1 + b_2 + \cdots + b_n$. The convex hull of the columns of A is a polytope of dimension $d - 1$, denoted by $\mathrm{conv}(A)$, and its normalized volume is denoted by $\mathrm{vol}(A)$. Gel'fand, Kapranov and Zelevinsky found in their original work that the holonomic rank of the GKZ-hypergeometric system $H_A(\beta)$ is generally equal to the volume $\mathrm{vol}(A)$.

Let us summarize what we have seen concerning the holonomic rank of $H_A(\beta)$. The explicit construction of logarithmic series solutions in Theorem 3.5.1 shows that $\mathrm{vol}(A)$ is a lower bound for $\mathrm{rank}(H_A(\beta))$ for all parameters:

$$\mathrm{rank}(H_A(\beta)) \quad \geq \quad \mathrm{vol}(A). \tag{4.1}$$

If $\mathrm{conv}(A)$ admits a unimodular regular triangulation Δ_w as in Section 3.6, then the holonomic rank is equal to the volume, by Theorem 3.6.6.

For generic parameters β, and generic weight vectors $w \in \mathbf{R}^n$,

$$\mathrm{rank}(\mathrm{in}_{(-w,w)}(H_A(\beta))) \quad = \quad \mathrm{rank}(\widetilde{\mathrm{in}}_{(-w,w)}(H_A(\beta))) \quad = \quad \mathrm{vol}(A).$$

The first equality is Theorem 2.3.9, and the second is Theorem 3.2.10. Since $H_A(\beta)$ is regular holonomic, Theorem 2.5.1 shows that, over the complex field $\mathbf{k} = \mathbf{C}$,

$$\mathrm{rank}(\mathrm{in}_{(-w,w)}(H_A(\beta))) \quad = \quad \mathrm{rank}(H_A(\beta)). \tag{4.2}$$

The same result holds over any field of characteristic zero, since the holonomic rank is invariant under field extensions, and it suffices to work in $\mathbf{Q}(\beta_1, \ldots, \beta_d)$ which can always be embedded into $\mathbf{C}$. We therefore conclude

$$\mathrm{rank}(H_A(\beta)) \quad = \quad \mathrm{vol}(A) \quad \text{for generic } \beta. \tag{4.3}$$

This identity need not hold for special β; see Example 4.2.7 below.

In this chapter we present three sufficient conditions (Theorems 4.3.8, 4.5.2 and 4.6.1) for the holonomic rank of $H_A(\beta)$ to coincide with the volume $\mathrm{vol}(A)$. These three conditions are different by nature. The first condition, the Cohen-Macaulayness of the toric ideal I_A, is parameter-free, depending only on the matrix A. The second one, the semi-nonresonance of the parameter β, is a condition on the matrix A and the parameter β, described geometrically in the parameter space. The third condition, w-flatness of the parameter β, really depends on the Gröbner deformation while the other two do not. Besides these three conditions for "rank = vol" this chapter contains a general upper bound for the holonomic rank (Theorem 4.1.1), and an exact formula for the case of dimension $d = 2$ (Theorem 4.2.4).

In this chapter we need some additional tools from commutative algebra, such as the Cohen-Macaulay property and Koszul complexes. For the readers' convenience we briefly review some relevant basics. General references on commutative algebra are [20, Chapter 1], [32], [68], [94]. Let S be the polynomial ring in $x_1, \ldots, x_n$ over $\mathbf{k}$, and I a homogeneous ideal. Total degree of polynomials induces a natural grading on the quotient ring $S/I = \oplus_{m=0}^{\infty}(S/I)_m$. The elements in $(S/I)_m$ are images of homogeneous polynomials of degree m. The *Hilbert series* $H_I(z)$ is a formal generating function in one variable:

$$H_I(z) \quad := \quad \sum_{m=0}^{\infty} \dim_{\mathbf{k}}(S/I)_m \cdot z^m. \tag{4.4}$$

A standard technique for computing Hilbert series is to utilize the identity

$$H_I(z) \quad = \quad H_{\mathrm{in}_{\prec}(I)}(z) \tag{4.5}$$

where $\prec$ is any term order. There are many algorithms for computing the Hilbert series of a monomial ideal $\mathrm{in}_{\prec}(I)$. Computer algebra systems for algebraic geometry such as CoCoA, Macaulay, Macaulay 2, and Singular ([22], [10], [42], [43]) have standard commands for this. They are respectively Poincare, hilb, hilbertSeries, and hilb. The *Krull dimension* of S/I is the dimension of the variety of the ideal I. This number is denoted by $\dim(I)$. It equals the length of longest chains of prime ideals of S/I, or the maximal number of elements of S/I algebraically independent over $\mathbf{k}$, or the order of pole at $z = 1$ of $H_I(z)$. One can evaluate $\dim(I)$ with a one-line command in the systems above. An ideal of Krull dimension 0 is called a *zero-dimensional ideal*. Its variety is a finite set of points. For $m \gg 0$, $\dim_{\mathbf{k}}(S/I)_m$ is a polynomial in m of degree $\dim(I)-1$, which is called the *Hilbert polynomial* (see, e.g., [12, 9.3, p.441], [27, Chapter 9]). The leading coefficient of the Hilbert polynomial multiplied by $(\dim(I) - 1)!$ is the *degree* of I, denoted by $\mathrm{degree}(I)$. The Hilbert series $H_I(z)$ can be written as $\frac{q(z)}{(1-z)^{\dim(I)}}$ for a polynomial $q(z)$ with $q(1) \neq 0$. Note that $q(1) = \mathrm{degree}(I)$. For example, if A is a $d \times n$-integer matrix of rank d with homogeneity condition (3.3) then the toric ideal I_A has dimension d, and $\mathrm{degree}(I_A) = \mathrm{vol}(A)$; see e.g. [96, Theorem 4.16].

Put $d = \dim(I)$. Then d homogeneous elements $h_1, \ldots, h_d$ are called a *homogeneous system of parameters* (h.s.o.p.) for S/I if $\dim_{\mathbf{k}}(S/I + \langle h_1, \ldots, h_d \rangle)$ is finite, or equivalently S/I is a finitely generated $\mathbf{k}[h_1, \ldots, h_d]$-module. A *linear system of parameters* (l.s.o.p.) is an h.s.o.p. consisting of elements of degree 1. *Noether normalization* guarantees the existence of an l.s.o.p. for S/I. Let $h_1, \ldots, h_d$ be an h.s.o.p. A subsequence $h_1, \ldots, h_r$ is called a homogeneous S/I-*regular sequence* if S/I is a free $\mathbf{k}[h_1, \ldots, h_r]$-module, or equivalently, if the following identity of the Hilbert series holds:

$$H_I(z) \quad = \quad \frac{H_{I+\langle h_1, \ldots, h_r \rangle}(z)}{\prod_{i=1}^r (1 - z^{\deg h_i})}. \tag{4.6}$$

A degree-preserving exact sequence of finitely generated graded S-modules

$$0 \longrightarrow F^l \longrightarrow \cdots \longrightarrow F^1 \longrightarrow F^0 = S \longrightarrow S/I \longrightarrow 0 \tag{4.7}$$

is called a *free resolution of S/I of length l* if every F^j is a free S-module. A free resolution is said to be *minimal* when the ranks of F^j are taken as small as possible. Since I is homogeneous, the minimal free resolution is essentially unique (see [32, §20.1]). Its length satisfies $l \leq n$ by *Hilbert's Syzygy Theorem* and $l \geq n - d$ by the *Auslander-Buchsbaum formula* (see [32, §19.3]).

Definition 4.0.1. Put $d = \dim(I)$. Then the following are equivalent:

(1) Some h.s.o.p. for S/I is a regular sequence, i.e., (4.6) holds with $r = d$.
(2) Every h.s.o.p. for S/I is a regular sequence, i.e., (4.6) holds with $r = d$.
(3) The length l of the minimal free resolution of S/I is equal to $n - d$.

When these conditions hold, then the ideal I is said to be *Cohen-Macaulay*.

The minimal free resolution can be computed with Schreyer's algorithm [32, §15.5] which is implemented in the computer algebra systems cited above. Hence the last definition of Cohen-Macaulayness can be checked by computer. Here is an alternative method for testing whether I is Cohen-Macaulay.

Theorem 4.0.2 ([20, Cor. 4.6.11]). *Let $l_1, \ldots, l_d$ be an l.s.o.p. of S/I. Then*

$$\deg(I) \quad \leq \quad \deg(I + \langle l_1, \ldots, l_d \rangle) \quad - \quad \dim_{\mathbf{k}} S/(I + \langle l_1, \ldots, l_d \rangle), \tag{4.8}$$

and equality holds if and only if the homogeneous ideal I is Cohen-Macaulay.

An important property of Cohen-Macaulay ideals is that they have no embedded primes ([32, Corollary 18.14]). But this is not enough for Cohen-Macaulayness as the example $I = \langle x_1, x_2 \rangle \cap \langle x_3, x_4 \rangle = \langle x_1 x_3, x_1 x_4, x_2 x_3, x_2 x_4 \rangle$ shows. The reader is invited to verify, using the above criteria, that I is not Cohen-Macaulay. Does removing one of the four minimal generators lead to a Cohen-Macaulay ideal ? The monomial case is important because a homogeneous ideal I is Cohen-Macaulay if any one of its initial monomial ideals $\mathrm{in}_\prec(I)$ is Cohen-Macaulay.

For the toric ideal I_A, *Hochster's Theorem* [94, Corollary I.7.6] gives a sufficient condition for Cohen-Macaulayness. It states that I_A is Cohen-Macaulay if the semigroup spanned by the columns of A is *normal*, i.e., $\mathbf{N}A = \mathbf{Z}A \cap \mathrm{pos}(A)$. This condition is not necessary; see e.g. Example 4.1.6. If the initial monomial ideal $\mathrm{in}_\prec(I_A)$ is square-free (as in Section 3.6), then A is normal (see e.g. [96, Proposition 13.15]) and hence I_A is Cohen-Macaulay.

4.1 The Fake Indicial Ideal and an Upper Bound

Fix any generic weight vector $w \in \mathbf{R}^n$ for the toric ideal I_A. Then $\mathrm{in}_w(I_A)$ is a monomial ideal in $\mathbf{k}[\partial_1, \ldots, \partial_n]$ and we can consider its distraction $\widetilde{\mathrm{in}}_w(I_A) := (R \cdot \mathrm{in}_w(I_A)) \cap \mathbf{k}[\theta] = (D \cdot \mathrm{in}_w(I_A)) \cap \mathbf{k}[\theta]$. Let $\beta \in \mathbf{k}^d$. The objective in this section is to study the structure of the *fake indicial ideal*

$$\widetilde{\mathrm{fin}}_w(H_A(\beta)) = \widetilde{\mathrm{in}}_w(I_A) + \langle A \cdot \theta - \beta \rangle \subset \mathbf{k}[\theta_1, \ldots, \theta_n] = \mathbf{k}[\theta].$$

This commutative ideal was introduced in (3.10). It is zero-dimensional as shown later in Lemma 4.1.3. Our goal is to give an upper bound on its rank, which is its $\mathbf{k}$-codimension in $\mathbf{k}[\theta]$.

Theorem 4.1.1. *For any matrix A with (3.3), generic w, and any $\beta \in \mathbf{k}^d$,*

$$\mathrm{rank}(\widetilde{\mathrm{fin}}_w(H_A(\beta))) \leq 2^{2d} \cdot \mathrm{vol}(A).$$

The fake indicial ideal $\widetilde{\mathrm{fin}}_w(H_A(\beta))$ is contained in the indicial ideal $\widetilde{\mathrm{in}}_{(-w,w)}(H_A(\beta))$ whose rank equals that of the initial ideal $\mathrm{in}_{(-w,w)}(H_A(\beta))$, by Theorem 2.3.9. In view of (4.2), we conclude that the rank of $\widetilde{\mathrm{fin}}_w(H_A(\beta))$ is always an upper bound on the rank of the hypergeometric D-ideal $H_A(\beta)$.

Corollary 4.1.2. *For any matrix A with (3.3) and any $\beta \in \mathbf{k}^d$,*

$$\mathrm{rank}(H_A(\beta)) \leq 2^{2d} \cdot \mathrm{vol}(A).$$

This inequality is the only general upper bound for the rank of the GKZ-system given in this book. Most likely it is far from optimal. Yet it remains an open problem to find an upper bound which is substantially better. For instance, it would be desirable to know whether the ratio $\mathrm{rank}(H_A(\beta))/\mathrm{vol}(A)$ can be bounded above by some polynomial function in d. The proof of Theorem 4.1.1 will be given at the end of this section. The material to follow is combinatorial commutative algebra, with no appearance of D-modules.

The roots of the zero-dimensional ideal $\widetilde{\mathrm{fin}}_w(H_A(\beta))$ are called the *fake exponents*. They are always $\mathbf{k}$-rational, and their number is finite and bounded above by the number of standard pairs of $\mathrm{in}_w(I_A)$. This follows from

Lemma 4.1.3. *For any parameter vector $\beta \in \mathbf{k}^d$, the radical of the fake indicial ideal $\widetilde{\mathrm{fin}}_w(H_A(\beta))$ is zero-dimensional, and it equals*

$$\mathrm{rad}(\widetilde{\mathrm{fin}}_w(H_A(\beta))) = \bigcap_{(\partial^a, \sigma) \in \mathcal{S}(\mathrm{in}_w(I_A))} (\langle \theta_i - a_i : i \notin \sigma \rangle + \langle A \cdot \theta - \beta \rangle). \quad (4.9)$$

Proof. The formula (3.17) represents the fake indicial ideal, even for non-generic parameters, and the radical of (3.17) is the right hand side of (4.9). $\qquad\square$

The number of fake exponents is at most the rank of the fake indicial ideal, and equality holds in this bound if and only if $\widetilde{\mathrm{fin}}_w(H_A(\beta))$ is a radical ideal. We also consider the following (possibly larger) zero-dimensional ideal:

$$\widetilde{\mathrm{topfin}}_w(H_A(\beta)) \quad := \quad \mathrm{top}(\widetilde{\mathrm{in}}_w(I_A)) + \langle A \cdot \theta - \beta \rangle \quad \subset \quad \mathbf{k}[\theta].$$

The radical of $\widetilde{\mathrm{topfin}}_w(H_A(\beta))$ equals (3.18) which may be larger than (4.9). The cardinality of the set $\mathcal{T}(\mathrm{in}_w(I_A))$ of top-dimensional standard pairs of the monomial ideal $\mathrm{in}_w(I_A) \subset \mathbf{k}[\partial_1, \ldots, \partial_n]$ equals $\mathrm{vol}(A)$. Theorem 3.2.10 implies

Corollary 4.1.4. *If $\beta \in \mathbf{k}^d$ is generic, then $\widetilde{\mathrm{fin}}_w(H_A(\beta)) = \widetilde{\mathrm{topfin}}_w(H_A(\beta))$, this ideal is radical, every fake exponent is an exponent, and the number of fake exponents equals $\mathrm{rank}(\widetilde{\mathrm{fin}}_w(H_A(\beta))) = \mathrm{vol}(A)$.*

All of the above assertions can fail for non-generic parameters $\beta \in \mathbf{k}^d$. In Example 3.1.4 this happens for $\beta_2 = 1$ and suitable β_1. In the next result we do not make any genericity assumption on $\beta \in \mathbf{k}^d$. It works for all β.

Theorem 4.1.5. *Let A and w as above and $\beta \in \mathbf{k}^d$ any parameter vector.*

(1) *Then $\mathrm{rank}(\widetilde{\mathrm{fin}}_w(H_A(\beta))) \geq \mathrm{rank}(\widetilde{\mathrm{topfin}}_w(H_A(\beta))) \geq \mathrm{vol}(A)$.*
(2) *If $\mathrm{in}_w(I_A)$ is Cohen-Macaulay then $\mathrm{rank}(\widetilde{\mathrm{fin}}_w(H_A(\beta))) = \mathrm{vol}(A)$.*

The inequality $\mathrm{rank}(\widetilde{\mathrm{fin}}_w(H_A(\beta))) \geq \mathrm{vol}(A)$ in part (1) can be derived from our lower bound for the rank of the GKZ-system in (4.1). In what follows, however, we give a self-contained proof using only commutative algebra.

Proof. The first inequality is trivial since $\widetilde{\mathrm{fin}}_w(H_A(\beta)) \subseteq \widetilde{\mathrm{topfin}}_w(H_A(\beta))$. To show the second inequality, we introduce a new variable θ_0 and we set $S := \mathbf{k}[\theta_0, \theta_1, \ldots, \theta_n]$. Let J be the ideal of S generated by the homogenizations of all elements in $\mathrm{top}(\widetilde{\mathrm{in}}_w(I_A))$. It follows from Theorem 3.2.2 that J is generated by the forms $\prod_{i=1}^n \prod_{j=0}^{a_i-1}(\theta_i - j\theta_0)$ where $\partial_1^{a_1} \cdots \partial_n^{a_n}$ runs over all minimal generators of $\mathrm{top}(\mathrm{in}_w(I_A))$. Moreover, these generators constitute the reduced Gröbner basis for J with respect to any term order that satisfies $\theta_0 \prec \theta_i$ for $i = 1, \ldots, n$, and thus the Hilbert series of S/J and $\mathbf{k}[\partial_0, \ldots, \partial_n]/\mathbf{k}[\partial_0, \ldots, \partial_n]\mathrm{top}(\mathrm{in}_w(I_A))$ are identical. Therefore the graded algebra S/J has Krull dimension $d+1$, and $\mathrm{degree}(J) = \mathrm{vol}(A)$.

Abbreviate $\ell_j := (A \cdot \theta)_j - \beta_j \theta_0$ and consider the ideal $J' := J + (\ell_1, \ldots, \ell_d)$ in S. We claim that the projective variety defined by J' consists of finitely many points in $\mathbf{P}^n$, and that none of them lies on the "hyperplane at infinity" $\{\theta_0 = 0\}$. For $\theta_0 = 1$ these points are the fake exponents, which form a finite set by Lemma 4.1.3. For $\theta_0 = 0$ we get the zero set of the monomial ideal $\mathrm{top}(\mathrm{in}_w(I_A))$ intersected with the kernel of A. This intersection is just the origin since the associated primes of $\mathrm{top}(\mathrm{in}_w(I_A))$ have the form $\langle \theta_i : i \notin \sigma \rangle$ where the $d \times d$-submatrix of A defined by σ is invertible. We conclude that the graded algebra S/J' has Krull dimension 1, and that θ_0 lies in no minimal prime of J'.

Since the Krull dimension of S/J' is 1, the only possible embedded prime of J' is the maximal ideal $\langle \theta_0, \ldots, \theta_n \rangle$. Write $J' = N \cap Q$ where N is the intersection of the primary components belonging to minimal primes, and Q is a primary component belonging to the maximal ideal $\langle \theta_0, \ldots, \theta_n \rangle$. Then each graded component S_m for $m \gg 0$ is contained in Q. Hence for $x \in (S/J')_m$ with $m \gg 0$ the annihilator $\mathrm{Ann}(x)$ equals $\{y \in S/J' : yx \in N/J'\}$. Since θ_0 does not lie in a minimal prime of J', multiplication by θ_0 defines an injection from $(S/J')_m$ to $(S/J')_{m+1}$. It is even a **k**-isomorphism because $(S/J')_m$ is a **k**-vector space of dimension degree(J') for $m \gg 0$. Hence the quotient ring $(S/J')/\langle \theta_0 - 1 \rangle$ is zero-dimensional and has **k**-vector space dimension equal to degree(J'). Note, however, that $(S/J')/\langle \theta_0 - 1 \rangle = \mathbf{k}[\theta]/\widetilde{\mathrm{topfin}}_w(H_A(\beta))$ and therefore $\mathrm{rank}(\widetilde{\mathrm{topfin}}_w(H_A(\beta))) = \mathrm{degree}(J')$. Since Theorem 4.0.2 says degree$(J') \geq$ degree(J), the proof of (1) is complete.

For part (2) assume the monomial ideal $\mathrm{in}_w(I_A)$ to be Cohen-Macaulay. Then $\mathrm{in}_w(I_A)$ has no embedded primes (cf. [32, Corollary 18.14]), and hence $\mathrm{in}_w(I_A) = \mathrm{top}(\mathrm{in}_w(I_A))$ and $\widetilde{\mathrm{fin}}_w(H_A(\beta)) = \widetilde{\mathrm{topfin}}_w(H_A(\beta))$. We have degree$(J') = \mathrm{rank}(\widetilde{\mathrm{fin}}_w(H_A(\beta)))$ and degree$(J) = \mathrm{vol}(A)$. Since $\mathrm{top}(\mathrm{in}_w(I_A))$ is an initial monomial ideal of J, we know that J is Cohen-Macaulay as well. The second part of Theorem 4.0.2 implies degree$(J) = $ degree(J'). $\square$

A necessary condition for $\mathrm{in}_w(I_A)$ to be Cohen-Macaulay is that the toric ideal I_A itself is Cohen-Macaulay; see [11]. By Hochster's Theorem [94, Corollary I.7.6], this happens, for instance, when the semigroup spanned by the columns of A is *normal*, i.e., $\mathbf{N}A = \mathbf{Z}A \cap \mathrm{pos}(A)$. It is an open problem whether every normal toric ideal I_A possesses a Cohen-Macaulay initial ideal. A widely used sufficient condition for $\mathrm{in}_w(I_A)$ to be Cohen-Macaulay is that the regular triangulation Δ_w is unimodular. This case was discussed in detail in Section 3.6. Recall that Δ_w is unimodular if and only if the initial monomial ideal $\mathrm{in}_w(I_A)$ is square-free. In this case $\mathrm{in}_w(I_A)$ equals the Stanley-Reisner ideal of Δ_w, which is Cohen-Macaulay since Δ_w is a shellable ball. Hence the rank statement in Theorem 3.6.6 is a special case of Theorem 4.1.5 (2).

We wish to emphasize that the class of toric ideals which are Cohen-Macaulay is much larger than the class of those which are normal. Here is a typical example of a Cohen-Macaulay toric ideal which is not normal:

Example 4.1.6. Let $d = 2, n = 5$ and $A = \begin{pmatrix} 1 & 1 & 1 & 1 & 1 \\ 0 & 8 & 16 & 21 & 26 \end{pmatrix}$. Then

$$I_A = \langle \underline{\partial_4^2} - \partial_3\partial_5, \ \underline{\partial_3^7} - \partial_1^2\partial_2\partial_5^4, \ \partial_2\partial_3^6 - \partial_1^3\partial_5^4, \ \underline{\partial_2^2} - \partial_1\partial_3 \rangle.$$

The underlined monomials generate $\mathrm{in}_{w'}(I_A)$ for any weight vector w' which represents the *reverse lexicographic* term order with $\partial_1 \succ \partial_2 \succ \partial_3 \succ \partial_4 \succ \partial_5$.

This monomial ideal is Cohen-Macaulay and thus I_A is Cohen-Macaulay. Theorem 4.1.5 (2) tells us that $\mathrm{rank}(\widetilde{\mathrm{fin}}_{w'}(H_A(\beta))) = \mathrm{vol}(A) = 26$.

Next choose $w = (0, 0, 8, 15, 28)$. For this particular weight vector, the initial ideal $\mathrm{in}_w(I_A)$ equals

$$\langle \partial_3\partial_5, \partial_2^5\partial_5^4, \partial_2^5\partial_2^2\partial_3^3, \partial_2^5\partial_4^4\partial_5^2, \partial_2^5\partial_4^6\partial_5, \partial_2^5\partial_4^8, \partial_1\partial_4^2, \partial_1\partial_3, \partial_1\partial_3^3\partial_4^4, \partial_1^2\partial_2\partial_5^4, \partial_1^3\partial_5^4 \rangle$$

$$= \quad \langle \partial_1, \partial_2^5, \partial_3 \rangle \cap \langle \partial_1, \partial_2^5, \partial_5 \rangle \cap \langle \partial_1, \partial_4^8, \partial_5 \rangle \cap \langle \partial_3, \partial_4^2, \partial_5^4 \rangle$$

$$\cap \langle \partial_1^3, \partial_2, \partial_3, \partial_4^2 \rangle \cap \langle \partial_1^2, \partial_2^3, \partial_3, \partial_4^2 \rangle \cap \langle \partial_1, \partial_3, \partial_4^4, \partial_5^3 \rangle \cap \langle \partial_1, \partial_3, \partial_4^6, \partial_5^2 \rangle.$$

What makes this initial ideal interesting is that I_A *is Cohen-Macaulay but* $\mathrm{top}(\mathrm{in}_w(I_A))$ *is not Cohen-Macaulay.* We will show how this can be seen.

The intersection of the first four primary components equals

$$\mathrm{top}(\mathrm{in}_w(I_A)) \quad = \quad \langle \partial_1\partial_3, \partial_3\partial_5, \partial_1\partial_4^2, \partial_1\partial_5^4, \partial_2^5\partial_4^2\partial_5, \partial_2^5\partial_5^4, \partial_2^5\partial_4^8 \rangle. \quad (4.10)$$

Take the ideal quotient with respect to the special monomial $m := \partial_2^4\partial_4\partial_5^3$:

$$\big(\mathrm{top}(\mathrm{in}_w(I_A)) : m\big) \quad = \quad \langle \partial_1, \partial_2, \partial_3 \rangle \cap \langle \partial_1, \partial_4, \partial_5 \rangle. \quad (4.11)$$

This ideal quotient is not Cohen-Macaulay. It is a general fact that the Cohen-Macaulay property of a monomial ideal is preserved under taking ideal quotients by monomials. We conclude that $\mathrm{top}(\mathrm{in}_w(I_A))$ is not Cohen-Macaulay.

The monomial m defines a parameter vector $\beta = A \cdot (0, 4, 0, 1, 3)^T = (8, 131)^T$ for which the second inequality in Theorem 4.1.5 (1) is strict:

$$\mathrm{rank}(\widetilde{\mathrm{fin}}_w(H_A(\beta))) \quad = \quad \mathrm{rank}(\widetilde{\mathrm{topfin}}_w(H_A(\beta))) \quad = \quad 27 \quad > \quad \mathrm{vol}(A) \quad = \quad 26.$$

This example will reappear in the end of this section.

Our next goal is to give a geometric upper bound on the number of fake exponents. Let Δ_w be the regular triangulation gotten from $\mathrm{in}_w(I_A)$ as in Section 3.2 and [96, Theorem 8.3]. We define the *arithmetic volume* of Δ_w to be the number

$$\mathrm{avol}(\Delta_w) \quad := \quad \sum_{\sigma \,:\, \text{face of } \Delta_w} [\mathrm{image}_\mathbf{R}(A_\sigma) \cap \mathbf{Z}^d : \mathrm{image}_\mathbf{Z}(A_\sigma)]. \quad (4.12)$$

We are summing the natural lattice indices associated with each face of Δ_w. This includes the empty face $\sigma = \emptyset$, which contributes 1. The relation to volume is as follows: if we sum only over facets σ of Δ_w then we get precisely the normalized volume $\mathrm{vol}(A)$. When $d = 2$, edges of Δ_w altogether contribute $\mathrm{vol}(A)$, each vertex contributes 1, and the empty face contributes 1. Hence if $d = 2$ and Δ_w has l vertices, then $\mathrm{avol}(\Delta_w) = \mathrm{vol}(A) + l + 1 \geq \mathrm{vol}(A) + 3$ and this inequality is tight for suitable w.

Example 4.1.7. Let $A = \begin{pmatrix} 1 & 1 & 1 & 1 \\ 0 & 1 & 2 & 3 \end{pmatrix}$, and $w = (0, 1, 1, 0)$. Then Δ_w is the one-dimensional simplex $\{\{1, 4\}\}$. Hence $\mathrm{avol}(\Delta_w) = \mathrm{vol}(A) + 2 + 1 = 6$.

Theorem 4.1.8. *Fix $w \in \mathbf{R}^n$ generic. For any parameter vector $\beta \in \mathbf{k}^d$, the number of fake exponents is at most the arithmetic volume of the underlying regular triangulation. In symbols,*

$$\operatorname{rank}\big(\operatorname{rad}(\widetilde{\operatorname{fin}}_w(H_A(\beta)))\big) \quad \leq \quad \operatorname{avol}(\Delta_w). \qquad (4.13)$$

Proof. Let $M = \operatorname{in}_w(I_A)$ and $V = V(\widetilde{\operatorname{fin}}_w(H_A(\beta))) \subset \mathbf{k}^n$ be the set of fake exponents. Every point $p \in V$ corresponds to a maximal ideal $\langle \theta - p \rangle = \langle \theta_1 - p_1, \dots, \theta_n - p_n \rangle$. Let $\mathcal{S}_p(M)$ be the set of standard pairs $(\partial^u, \sigma) \in \mathcal{S}(M)$ such that

$$\langle \theta_i - u_i : i \notin \sigma \rangle + \langle A \cdot \theta - \beta \rangle \quad = \quad \langle \theta - p \rangle. \qquad (4.14)$$

We will prove that the set $\mathcal{S}_\beta(M) := \bigcup_{p \in V} \mathcal{S}_p(M)$ has cardinality at most $\operatorname{avol}(\Delta_w)$. This implies our claim because

$$\operatorname{rank}\big(\operatorname{rad}(\widetilde{\operatorname{fin}}_w(H_A(\beta)))\big) = \#V \leq \sum_{p \in V} \#\mathcal{S}_p(M) = \#\mathcal{S}_\beta(M).$$

Note that

$$(\partial^u, \sigma) \in \mathcal{S}_\beta(M) \quad \text{if and only if} \quad \beta \text{ lies in } A \cdot u + \operatorname{image}_{\mathbf{k}}(A_\sigma). \qquad (4.15)$$

Let (∂^u, σ) and (∂^v, σ) be two distinct standard pairs in $\mathcal{S}_\beta(M)$. Then

$$A \cdot (u - v) \quad \text{lies in} \quad \operatorname{image}_{\mathbf{k}}(A_\sigma). \qquad (4.16)$$

On the other hand, the definition of standard pairs implies

$$\big(A \cdot u + \operatorname{image}_{\mathbf{N}}(A_\sigma)\big) \quad \cap \quad \big(A \cdot v + \operatorname{image}_{\mathbf{N}}(A_\sigma)\big) \quad = \quad \emptyset.$$

This implies

$$A \cdot (u - v) \quad \text{does not lie in} \quad \operatorname{image}_{\mathbf{Z}}(A_\sigma). \qquad (4.17)$$

The conditions (4.16) and (4.17) state that the vector $u - v$ represents a non-zero element in the finite abelian group $\big(\operatorname{image}_{\mathbf{R}}(A_\sigma) \cap \mathbf{Z}^d\big)/\operatorname{image}_{\mathbf{Z}}(A_\sigma)$. Therefore the order of this group is an upper bound on the number of standard pairs $(\bullet, \sigma)$ appearing in $\mathcal{S}_\beta(M)$. The desired inequality $\#\mathcal{S}_\beta(M) \leq \operatorname{avol}(\Delta_w)$ follows by summing over all faces σ of Δ_w. $\qquad \Box$

In what follows we retain the notation introduced in the proof above. We first record the following fact about the sets $\mathcal{S}_p(M)$ for future use.

Lemma 4.1.9. *Let $(\partial^u, \sigma), (\partial^v, \tau)$ be distinct elements of $\mathcal{S}_p(M)$. Then $\sigma \not\subset \tau$ and $\tau \not\subset \sigma$.*

Proof. Suppose $\sigma \subset \tau$. For all indices $i \notin \tau$, we have $p_i = u_i = v_i$. This implies $A \cdot u + \operatorname{image}_{\mathbf{N}}(A_\sigma) \subset A \cdot v + \operatorname{image}_{\mathbf{N}}(A_\tau)$, which is impossible for two distinct standard pairs (∂^u, σ) and (∂^v, τ) of M. $\qquad \Box$

In Theorem 4.1.8 we showed that $\#\mathcal{S}_\beta(M) \leq \mathrm{avol}(\Delta_w)$. Often there is a big gap in this inequality, and we wish to improve it. Call a face σ of Δ_w *associated* to $M = \mathrm{in}_w(I_A)$ if there exists a standard pair of the form $(\bullet, \sigma)$. Define $\mathrm{avol}(M)$ to be the subsum of (4.12) gotten by summing only over associated faces σ. Hence $\mathrm{vol}(A) \leq \mathrm{avol}(M) \leq \mathrm{avol}(\Delta_w)$. Note that $\mathrm{vol}(A) = \mathrm{avol}(M)$ if and only if M has no embedded components.

Corollary 4.1.10. *The number of fake exponents is at most* $\mathrm{avol}(M)$.

Proof. In the proof of Theorem 4.1.8, it suffices to consider only those faces σ of Δ_w which are associated to M. $\qquad\square$

Corollary 4.1.10 gives a better bound than Theorem 4.1.8. For instance, in Example 4.1.6 we have $\mathrm{vol}(A) = \#\mathcal{S}_\beta(M) = 26$, $\mathrm{avol}(M) = 28$, $\mathrm{avol}(\Delta_w) = 32$, and $\#\mathcal{S}(M) = 40$.

The primary decomposition of the fake indicial ideal has the form

$$\widetilde{\mathrm{fin}}_w(H_A(\beta)) = \bigcap_{p \in V} Q_{A,p,w}(\theta - p),$$

where $Q_{A,p,w}$ is a zero-dimensional ideal in $\mathbf{k}[\theta]$ which is primary to the maximal ideal $\langle \theta_1, \ldots, \theta_n \rangle$. The shifted ideal $Q_{A,p,w}(\theta - p)$ is primary to the maximal ideal $\langle \theta - p \rangle$. In view of Theorem 2.3.11, it is important to describe the ideals $Q_{A,p,w}$ and to derive an upper bound for the *fake rank*

$$\mathrm{rank}(\widetilde{\mathrm{fin}}_w(H_A(\beta))) = \sum_{p \in V} \mathrm{rank}(Q_{A,p,w}). \tag{4.18}$$

Here is a first crude estimate for the ranks of the primary components:

Proposition 4.1.11. *Let* $M = \mathrm{in}_w(I_A)$ *and* $m(p) = \#\mathcal{S}_p(M)$. *Then*

$$\mathrm{rank}(Q_{A,p,w}) \leq \binom{n - d + m(p) - 1}{n - d}. \tag{4.19}$$

Proof. By localizing the formula (3.17) at the point p we get

$$Q_{A,p,w}(\theta - p) = \langle A \cdot \theta - \beta \rangle + \bigcap_{(\partial^a, \sigma) \in \mathcal{S}_p(M)} \langle \theta_i - a_i : i \notin \sigma \rangle,$$

or, equivalently, $\quad Q_{A,p,w} = \langle A \cdot \theta \rangle + \bigcap_{(\partial^a, \sigma) \in \mathcal{S}_p(M)} \langle \theta_i : i \notin \sigma \rangle. \tag{4.20}$

From (4.20) we see that our primary ideal $Q_{A,p,w}$ contains the $m(p)$-th power of the maximal ideal $\langle \theta_1, \ldots, \theta_n \rangle$:

$$\langle \theta \rangle^{m(p)} = \langle \theta_1, \ldots, \theta_n \rangle^{m(p)} = \prod_{(\partial^a, \sigma) \in \mathcal{S}_p(M)} (\langle A \cdot \theta \rangle + \langle \theta_i : i \notin \sigma \rangle).$$

This implies that $\langle A \cdot \theta \rangle + \langle \theta \rangle^{m(p)}$ is a subideal of $Q_{A,p,w}$. The right side of (4.19) is the rank of this subideal. It bounds the rank of $Q_{A,p,w}$. $\qquad\square$

Let $\mathcal{U}_p(M)$ denote the set of all faces σ of Δ_w such that $(\partial^a, \sigma) \in \mathcal{S}_p(M)$ for some $a \in \mathbf{N}^n$ (which is unique). Lemma 4.1.9 states that $\mathcal{U}_p(M)$ is the set of facets of a simplicial complex on $\{1, 2, \ldots, n\}$. We identify $\mathcal{U}_p(M)$ with this complex, which is a subcomplex of the triangulation Δ_w. The number of facets of $\mathcal{U}_p(M)$ equals $m(p) = \#\mathcal{S}_p(M)$. We also abbreviate $\mathcal{T}_p(M) := \mathcal{S}_p(M) \cap T(M)$ and $m_{\text{top}}(p) := \#\mathcal{T}_p(M)$. Thus $m_{\text{top}}(p)$ is the number of $(d-1)$-dimensional facets of $\mathcal{U}_p(M)$.

Example 4.1.12. (Continuation of Examples 3.1.4, 3.2.6)
Let $A = \begin{pmatrix} 1 & 1 & 1 & 1 \\ 0 & 1 & 2 & 3 \end{pmatrix}$, and $w = (1, 3, 0, 0)$. Then the monomial ideal $M = \text{in}_w(I_A)$ has four standard pairs: $(1, \{3, 4\})$, $(1, \{1, 3\})$, $(\partial_4, \{1, 3\})$, and $(\partial_2, \{1\})$. The following are typical examples of $\mathcal{S}_p(M)$ and $\mathcal{U}_p(M)$.

$$\mathcal{S}_{(1,0,0,0)}(M) = \{(1, \{1, 3\})\}, \quad \mathcal{U}_{(1,0,0,0)}(M) = \{\{1, 3\}\},$$

$$\mathcal{S}_{(0,1,0,0)}(M) = \{(\partial_2, \{1\})\}, \quad \mathcal{U}_{(0,1,0,0)}(M) = \{\{1\}\},$$

$$\mathcal{S}_{(0,0,1,0)}(M) = \{(1, \{1, 3\}), (1, \{3, 4\})\}, \quad \mathcal{U}_{(0,0,1,0)}(M) = \{\{1, 3\}, \{3, 4\}\}.$$

The Stanley-Reisner ideal of the simplicial complex $\mathcal{U}_p(M)$ is the ideal $\bigcap_{\sigma \in \mathcal{U}_p(M)} \langle \theta_i : i \notin \sigma \rangle$, which appears on the right hand side in (4.20). The generators of this ideal are square-free monomials representing the minimal non-faces of $\mathcal{U}_p(M)$. Thus our primary component $Q_{A,p,w}$ equals the Stanley-Reisner ideal of $\mathcal{U}_p(M)$ plus a linear system of parameters. This implies the following bound which is better than Proposition 4.1.11 if $n - d \gg 0$.

Theorem 4.1.13. *Let A and w be as above and $p \in V$ any fake exponent. The rank of $Q_{A,p,w}$ is less than or equal to the number of faces of $\mathcal{U}_p(M)$, and it is greater than or equal to the number of $(d-1)$-facets of $\mathcal{U}_p(M)$. In particular,*

$$m_{\text{top}}(p) \leq \text{rank}(Q_{A,p,w}) \leq m(p) \cdot 2^d.$$

Proof. The upper bound is a direct application of [94, Lemma III.2.4 (b), page 81]. The lower bound is derived from Theorem 4.0.2. This is analogous to Theorem 4.1.5 (1). □

Recall from Stanley's book [94] that the simplicial complex $\mathcal{U}_p(M)$ is called a *Cohen-Macaulay complex* if its Stanley-Reisner ideal $\bigcap_{\sigma \in \mathcal{U}_p(M)} \langle \theta_i : i \notin \sigma \rangle$ is Cohen-Macaulay. For instance, if $\beta = 0$, then the complex associated with this exponent is the triangulation $\mathcal{U}_0(M) = \Delta_w$ itself, which is known to be shellable and hence Cohen-Macaulay [94, Theorem III.2.5].

Lemma 4.1.14. *Let $p \in V$ be a fake exponent. If the simplicial complex $\mathcal{U}_p(M)$ is Cohen-Macaulay then $\text{rank}(Q_{A,p,w}) = m_{\text{top}} = m(p)$.*

Proof. This follows from the second part of Theorem 4.0.2. □

Fix parameters $\beta \in \mathbf{k}^d$. Recall that $\mathcal{S}_\beta(M)$ is the set of standard pairs $(\partial^u, \sigma) \in \mathcal{S}(M)$ such that $\langle \theta_i - u_i : i \notin \sigma \rangle + \langle A \cdot \theta - \beta \rangle$ is a proper ideal. We also abbreviate $\mathcal{T}_\beta(M) := \mathcal{S}_\beta(M) \cap T(M)$. The cardinalities of these two sets determine the following bounds on the fake rank.

Corollary 4.1.15. *For all $\beta \in \mathbf{k}^d$ we have*

$$\#\mathcal{T}_\beta(M) \leq \mathrm{rank}(\widetilde{\mathrm{fin}}_w(H_A(\beta))) \leq \#\mathcal{S}_\beta(M) \cdot 2^d.$$

The lower bound is attained if $\mathcal{U}_p(M)$ is Cohen-Macaulay for every $p \in V$.

Proof. Note that $\sum_{p \in V} \#\mathcal{T}_p(M) = \#\mathcal{T}_\beta(M)$ and use equation (4.18). Then Corollary 4.1.15 follows directly from Theorem 4.1.13 and Lemma 4.1.14. $\square$

Example 4.1.16. (Continuation of Example 4.1.6) The lower bound in Corollary 4.1.15 is strict for $\beta = (8, 131)$. Namely, $\mathrm{rank}(\widetilde{\mathrm{fin}}_w(H_A(\beta))) = 27$ but $\#\mathcal{T}_\beta(M) = 26$. Hence also the inequality in Theorem 4.1.13 must be strict for some fake exponent p. Indeed, for $p = (0, 4, 0, 1, 3)$, the simplicial complex $\mathcal{U}_p(M) = \{\{2, 3\}, \{4, 5\}\}$ is *disconnected* and hence not Cohen-Macaulay. Its Stanley-Reisner ideal equals (4.11), with ∂_i replaced by θ_i. Therefore

$$Q_{A,p,w} = \langle \theta_1 + \theta_2 + \theta_3 + \theta_4 + \theta_5, 8\theta_2 + 16\theta_3 + 21\theta_4 + 26\theta_5, \theta_1, \theta_2\theta_5, \theta_2\theta_4, \theta_3\theta_5, \theta_3\theta_4 \rangle.$$

This ideal has rank 3 but the number of facets of $\mathcal{U}_p(M)$ is only $m(p) = 2$.

We close this section by proving the rank bound stated in the beginning.

Proof (of Theorem 4.1.1). In the proof of Theorem 4.1.8, we showed that $\#\mathcal{S}_\beta(M) \leq \mathrm{avol}(\Delta_w)$. From the upper bound in Corollary 4.1.15 we find

$$\mathrm{rank}(\widetilde{\mathrm{fin}}_w(H_A(\beta))) \leq 2^d \cdot \mathrm{avol}(\Delta_w).$$

On the other hand, volume and arithmetic volume are related as follows:

$$\mathrm{avol}(\Delta_w) \leq 2^d \cdot \mathrm{vol}(\Delta_w).$$

This holds because the lattice index $[\mathrm{image}_{\mathbf{R}}(A_\sigma) \cap \mathbf{Z}^d : \mathrm{image}_{\mathbf{Z}}(A_\sigma)]$ weakly decreases when the face σ of Δ_w is replaced by a proper subface of σ. Theorem 4.1.1 now follows by combining the previous two displayed inequalities.

4.2 Hypergeometric Functions from Toric Curves

The purpose of this section is to tighten our bounds and other results on GKZ-hypergeometric systems in the cases when the dimension d is 2. This means that I_A is the homogeneous prime ideal of a toric curve (or *monomial curve*) in projective $(n-1)$-space. The results in this section were obtained by different methods also by Cattani, Dickenstein and D'Andrea [24].

We start out with the observation that the case $d = 1$ is trivial for homogeneous toric ideals I_A, so the first case of interest is $d = 2$. In this case we may assume without loss of generality that the given matrix has the form

$$A \;=\; \begin{pmatrix} 1 & 1 & 1 & \cdots & 1 & 1 \\ 0 & i_2 & i_3 & \cdots & i_{n-1} & i_n \end{pmatrix}$$

with $0 < i_2 < i_3 < \cdots < i_n$ relatively prime integers, i.e., the greatest common divisor of $i_2, i_3, \ldots, i_n$ is 1. We write $a_1, \ldots, a_n$ for the columns of A. Their convex hull is the 1-dimensional polytope with vertices $a_1 = (1,0)$ and $a_n = (1, i_n)$. Its normalized volume equals $\mathrm{vol}(A) = i_n$.

Let $\mathbf{N}A$ be the monoid spanned by $a_1, \ldots, a_n$, and set $\mathbf{Z}a_i := \{\lambda a_i : \lambda \in \mathbf{Z}\}$. Clearly, $\mathbf{N}A$ is contained in the possibly larger monoid $(\mathbf{N}A + \mathbf{Z}a_1) \cap (\mathbf{N}A + \mathbf{Z}a_n)$. Both are submonoids of $\mathbf{N}^2$. Consider their set difference

$$\mathcal{E}(A) \;:=\; \big((\mathbf{N}A + \mathbf{Z}a_1) \cap (\mathbf{N}A + \mathbf{Z}a_n)\big) \setminus \mathbf{N}A.$$

The elements of $\mathcal{E}(A)$ are called the *holes* of A.

Lemma 4.2.1. *The set $\mathcal{E}(A)$ of holes of A is finite.*

Proof. Each element u in $\mathcal{E}(A)$ has two different representations,

$$u \;=\; (-1-\lambda_1)\cdot a_1 + \sum_{j=2}^{n}\lambda_j \cdot a_j \;=\; (-1-\mu_n)\cdot a_n + \sum_{j=1}^{n-1}\mu_j \cdot a_j, \quad (4.21)$$

where all λ_j and μ_j are non-negative integers. These representations imply

$$\det(u, a_n) \;=\; \sum_{j=1}^{n-1}\mu_j \cdot \det(a_j, a_n) \;\geq\; 0,$$

$$\det(a_1, u) \;=\; \sum_{j=2}^{n}\lambda_j \cdot \det(a_1, a_j) \;\geq\; 0.$$

These two inequalities are equivalent to $u \in \mathbf{R}_+A$. We thus conclude that the set $\mathcal{E}(A)$ of holes is contained in the cone spanned by $a_1, \ldots, a_n$.

We shall now prove that both of the determinants $\det(u, a_n)$ and $\det(a_1, u)$ are less than or equal to $-i_n + (n-2)i_n^2$. This will imply the finiteness of $\mathcal{E}(A)$. First, after adding a_1 and a_n several times if necessary, we may assume that $\lambda_1 = \mu_n = 0$. This means that $u + a_1$ and $u + a_n$ both lie in $\mathbf{N}A$ but u does not. If $2 \leq j \leq n-1$ then we have $\lambda_j < i_n$ because otherwise we could replace $\lambda_j \cdot a_j$ by $(\lambda_j - i_n)a_j + (i_n - i_j)\cdot a_1 + i_j \cdot a_n$ in (4.21) to infer $u \in \mathbf{N}A$, and by the same reasoning we have $\mu_j < i_n$. Also the determinants $\det(a_1, a_j)$ and $\det(a_j, a_n)$ are bounded above by i_n, and hence by the two representations of u, $\det(a_1, u) \leq -i_n + (n-2)i_n^2$ and $\det(u, a_n) \leq -i_n + (n-2)i_n^2$. This completes our finiteness proof for the set $\mathcal{E}(A)$ of holes. $\square$

The set of holes must not be confused with $(\mathbf{Z}A \cap \mathbf{R}_+ A) \backslash \mathbf{N}A$, the set of elements which lie in the *normalization* of $\mathbf{N}A$ but not in $\mathbf{N}A$. The latter set contains $\mathcal{E}(A)$ but it is not finite unless $a_2 = (1,1)$ and $a_{n-1} = (1, i_n - 1)$.

Example 4.2.2. Let $n = 5$ and $A = \begin{pmatrix} 1 & 1 & 1 & 1 & 1 \\ 0 & 2 & 4 & 7 & 9 \end{pmatrix}$. The normalization $(\mathbf{Z}A \cap \mathbf{R}_+ A)$ of $\mathbf{N}A$ consists of all integer vectors (i,j) with $0 \leq j \leq 9i$. There are precisely three holes in this example:

$$\mathcal{E}(A) = \{ (2,10), (2,12), (3,19) \}.$$

The following lemma shows the algebraic significance of the set of holes.

Lemma 4.2.3. *The toric ideal I_A is Cohen-Macaulay if and only if $\mathcal{E}(A) = \emptyset$.*

Proof. The two variables ∂_1 and ∂_n form a linear system of parameters of $\mathbf{k}[\mathbf{N}A] = \mathbf{k}[\partial]/I_A$. The toric ideal I_A is Cohen-Macaulay if and only if ∂_1, ∂_n forms a regular sequence modulo I_A. This fails to hold if and only if I_A has a minimal generator of the form $\partial_1^a \partial^\lambda - \partial_n^b \partial^\mu$ with $a, b > 0$. On the other hand, such a minimal generator exists in I_A if and only if $\mathcal{E}(A) \neq \emptyset$. $\square$

We now return to the study of the GKZ-hypergeometric system $H_A(\beta)$. The following result completely determines its holonomic rank for $d = 2$.

Theorem 4.2.4. *Let $d = 2$. Then*

$$\mathrm{rank}(H_A(\beta)) = \begin{cases} \mathrm{vol}(A) & \textit{if } \beta \notin \mathcal{E}(A), \\ \mathrm{vol}(A) + 1 & \textit{if } \beta \in \mathcal{E}(A). \end{cases}$$

In view of Lemma 4.2.3 and our general lower bound $\mathrm{rank}(H_A(\beta)) \geq \mathrm{vol}(A)$, this theorem implies the following result. For $d = 2$, the toric ideal I_A is Cohen-Macaulay if and only if $\mathrm{rank}(H_A(\beta)) = \mathrm{vol}(A)$ for all parameters β. It is an open problem whether the same statement is true for $d \geq 3$.

Proof (of Theorem 4.2.4). The lower bound $\mathrm{rank}(H_A(\beta)) \geq \mathrm{vol}(A)$ has been established for general d in Section 3.5. We shall first prove the upper bound $\mathrm{rank}(H_A(\beta)) \leq \mathrm{vol}(A) + 1$. Choose $w \in \mathbf{R}^n$ to represent the reverse lexicographic term order for $\partial_2 \succ \partial_3 \succ \cdots \succ \partial_{n-1} \succ \partial_n \succ \partial_1$. Set $M = \mathrm{in}_w(I_A)$ and $P := \langle \partial_2, \partial_3, \ldots, \partial_{n-1} \rangle$. We have $\mathrm{rad}(M) = P$, because of the relations $\partial_j^{i_n} - \partial_1^{i_n - i_j} \partial_n^{i_j} \in I_A$ for $j = 2, 3, \ldots, n - 1$. Hence the regular triangulation Δ_w consists only of one segment, its two endpoints, and the empty set. This implies that the set $\mathrm{Ass}(M)$ of associated primes of M is a subset of $\{ P, P + \langle \partial_1 \rangle, P + \langle \partial_n \rangle, P + \langle \partial_1, \partial_n \rangle \}$. Since I_A is prime, by the property of the reverse lexicographic term order, the lowest variable ∂_1 does not appear in the generators of M, and hence is not in any associated prime of M. Therefore

$$P \in \mathrm{Ass}(M) \subseteq \{ P, P + \langle \partial_n \rangle \}.$$

Lemma 4.1.9 shows that the cardinality $m(p)$ of $\mathcal{S}_p(M)$ equals 1 for all exponents p. Hence $\widetilde{\mathrm{fin}}_w(H_A(\beta))$ is a radical ideal by (4.20). Therefore $\mathrm{rank}(H_A(\beta))$ is bounded above by the number of fake exponents, which is in turn bounded above by the arithmetic volume $\mathrm{avol}(M)$. We finally note that $\mathrm{avol}(M) \leq \mathrm{vol}(A) + 1$ because the face σ corresponding to $P + \langle \partial_n \rangle$ is a point, which has normalized volume 1. This completes the proof of $\mathrm{rank}(H_A(\beta)) \leq \mathrm{vol}(A) + 1$.

Suppose now that β has the property that there exist $\mathrm{vol}(A)+1$ many fake exponents. Then there is a unique lower-dimensional standard pair $(\partial^b, \{1\})$ associated with β, i.e., $\beta = A \cdot b + \lambda a_1$ for some $\lambda \in \mathbf{k}$. The corresponding fake exponent $v = b + \lambda e_1$ is always an exponent, even if λ is a negative integer. Indeed, the hypergeometric series

$$\phi_v \quad := \quad \sum_{u \in C_w(H_A(\beta))^*} \frac{[v]_{u_-}}{[v+u]_{u_+}} \cdot x^{v+u} \tag{4.22}$$

is well-defined because every vector u in the polar Gröbner cone $C_w(H_A(\beta))^*$ satisfies $u_1 \leq 0$, by the property of the reverse lexicographic term order. This shows that the denominator of any term of ϕ_v is not zero.

There exists a second standard pair $(\partial^a, \{1, n\})$ with the property that $Aa \equiv Ab \mod \mathbf{Z}\{a_1, a_n\}$. This congruence translates into a binomial

$$\partial^b \partial_n^i - \partial^a \partial_1^j \ \in \ I_A \tag{4.23}$$

where i is a positive integer and j is a non-negative integer.

Consider any standard pair $(\partial^c, \{1, n\})$ with $c \neq a$. There are $\mathrm{vol}(A) - 1$ many of these. The corresponding fake exponent $c + ue_1 + ve_n$ is always an exponent, because $Ac + ua_1 + va_n = Ab + \lambda a_1$ and $Ac \not\equiv Ab \mod \mathbf{Z}\{a_1, a_n\}$ imply that v cannot be an integer. It is possible that u is a negative integer, but this is not a problem by the same reasoning as above. We have thus identified $\mathrm{vol}(A)$ many distinct exponents; which amounts to an independent proof of the inequality $\mathrm{rank}(H_A(\beta)) \geq \mathrm{vol}(A)$.

We shall next examine the fake exponent $a + (j+\lambda)e_1 - ie_n$, which corresponds to the standard pair $(\partial^a, \{1, n\})$. We shall prove the following claim:

(1) *The fake exponent $a+(j+\lambda)e_1 - ie_n$ is an exponent if and only if $\beta \in \mathcal{E}(A)$.*

We first prove the following easier fact:

(2) $\beta \in \mathcal{E}(A)$ *if and only if* $\lambda \in \{-j, -j+1, \ldots, -2, -1\}$.

For the if-direction of (2) note that $\beta = Ab + \lambda a_1 = Aa + (j+\lambda)a_1 - ia_n$ implies $\beta \in (\mathbf{N}A + \mathbf{Z}a_1) \cap (\mathbf{N}A + \mathbf{Z}a_n)$. But $\beta \in \mathbf{N}A$ is impossible since an identity $Ab + \lambda a_1 = Ad$ with $d \in \mathbf{N}^n$ would violate the standardness of ∂^b.

To prove the only-if direction of (2) suppose that $\beta \in \mathcal{E}(A)$. The hypothesis $\beta = Ab + \lambda a_1 \in (\mathbf{N}A + \mathbf{Z}a_1) \backslash \mathbf{N}A$ implies that λ is a negative integer. Suppose that $j + \lambda$ is a negative integer. The relation $\beta = Aa + (j + \lambda)a_1 - ia_n \in$

$\mathbf{N}A + \mathbf{Z}a_n$ violates the standardness of $\partial^a \partial_n^l$ for large $l \gg 0$. Therefore λ is an integer between $-j$ and -1, and (2) follows.

It now remains to establish the following claim:

(3) $a + (j{+}\lambda)e_1 - ie_n$ *is an exponent if and only if* $\lambda \in \{-j, -j{+}1, \ldots, -1\}$.

We first prove the only-if direction. Suppose that $\lambda \notin \{-j, -j{+}1, \ldots, -1\}$. We must show that $a + (j{+}\lambda)e_1 - ie_n$ is not an exponent. Consider any logarithmic series solution ϕ whose w-starting term equals $x^a x_1^{j+\lambda} x_n^{-i}$. This monomial is annihilated by $\partial^b \partial_n^i$, the leading term of our binomial (4.23). However, $x^a x_1^{j+\lambda} x_n^{-i}$ is not annihilated by the trailing term $\partial^a \partial_1^j$. Hence the series ϕ contains a component $p(\log(x)) \cdot x^b x_1^\lambda$ which is mapped to $x_1^\lambda x_n^{-i}$ by the operator $\partial^b \partial_n^i$. This implies $p(\log(x)) = \gamma \cdot \log(x_n)$ for some $\gamma \in \mathbf{k}^*$. However, $\log(x_n) x^b x_1^\lambda$ is not annihilated by the Euler operator $x_1 \partial_1 + x_2 \partial_2 + \cdots + x_n \partial_n - \beta_1$, which means it is impossible to choose $p(\log(x))$. Consequently there is no logarithmic series solution ϕ to $H_A(\beta)$ with $\mathrm{in}_w(\phi) = x^a x_1^{j+\lambda} x_n^{-i}$. This proves that $a + (j{+}\lambda)e_1 - ie_n$ is not an exponent.

For the if-direction suppose that $\beta = Ab + \lambda a_1 = Aa + (j + \lambda)a_1 - ia_n \in \mathcal{E}(A)$. We will prove that $v = a + (j + \lambda)e_1 - ie_n$ is an exponent. The negative support of v is $\{n\}$. This is minimal, since otherwise there exists $u \in L = \ker_{\mathbf{Z}}(A)$ such that $v + u$ has no negative support, which contradicts $\beta \notin \mathbf{N}A$. Hence by Theorem 3.4.14, v is an exponent, and ϕ_v in (3.34) is a well-defined canonical solution.

It remains to be shown that if β has only vol(A) fake exponents then $\beta \notin \mathcal{E}(A)$. We prove the contrapositive. Suppose $\beta \in \mathcal{E}(A)$. Then $\beta = A \cdot b - ja_1 = A \cdot a - ia_n$ where i, j are positive integers and $a, b \in \mathbf{N}^n$. We may further assume that b has zero first coordinate and that $\partial^b \notin M$. The binomial (4.23) shows that $\partial^b \partial_n^i \in M$. This implies that $(\partial^b, \{1\})$ is a standard pair. This is a lower-dimensional standard pair associated with β. We conclude that β has vol$(A) + 1$ many fake exponents. □

Our proof of Theorem 4.2.4 has the following remarkable corollary.

Corollary 4.2.5. *Let* $d = 2$. *There exists a generic vector* $w \in \mathbf{R}^n$ *such that the canonical solutions of* $H_A(\beta)$ *with respect to* w *are free of logarithms, i.e, they have the form* $x^v \cdot f$ *where* $f \in \mathbf{C}[[C_w(H_A(\beta))_{\mathbf{Z}}^*]]$.

We next show that our results for $d = 2$ are no longer true for $d = 3$.

Example 4.2.6. Corollary 4.2.5 does not hold for $d \geq 3$.

Let $d = 3, n = 4$ and $A = \begin{pmatrix} 1 & 1 & 1 & 1 \\ 0 & 1 & 0 & 1 \\ 0 & 0 & 1 & 1 \end{pmatrix}$. Thus $I_A = \langle \partial_1 \partial_4 - \partial_2 \partial_3 \rangle$ and $H_A(\beta)$ is the system defining the classical Gauss hypergeometric function. Take $\beta = (0, 0, 0)$. The two possible initial ideals are $\mathrm{in}_w(I_A) = \langle \partial_1 \partial_4 \rangle$ or $\mathrm{in}_w(I_A) = \langle \partial_2 \partial_3 \rangle$. In either case, there is only one exponent, namely

$(0, 0, 0, 0)$, and this exponent has multiplicity 2. In fact, the solution space to $H_A(\beta)$ is spanned by $\log(x_1 x_4 / x_2 x_3)$ and the constant function 1.

Example 4.2.7. The "pinched hexagon" has rank = vol + 2.

Let $d = 3, n = 6$, $A = \begin{pmatrix} 1 & 1 & 1 & 1 & 1 & 1 \\ 0 & 1 & 1 & 0 & -1 & -1 \\ -1 & -1 & 0 & 1 & 1 & 0 \end{pmatrix}$ and $\beta = (1, 0, 0)$. The toric ideal I_A has the following reverse lexicographic Gröbner basis:

$$\partial_2 \partial_5 - \partial_3 \partial_6, \; \underline{\partial_1 \partial_4} - \partial_3 \partial_6, \; \partial_3 \partial_5^2 - \partial_4^2 \partial_6, \; \partial_2 \partial_4^2 - \partial_3^2 \partial_5, \; \partial_1 \partial_5^2 - \partial_4 \partial_6^2,$$
$$\underline{\partial_1 \partial_3 \partial_5} - \partial_2 \partial_4 \partial_6, \; \underline{\partial_1 \partial_3^2} - \partial_2^2 \partial_4, \; \underline{\partial_1^2 \partial_5} - \partial_2 \partial_6^2, \; \underline{\partial_1^2 \partial_3} - \partial_2^2 \partial_6.$$

The underlined monomials generate the initial ideal $\text{in}_w(I_A)$. It has eight standard pairs, six of which are top-dimensional. This implies $\text{vol}(A) = 6$ and $\text{rank}(H_A(\beta)) \leq 8$. Each standard pair gives a hypergeometric function:

standard pair	exponent	solution to $H_A(\beta)$
$(\partial_1 \partial_3, \{2, 6\})$	$(1, -1, 1, 0, 0, 0)$	$x_1 x_3 / x_2$
$(\partial_4, \{2, 3, 6\})$	$(0, 1, -1, 1, 0, 0)$	$x_2 x_4 / x_3$
$(\partial_5, \{3, 4, 6\})$	$(0, 0, 1, -1, 1, 0)$	$x_3 x_5 / x_4$
$(1, \{4, 5, 6\})$	$(0, 0, 0, 1, -1, 1)$	$x_4 x_6 / x_5$
$(\partial_1 \partial_5, \{6\})$	$(1, 0, 0, 0, 1, -1)$	$x_5 x_1 / x_6$
$(1, \{1, 2, 6\})$	$(-1, 1, 0, 0, 0, 1)$	$x_6 x_2 / x_1$
$(1, \{2, 3, 6\})$	$(0, 0, \frac{1}{2}, 0, 0, \frac{1}{2})$	$x_3^{\frac{1}{2}} x_6^{\frac{1}{2}} + \cdots$
$(1, \{3, 4, 6\})$	$(0, 0, \frac{1}{2}, 0, 0, \frac{1}{2})$	$x_3^{\frac{1}{2}} x_6^{\frac{1}{2}} \log(\frac{x_3^3 x_6}{x_2^2 x_4^2}) + \cdots$

The indicial ideal has seven roots but it has multiplicity eight. We conclude

$$\text{vol}(A) = 6 \quad \text{and} \quad \text{rank}(H_A(\beta)) = 8.$$

Hence the upper bound in Theorem 4.2.4 does not hold here. Note that $H_A(\beta)$ has six linearly independent rational solutions, which are Laurent monomials. The other two solutions are honest series derived from the same exponent, hence one of them has a logarithm in the starting monomial. Note also that the logarithmic solution arises from two top-dimensional standard pairs.

We close this section with an important example of a hypergeometric function arising from the toric ideal I_A associated with a curve as above. Consider the following sparse generic polynomial in a single variable t:

$$f(t) \;\; = \;\; x_1 + x_2 t^{i_2} + x_3 t^{i_3} + \cdots + x_{n-1} t^{i_{n-1}} + x_n t^{i_n}.$$

The equation $f(t) = 0$ has i_n distinct roots, and each of the roots is an algebraic function of the coefficients, say, $t = t(x_1, x_2, \ldots, x_n)$. For instance, over the field of complex numbers, we may use Cauchy's formula to write

$$t = \frac{1}{2\pi i} \int_\Gamma \frac{z f'(z)}{f(z)} dz.$$

where Γ is a suitable loop in the complex plane. From this integral representation of the roots, it follows that the algebraic function t is annihilated by the hypergeometric system $H_A(\beta)$ for $\beta = (0, -1)$. In fact, the i_n distinct roots locally form a solution basis for the D-ideal $H_A(\beta)$. This is consistent with Theorem 4.2.4 since $(0, -1) \notin \mathcal{E}(A)$. A detailed discussion of these universal algebraic hypergeometric functions and their series expansions is given in [97]. A solution basis for $H_A(\beta)$ with β arbitrary was constructed by Cattani, Dickenstein and D'Andrea in [24]. Their solution basis involves residues and powers of roots. Hypergeometric integrals similar to Cauchy's integral above will be studied by algebraic methods in Chapter 5.

4.3 Koszul Complexes and Cohen-Macaulay Property

In this section we prove that if the toric ideal is Cohen-Macaulay then $\mathrm{rank}(H_A(\beta)) = \mathrm{vol}(A)$ for all parameters $\beta \in \mathbf{k}^d$. For $d = 2$ this was seen in the previous section. The proof for general d is based on a different characterization of the Cohen-Macaulay property, using homological methods in commutative algebra, specifically, the Koszul complex on an l.s.o.p. We first review the necessary background, and we then introduce fake initial ideals of the GKZ-system. The main point, to be shown in Theorem 4.3.8, is that the fake characteristic ideal equals the characteristic ideal if I_A is Cohen-Macaulay. This result is due to Gel'fand, Kapranov and Zelevinsky [38].

Let S be a commutative polynomial ring, I a homogeneous ideal in S and $h_1, \ldots, h_r \in S$ homogeneous polynomials of positive degree. Then the *Koszul complex* $K_\bullet(h_1, \ldots, h_r; S/I)$ is the following complex of free S/I-modules:

$$0 \longrightarrow K_r(h_1, \ldots, h_r; S/I) \xrightarrow{d_r} \cdots$$
$$\xrightarrow{d_2} K_1(h_1, \ldots, h_r; S/I) \xrightarrow{d_1} K_0(h_1, \ldots, h_r; S/I) \longrightarrow 0 \qquad (4.24)$$

where

$$K_p(h_1, \ldots, h_r; S/I) \quad := \quad \bigoplus_{1 \le i_1 < \cdots < i_p \le r} (S/I)\, e_{i_1, \ldots, i_p} \quad \simeq \quad (S/I)^{\binom{r}{p}}$$

and the differential is defined by its image on the basis vectors as follows:

$$d_p(e_{i_1, \ldots, i_p}) \quad := \quad \sum_{k=1}^p (-1)^{k-1} h_{i_k}\, e_{i_1, \ldots, \widehat{i_k}, \ldots, i_p}.$$

This implies, for example, $d_1(\sum_{i=1}^r g_i\, e_i) = \sum_{i=1}^r g_i h_i$ and

$$d_2(\sum_{i<j} g_{ij}\, e_{ij}) \;=\; \sum_{i<j} g_{ij}(h_i\, e_j - h_j\, e_i).$$

The p-th homology of the Koszul complex $K_\bullet(h_1,\ldots,h_r;S/I)$ is denoted by $H_p(h_1,\ldots,h_r;S/I)$. This is an S/I-module. The first homology module $H_1(h_1,\ldots,h_r;S/I) = \mathrm{kernel}(d_1)/\mathrm{image}(d_2)$ plays the decisive role in the following characterization of regular sequences via the Koszul complex.

Theorem 4.3.1. ([68, Theorem 16.5]) *If* $h_1,\ldots,h_r$ *is a regular sequence in the quotient ring* S/I, *then* $H_p(h_1,\ldots,h_r;S/I) = 0$ *for* $p > 0$ *and* $H_0(h_1,\ldots,h_r;S/I) = S/(I+\langle h_1,\ldots,h_r\rangle)$. *Conversely if* $H_1(h_1,\ldots,h_r;S/I)$ *is the zero module, then* $h_1,\ldots,h_r$ *is a regular sequence in* S/I.

Later in this section we consider mildly non-commutative versions of the Koszul complex, and, in this context, we apply Theorem 4.3.1 to the toric ideal $I = I_A$. The following proposition exhibits the close relationship between the notion of regular sequence or Koszul complex and Gröbner theory.

Proposition 4.3.2. *Let* $f_1,\ldots,f_r \in S$ *be homogeneous and* $u \in \mathbf{R}^n$. *If the initial forms* $\mathrm{in}_u(f_1),\ldots,\mathrm{in}_u(f_r)$ *are a regular sequence in* S, *then* $\{f_1,\ldots,f_r\}$ *is a Gröbner basis of the ideal* $\langle f_1,\ldots,f_r\rangle$ *with respect to the weight* u.

Proof. Since the Gröbner cones are rational, we may assume $u \in \mathbf{Z}^n$. Assume the contrary of the proposition, and suppose $f \in \langle f_1,\ldots f_r\rangle$ does not have a standard representation with respect to $\{f_1,\ldots f_r\}$. Take an expression

$$f = g_1 f_1 + \cdots + g_r f_r \tag{4.25}$$

so that $\max\{\mathrm{ord}_u(g_i) + \mathrm{ord}_u(f_i) : i = 1,\ldots,r\}$ takes its minimal value q. Since f does not have a standard representation, we have $q > \mathrm{ord}_u(f)$, which means the subsum $\sum_{\mathrm{ord}_u(g_i)+\mathrm{ord}_u(f_i)=q} \mathrm{in}_u(g_i)\mathrm{in}_u(f_i)$ is zero. Hence the vector $\sum_{\mathrm{ord}_u(g_i)+\mathrm{ord}_u(f_i)=q} \mathrm{in}_u(g_i)\, e_i$ lies in $\mathrm{kernel}(d_1)$ for the Koszul complex $K_\bullet(\mathrm{in}_u(f_1),\ldots,\mathrm{in}_u(f_r);S)$. It lies in $\mathrm{image}(d_2)$ since the module $H_1(\mathrm{in}_u(f_1),\ldots,\mathrm{in}_u(f_r);S)$ vanishes by Theorem 4.3.1. Hence there exist $h_{jk} \in S$ such that

$$\sum_{\mathrm{ord}_u(g_i)+\mathrm{ord}_u(f_i)=q} \mathrm{in}_u(g_i)\, e_i = d_2(\sum_{j<k} h_{jk}\, e_{jk})$$

$$= \sum_{j<k} h_{jk}(\mathrm{in}_u(f_j)\, e_k - \mathrm{in}_u(f_k)\, e_j)$$

$$= \sum_i \left(-\sum_{i<j} h_{ij}\mathrm{in}_u(f_j) + \sum_{k<i} h_{ki}\mathrm{in}_u(f_k)\right) e_i.$$

Hence,

$$\mathrm{ord}_u\left(g_i - \left(-\sum_{i<j} h_{ij}f_j + \sum_{k<i} h_{ki}f_k\right)\right) < \mathrm{ord}_u(g_i)$$

for i such that $\mathrm{ord}_u(g_i) + \mathrm{ord}_u(f_i) = q$. Replace g_i by $g_i + \sum_{i<j} h_{ij}f_j - \sum_{k<i} h_{ki}f_k$ in the expression (4.25). Then we have a new expression for f which violates the minimality of q. Hence any polynomial in the ideal $\langle f_1, \ldots f_r \rangle$ has a standard representation with respect to $\{f_1, \ldots, f_r\}$. $\square$

We now return to our analysis of the hypergeometric D-ideal $H_A(\beta)$, for arbitrary d and arbitrary β. Let $u, v \in \mathbf{R}^n$ satisfy $u + v \geq 0$. The vector (u, v) defines a filtration $F_p^{(u,v)}(D)$ of the Weyl algebra D by $\mathbf{k}$-vector subspaces:

$$F_p^{(u,v)}(D) \quad := \quad \{\, P \in D : \mathrm{ord}_{(u,v)}(P) \leq p \,\}$$

where $p \in \mathbf{R}$ and $\mathrm{ord}_{(u,v)}(P)$ is the maximal value of $u \cdot a + v \cdot b$ among $x^a \partial^b$ appearing in P. For instance, for $(u, v) = (e, e)$ we get the *Bernstein filtration* whose components are finite-dimensional over $\mathbf{k}$, and for $(u, v) = (0, e)$ we get the *order filtration* whose components are not finite-dimensional over $\mathbf{k}$. Another important special case is $(u, v) = (-e_i, e_i)$, which is called the *V-filtration*, or more generally, the case $u + v = 0$. Put

$$F_{<p}^{(u,v)}(D) \quad := \quad \bigcup_{q<p} F_q^{(u,v)}(D) \quad = \quad \{\, P \in D : \mathrm{ord}_{(u,v)}(P) < p \,\}$$

and $$\mathrm{gr}_p^{(u,v)}(D) \quad := \quad F_p^{(u,v)}(D)/F_{<p}^{(u,v)}(D).$$

Then the graded algebra $\mathrm{gr}^{(u,v)}(D) := \bigoplus_p \mathrm{gr}_p^{(u,v)}(D)$ is equal to D itself if $u + v = 0$, and it equals the commutative polynomial ring $\mathbf{k}[x, \xi]$ if $u + v > 0$.

Let I be any D-ideal. The filtration $\{F_p^{(u,v)}(D)\}$ of D induces filtrations of the D-ideal I and of the left D-module D/I:

$$F_p^{(u,v)}(I) \quad := \quad F_p^{(u,v)}(D) \cap I, \quad F_{<p}^{(u,v)}(I) := \bigcup_{q<p} F_q^{(u,v)}(I),$$

$$F_p^{(u,v)}(D/I) \quad := \quad F_p^{(u,v)}(D)/F_p^{(u,v)}(D) \cap I,$$

$$F_{<p}^{(u,v)}(D/I) := \bigcup_{q<p} F_q^{(u,v)}(D/I).$$

Their graded objects are:

$$\mathrm{gr}_p^{(u,v)}(I) := F_p^{(u,v)}(I)/F_{<p}^{(u,v)}(I),$$

$$
\begin{aligned}
\mathrm{gr}_p^{(u,v)}(D/I) &:= F_p^{(u,v)}(D/I)/F_{<p}^{(u,v)}(D/I) \\
&= \frac{F_p^{(u,v)}(D)/F_p^{(u,v)}(D) \cap I}{F_{<p}^{(u,v)}(D)/F_{<p}^{(u,v)}(D) \cap I} \\
&= \frac{F_p^{(u,v)}(D)/F_{<p}^{(u,v)}(D)}{F_p^{(u,v)}(D) \cap I/F_{<p}^{(u,v)}(D) \cap I} \\
&= \frac{\mathrm{gr}_p^{(u,v)}(D)}{\mathrm{gr}_p^{(u,v)}(I)}.
\end{aligned}
\tag{4.26}
$$

Since $\mathrm{gr}^{(u,v)}(I) = \bigoplus_p \mathrm{gr}_p^{(u,v)}(I) = \mathrm{in}_{(u,v)}(I)$ by definition, we have

$$
\mathrm{gr}^{(u,v)}(D/I) = \mathrm{gr}^{(u,v)}(D)/\mathrm{gr}^{(u,v)}(I) = \mathrm{gr}^{(u,v)}(D)/\mathrm{in}_{(u,v)}(I).
\tag{4.27}
$$

From now on fix $u, v \in \mathbf{R}^n$ such that $u + v > 0$ or $u + v = 0$.

Definition 4.3.3. The *fake initial ideal* of the GKZ ideal $H_A(\beta)$ is the left ideal of $\mathrm{gr}^{(u,v)}(D)$ generated by the commutative polynomial ideal $\mathrm{in}_v(I_A)$ together with the d coordinates of the vector $\mathrm{in}_{(u,v)}(A\theta - \beta)$. In symbols,

$$
\mathrm{fin}_{(u,v)}(H_A(\beta)) \quad := \quad \mathrm{gr}^{(u,v)}(D) \cdot \mathrm{in}_v(I_A) + \mathrm{gr}^{(u,v)}(D) \cdot \mathrm{in}_{(u,v)}(A\theta - \beta).
$$

Hence we have the following exact sequence of (left) modules over the algebra $\mathrm{gr}^{(u,v)}(D)/\mathrm{gr}^{(u,v)}(D)\mathrm{in}_v(I_A)$:

$$
\bigoplus_{i=1}^d \left(\mathrm{gr}^{(u,v)}(D)/\mathrm{gr}^{(u,v)}(D)\mathrm{in}_v(I_A) \right) \cdot e_i
$$
$$
\xrightarrow{\bar{d}_1} \mathrm{gr}^{(u,v)}(D)/\mathrm{gr}^{(u,v)}(D)\mathrm{in}_v(I_A)
\tag{4.28}
$$
$$
\longrightarrow \mathrm{gr}^{(u,v)}(D)/\mathrm{fin}_{(u,v)}(H_A(\beta)) \longrightarrow 0.
$$

Here $\bar{d}_1(\sum_{i=1}^d P_i e_i) = \sum_{i=1}^d P_i \mathrm{in}_{(u,v)}(A\theta - \beta)_i$. Observe $\mathrm{in}_{(u,v)}(DI_A) = \mathrm{gr}^{(u,v)}(D) \mathrm{in}_v(I_A)$. By (4.27), the exact sequence (4.28) is the same as

$$
\bigoplus_{i=1}^d \mathrm{gr}^{(u,v)}(D/DI_A) e_i \xrightarrow{\bar{d}_1} \mathrm{gr}^{(u,v)}(D/DI_A) \longrightarrow
$$

$$
\mathrm{gr}^{(u,v)}(D)/\mathrm{fin}_{(u,v)}(H_A(\beta)) \longrightarrow 0.
\tag{4.29}
$$

Since $\mathrm{in}_{(u,v)}(A\theta - \beta)_j\, e_i - \mathrm{in}_{(u,v)}(A\theta - \beta)_i\, e_j$ clearly belongs to the kernel of $\bar{d}_1$, we can extend the exact sequence to a Koszul complex $K_\bullet^\beta(\mathrm{gr}^{(u,v)}(D/DI_A))$:

$$
\cdots \xrightarrow{\bar{d}_2} K_1^\beta(\mathrm{gr}^{(u,v)}(D/DI_A)) \xrightarrow{\bar{d}_1} K_0^\beta(\mathrm{gr}^{(u,v)}(D/DI_A)) \longrightarrow 0
$$

where

$$K_p^\beta(\mathrm{gr}^{(u,v)}(D/DI_A)) \;=\; \bigoplus_{1\le i_1<\cdots<i_p\le d} \mathrm{gr}^{(u,v)}(D/DI_A)\,e_{i_1\cdots i_p},$$

and

$$\bar d_p(e_{i_1\cdots i_p}) \;=\; \sum_{r=1}^{p}(-1)^{r-1}\,\mathrm{in}_{(u,v)}((A\theta-\beta)_{i_r})\,e_{i_1\cdots\widehat{i_r}\cdots i_p}. \tag{4.30}$$

If $u+v>0$ then $\mathrm{gr}^{(u,v)}(D)$ is a commutative polynomial ring and the Koszul complex is defined as in (4.24). The Koszul complex $K_\bullet^\beta(\mathrm{gr}^{(u,v)}(D/DI_A))$ is well-defined even for $u+v=0$. This is proved by the following lemma:

Lemma 4.3.4. *We have*

$$\partial^u(A\theta)_i \;=\; ((A\theta)_i+(Au)_i)\partial^u.$$

Proof. This follows from direct computation. □

Lemma 4.3.4 also assures that the following Koszul complex $K_\bullet^\beta(D/DI_A)$ is well-defined as well:

$$\cdots \longrightarrow K_2^\beta(D/DI_A) \xrightarrow{d_2} K_1^\beta(D/DI_A) \xrightarrow{d_1} K_0^\beta(D/DI_A) \longrightarrow 0$$

where

$$K_p^\beta(D/DI_A) = \bigoplus_{1\le i_1<\cdots<i_p\le d} D/DI_A\,e_{i_1\cdots i_p},$$

and the differential is given by the formula (4.30) with $\mathrm{in}_{(u,v)}$ erased.

Define a filtration $\{F_q^{(u,v)}K_\bullet^\beta(D/DI_A)\}$ of the complex $K_\bullet^\beta(D/DI_A)$ by

$$F_q^{(u,v)}K_p^\beta(D/DI_A) \;:=\; \bigoplus_{1\le i_1<\cdots<i_p\le d} F_{q-\sum_{k=1}^p c_{i_k}}^{(u,v)}(D/DI_A)\,e_{i_1\cdots i_p} \tag{4.31}$$

where $c_i=\mathrm{ord}_{(u,v)}((A\theta-\beta)_i)$. Clearly we have

$$K_\bullet^\beta(\mathrm{gr}^{(u,v)}(D/DI_A)) = \mathrm{gr}^{(u,v)}K_\bullet^\beta(D/DI_A). \tag{4.32}$$

The following theorem has the same spirit as Proposition 4.3.2.

Theorem 4.3.5. *If the first homology $H_1(K_\bullet^\beta(\mathrm{gr}^{(u,v)}(D/DI_A)))$ vanishes, then the initial ideal $\mathrm{in}_{(u,v)}(H_A(\beta))$ equals the fake initial ideal $\mathrm{fin}_{(u,v)}(H_A(\beta))$.*

Proof. Since the equivalence class containing (u,v) is a finite union of rational cones (Theorem 2.1.3), we can find an integral weight vector in the same equivalence class. Hence we may assume that (u,v) is integral.

We first prove that for any q the sequence

$$F_q K_1^\beta (D/DI_A) \xrightarrow{d_1} F_q K_0^\beta (D/DI_A) \longrightarrow F_q(D/H_A(\beta)) \longrightarrow 0 \qquad (4.33)$$

is exact. Let $P \in F_q(D) \cap H_A(\beta)$. Then P defines an element in $\mathrm{Im}\, d_1$, which we also denote by P. Let $P \in d_1(F_r K_1^\beta (D/DI_A))$. If $r \leq q$, then we are done. Suppose $r > q$, and take the smallest r with $P \in d_1(F_r K_1^\beta (D/DI_A))$. Such r exists because the weights are all integers. Let $P = d_1(Q)$ with $Q \in F_r K_1^\beta (D/DI_A)$. Then Q defines an element $\mathrm{gr}\, Q$ in $\mathrm{Ker}\, \bar{d}_1$, and thus in $\mathrm{Im}\, \bar{d}_2$ by the assumption. Hence there exists $Q' \in F_r K_2^\beta (D/DI_A)$ such that $\mathrm{gr}\, Q = \bar{d}_2(\mathrm{gr}\, Q')$. Then $Q - d_2(Q') \in F_{<r} K_1^\beta (D/DI_A)$. Moreover $d_1(Q - d_2(Q')) = d_1(Q) = P$, which contradicts the minimality of r. Therefore $r \leq q$.

Next we see that the sequence

$$K_1^\beta (\mathrm{gr}^{(u,v)}(D/DI_A)) \longrightarrow K_0^\beta (\mathrm{gr}^{(u,v)}(D/DI_A)) \longrightarrow \mathrm{gr}^{(u,v)}(D/H_A(\beta)) \longrightarrow 0$$

is exact from the exactness of (4.33). This establishes the theorem. $\qquad\square$

In the special case $(u,v) = (0,e)$ we call the fake initial ideal the *fake characteristic ideal*. The following is thus a special case of Theorem 4.3.5.

Corollary 4.3.6. *If the first homology $H_1(K_\bullet^\beta (\mathrm{gr}^{(0,e)}(D/DI_A)))$ vanishes then the characteristic ideal $\mathrm{in}_{(0,e)}(H_A(\beta))$ is independent of the parameter vector β and coincides with the fake characteristic ideal $\mathrm{fin}_{(0,e)}(H_A(\beta))$.*

Lemma 4.3.7. *The following six conditions are equivalent for a matrix A:*

(1) I_A *is Cohen-Macaulay;*
(2) $\mathbf{k}(x)[\mathbf{N}A]$ *is Cohen-Macaulay;*
(3) $\mathbf{k}[x][\mathbf{N}A]$ *is Cohen-Macaulay;*
(4) *The d coordinates of $Ax\xi$ are a regular sequence in $\mathbf{k}[x][\mathbf{N}A]$;*
(5) $H_1(K_\bullet^\beta (\mathrm{gr}^{(0,e)}(D/DI_A))) = 0$;
(6) $\dim_{\mathbf{k}(x)} \mathbf{k}(x)[\mathbf{N}A]/\langle Ax\xi \rangle = \mathrm{vol}(A)$.

Here $\mathbf{k}[x][\mathbf{N}A]$ is the quotient ring of $\mathbf{k}[x][\xi]$ factored by the toric ideal I_A with variables in ξ, and $Ax\xi$ is a vector $\sum_{j=1}^n x_j \xi_j a_j$, and $\langle Ax\xi \rangle$ is the ideal generated by the d-coordinates $(Ax\xi)_1, \ldots, (Ax\xi)_d$.

Proof. First note that $\mathbf{k}[x][\mathbf{N}A] = \mathbf{k}[x][\xi]/I_A$ is a graded algebra since A is homogeneous. Since $\mathbf{k}(x)[\mathbf{N}A]/\langle Ax\xi \rangle$ is supported at the origin, the $\mathbf{k}(x)$-dimension of $\mathbf{k}(x)[\mathbf{N}A]/\langle Ax\xi \rangle$ is finite. Hence the d coordinates $(Ax\xi)_1, \ldots, (Ax\xi)_d$ form a linear system of parameters for the graded algebra $\mathbf{k}(x)[\mathbf{N}A]$.

It follows from [20, Theorems 2.1.9, 2.1.10] that the Cohen-Macaulay property is preserved under replacing the field $\mathbf{k}$ by a ring extension which is an integral domain. This shows that (1), (2) and (3) are equivalent. From the first paragraph, the d coordinates $(Ax\xi)_1, \ldots, (Ax\xi)_d$ are part of an l.s.o.p. for $\mathbf{k}[x][\mathbf{N}A]$. Hence (3) implies (4).

Since $\mathrm{gr}^{(0,e)}(D/DI_A) = \mathbf{k}[x][\mathbf{N}A]$, the condition (4) is equivalent to the Koszul complex of that commutative algebra, which is $K_\bullet^\beta (\mathrm{gr}^{(0,e)}(D/DI_A))$, being acyclic at degree 0. This latter condition is equivalent to the seemingly weaker condition (5) (see Theorem 4.3.1).

Since localization preserves the exactness, (5) implies that the first homology $H_1((Ax\xi)_1, \ldots, (Ax\xi)_d; \mathbf{k}(x)[\mathbf{N}A])$ vanishes. Thus (2) holds by the first paragraph and Theorem 4.3.1.

The degree of I_A is the volume $\mathrm{vol}(A)$ (see [96, Theorem 4.16]). Hence the left hand side of (6) is always greater than or equal to $\mathrm{vol}(A)$ and the equality holds if and only if the condition (2) is satisfied (see Theorem 4.0.2). This completes the proof of Lemma 4.3.7   □

The following theorem is the main result in this section. The first part is due to Gel'fand-Kapranov-Zelevinsky and Adolphson.

Theorem 4.3.8.

(1) ([2], [38]) *If the toric ideal I_A is Cohen-Macaulay, then the characteristic ideal is independent of the parameter β and coincides with the fake characteristic ideal, and* $\mathrm{rank}(H_A(\beta))$ *is equal to* $\mathrm{vol}(A)$.

(2) *Conversely, if the characteristic ideal coincides with the fake characteristic ideal for generic β, then I_A is Cohen-Macaulay.*

Proof. (1) is immediate from Corollary 4.3.6 and Lemma 4.3.7. The left hand side of Lemma 4.3.7 (6) is the fake rank. The assumption of (2) implies that the holonomic rank of $H_A(\beta)$ equals the fake rank for generic β. On the other hand, the holonomic rank of $H_A(\beta)$ for generic β equals $\mathrm{vol}(A)$ by Theorem 2.5.1 and Theorem 3.2.10. Hence the equality in Lemma 4.3.7 (6) holds. Therefore I_A is Cohen-Macaulay.   □

Here is the smallest example where the characteristic ideal depends on the parameters β and hence differs from the fake characteristic ideal.

Example 4.3.9. Let $A = \begin{pmatrix} 1 & 1 & 1 & 1 \\ 0 & 1 & 3 & 4 \end{pmatrix}$. Then I_A is not Cohen-Macaulay. The characteristic ideal $\mathrm{in}_{(0,e)}(H_A(\beta))$ is minimally generated by

$$\xi_2\xi_3 - \xi_1\xi_4, \quad \xi_2^3 - \xi_1^2\xi_3, \quad \xi_2^2\xi_4 - \xi_1\xi_3^2, \quad \xi_3^3 - \xi_2\xi_4^2,$$

$$x_1\xi_1 + x_2\xi_2 + x_3\xi_3 + x_4\xi_4, \quad x_2\xi_2 + 3x_3\xi_3 + 4x_4\xi_4,$$

$$(\beta_2 - 2)x_1\xi_2^2 + (\beta_2 - \beta_1 - 1)x_2\xi_1\xi_3 + (\beta_2 - 3\beta_1 + 1)x_3\xi_2\xi_4 + (\beta_2 - 4\beta_1 + 2)x_4\xi_3^2.$$

Hence it depends on the parameter β. It coincides with the fake characteristic ideal $\langle I_A, Ax\xi \rangle$ if and only if $\beta = \begin{pmatrix} 1 \\ 2 \end{pmatrix}$. The holonomic rank $\mathrm{rank}(H_A(\beta))$ is 5 if $\beta = \begin{pmatrix} 1 \\ 2 \end{pmatrix}$, and 4 otherwise. This was proved in Theorem 4.2.4.

Remark 4.3.10. Assume that the matrix A is normal, i.e., it satisfies $\mathbf{N}A = \text{pos}(A) \cap \mathbf{Z}A$. Then by Hochster's theorem ([94, Corollary I,7.6]) the toric ideal I_A is Cohen-Macaulay, and we conclude that $\text{rank}(H_A(\beta))$ equals $\text{vol}(A)$.

We are now prepared to pay back some old debts from Section 3.1.

Proof (of Proposition 3.1.9). We shall prove that $H_A(\beta)^{(h)} = \overline{H}_A(\beta)$ if and only if $\text{in}_{(e,e)}(H_A(\beta)) = \langle I_A, Ax\xi \rangle$. Since Lemma 4.3.7 is valid for the weight vector (e,e) as well, this will imply that $H_A(\beta)^{(h)} = \overline{H}_A(\beta)$ if I_A is Cohen-Macaulay by Theorem 4.3.5.

First assume $H_A(\beta)^{(h)} = \overline{H}_A(\beta)$. Then for $P \in H_A(\beta)$ the homogenization $P^{(h)}$ belongs to $\overline{H}_A(\beta) = D^{(h)}I_A + D^{(h)}\langle A\theta - h^2\beta \rangle$. Take the h-constant terms to see that $\text{in}_{(e,e)}(P) \in \langle I_A, Ax\xi \rangle$.

Next assume $\text{in}_{(e,e)}(H_A(\beta)) = \langle I_A, Ax\xi \rangle$, and suppose $H_A(\beta)^{(h)} \neq \overline{H}_A(\beta)$. Then there exists $P \in H_A(\beta)$ such that $P^{(h)} \notin \overline{H}_A(\beta)$. Let P have the minimal order among such. Since $\text{in}_{(e,e)}(P)$ belongs to $\langle I_A, Ax\xi \rangle$, there exists $Q \in H_A(\beta)$ such that $\text{in}_{(e,e)}(P) = \text{in}_{(e,e)}(Q)$ and $Q^{(h)} \in \overline{H}_A(\beta)$. The minimality of the order of P implies $(P - Q)^{(h)} \in \overline{H}_A(\beta)$ as well. Hence we see $P^{(h)} \in \overline{H}_A(\beta)$, which contradicts the choice of P. □

4.4 Integer Programming and Parametric *b*-functions

We develop a connection between integer programming and GKZ hypergeometric functions. This was the subject of our paper [89], but there are several new ideas and results in this section. In Section 4.5 we shall see that the optimal value function in integer programming is crucial for understanding the set of parameters $\beta \in \mathbf{k}^d$ for which $\text{rank}(H_A(\beta)) = \text{vol}(A)$ might fail.

We fix a $d \times n$-matrix A as before. For integral parameter vectors $\beta \in \mathbf{Z}^d$ and real cost vectors $w \in \mathbf{R}^n$ we consider the *integer programming problem*

$$\text{Minimize } w \cdot u \text{ subject to } A \cdot u = \beta \text{ and } u \in \mathbf{N}^n. \tag{4.34}$$

If β lies in the semigroup $\mathbf{N}A$ then this integer program has a solution. Our hypothesis (3.3) guarantees that the set of feasible solutions to (4.34) is finite. Hence there exists an *optimal solution* u which minimizes the cost function value $w \cdot u$. The optimal solution is unique if $w \in \mathbf{R}^n$ is generic but it need not be unique for special w. In this section we allow any cost vector w, so the integer program (4.34) may have many optimal solutions. All of these optimal solutions $u \in \mathbf{N}^n$ determine the same *optimal value* $v = w \cdot u \in \mathbf{R}$.

Let $\mathbf{k}$ be a subfield of the complex numbers and assume from now on that $w \in (\mathbf{k} \cap \mathbf{R})^n$. The semigroup $\mathbf{N}A$ is a subset of $\mathbf{Z}^d$ and hence a subset of $\mathbf{k}^d$. For each $\beta \in \mathbf{N}A$ the optimal value v of the integer program (4.34) lies in $\mathbf{k} \cap \mathbf{R}$ and is hence an element of the field $\mathbf{k}$. We consider the set of all pairs of feasible integer programs together with their optimal value:

$\text{Val}_{A,w} := \{(\beta, v) \in \mathbf{k}^{d+1} : \beta \in \mathbf{N}A \text{ and } v \text{ is the optimal value of } (4.34)\}.$

This is a discrete, infinite subset of $\mathbf{k}^{d+1}$. Let $\mathbf{k}[s, t] = \mathbf{k}[s_1, s_2, \ldots, s_d, t]$ denote the ring of polynomial functions on $\mathbf{k}^{d+1}$. We write $I(\text{Val}_{A,w})$ for the radical ideal of all polynomials in $\mathbf{k}[s, t]$ which vanish on the set $\text{Val}_{A,w}$. Thus the affine subvariety of $\mathbf{k}^{d+1}$ defined by $I(\text{Val}_{A,w})$ is the Zariski closure of the set $\text{Val}_{A,w}$ of pairs of feasible right hand sides with their optimal value.

We now explain the relationship to hypergeometric functions. Replace the parameters β_i by the indeterminates s_i and consider the hypergeometric ideal $H_A[s]$ in $D[s_1, \ldots, s_d]$. The cost vector $w \in \mathbf{R}^n$ defines the weight vector $(-w, w, 0) \in \mathbf{R}^{2n+d}$ for computing Gröbner bases in $D[s_1, \ldots, s_d]$. We denote the element $w_1 \theta_1 + \cdots + w_n \theta_n$ of D by the symbol t. This element represents the cost function in (4.34). The $(d + 1)$-dimensional polynomial ring $\mathbf{k}[s_1, \ldots, s_d, t] = \mathbf{k}[s, t] = \mathbf{k}[s, \sum_{i=1}^{n} w_i \theta_i]$ is thus a subring of $D[s_1, \ldots, s_d]$. We intersect this subring with the initial ideal $\text{in}_{(-w,w,0)}(H_A[s])$ of the hypergeometric ideal. The result is precisely the optimal value ideal $I(\text{Val}_{A,w})$ constructed above from integer programming.

Theorem 4.4.1. $I(\text{Val}_{A,w}) = \text{in}_{(-w,w,0)}(H_A[s]) \cap \mathbf{k}[s_1, \ldots, s_d, t]$.

In order to prove this theorem, we need the following construction of commutative algebra. Let J be any ideal in $\mathbf{k}[\partial_1, \ldots, \partial_n] = \mathbf{k}[\partial]$. Define $\text{mono}(J)$ to be the subideal generated by all monomials in J. In other words, $\text{mono}(J)$ is the largest monomial ideal in $\mathbf{k}[\partial]$ which is contained in J. The following algorithm can be used to compute $\text{mono}(J)$ from the generators of J.

Algorithm 4.4.2 (Computing all monomials in a given polynomial ideal)
Input: A polynomial ideal $J = \langle f_1, \ldots, f_r \rangle$ in $\mathbf{k}[\partial]$.
Output: Generators for $\text{mono}(J)$, the largest monomial ideal in J.

(1) Form the complete multi-homogenization of the r given generators:

$$g_j(\partial_1, \ldots, \partial_n, u_1, \ldots, u_n) := f_j\left(\frac{\partial_1}{u_1}, \ldots, \frac{\partial_n}{u_n}\right) \cdot u_1^{\deg_{\partial_1}(f_j)} \cdots u_n^{\deg_{\partial_n}(f_j)}.$$

Here $u_1, \ldots, u_n$ are new variables.

(2) Fix an elimination term order $\{u_1, \ldots, u_n\} \succ \{\partial_1, \ldots, \partial_n\}$ on $\mathbf{k}[\partial, u]$, and compute the reduced Gröbner basis $\mathcal{G}$ for the saturation ideal

$$L := \langle g_1, g_2, \ldots, g_r \rangle : \langle u_1 u_2 \cdots u_n \rangle^{\infty}$$

(3) Output all the monomials in the set $\mathcal{G}$ above. They generate $\text{mono}(J)$.

Note that none of the elements in $\mathcal{G}$ is divisible by any of $u_1, \ldots, u_n$. Hence all monomials in $\mathcal{G}$ actually lie in $\mathbf{k}[\partial]$. The correctness of Algorithm 4.4.2 is seen as follows. The ideal L is homogeneous with respect to the $\mathbf{N}^n$-grading $\deg(\partial_i) = \deg(u_i) = e_i$. Thus a u-free polynomial $p(\partial_1, \ldots, \partial_n)$ can lie in $\mathcal{G}$

only if it is a monomial $m = \partial_1^{a_1} \cdots \partial_n^{a_n}$. If such a monomial lies in J then it lies in L and its normal form modulo $\mathcal{G}$ is zero. Let $g \in \mathcal{G}$ such that $\mathrm{in}_{\prec}(g)$ divides m. Since the term order is an elimination order, we conclude that $g = \mathrm{in}_{\prec}(g)$, and hence m is divisible by a monomial in $\mathcal{G}$.

Example 4.4.3. We apply Algorithm 4.4.2 to the ideal

$$J = \langle \partial_1 + \partial_2 + \partial_3, \ \partial_1^2 + \partial_2^2 + \partial_3^2, \ \partial_1^3 + \partial_2^3 + \partial_3^3 \rangle$$

of symmetric polynomials without constant term. The homogenizations of the three given power sums are $\partial_1^i u_2^i u_3^i + \partial_2^i u_1^i u_3^i + \partial_3^i u_1^i u_2^i$ for $i = 1, 2, 3$. Saturating the ideal generated by these three with respect to $u_1 u_2 u_3$ gives

$$L = \langle \partial_1^3, \partial_2^3, \partial_3^3, \partial_2^2 \partial_3^2, \partial_1^2 \partial_2^2, \partial_1^2 \partial_3^2, \partial_1 \partial_2 \partial_3, \ldots \text{ and eight non-monomials} \rangle.$$

Thus $\mathrm{mono}(J)$ is the Artinian ideal generated by the seven monomials above.

If we think of J as a linear system of partial differential equations with constant coefficients, then the operation of passing from J to $\mathrm{mono}(J)$ amounts to intersection with the torus-invariant subring. The ideal J in the example above represents the *harmonic functions* in three variables. We have

$$D \cdot J \cap \mathbf{k}[\theta_1, \theta_2, \theta_3] = \langle \theta_1 \theta_2 \theta_3, \ \theta_i(\theta_i - 1)\theta_j(\theta_j - 1), \ \theta_i(\theta_i - 1)(\theta_i - 2) \text{ for } i \neq j \rangle.$$

The next lemma states this result in general.

Lemma 4.4.4. *Let J be an ideal in $\mathbf{k}[\partial]$. Then $D \cdot J \cap \mathbf{k}[\theta]$ equals the distraction $\widetilde{M}$ of the monomial ideal $M = \mathrm{mono}(J) \subset \mathbf{k}[\partial]$.*

Proof. Clearly, $\widetilde{M}$ is contained in $D \cdot J \cap \mathbf{k}[\theta]$. Conversely, suppose $f \in D \cdot J \cap \mathbf{k}[\theta]$. Then we write

$$f(\theta) = \sum_{u \in \mathbf{N}^n} c_u \cdot x^u \cdot p_u(\partial), \qquad \text{where } p_u \in J \text{ and } c_u \in \mathbf{k}.$$

The left hand side is homogeneous with respect to the $\mathbf{Z}^n$-grading and hence so is the right hand side. This implies that $p_u(\partial)$ is a scalar multiple of ∂^u whenever $c_u \neq 0$, so that $f(\theta)$ lies in the distraction of $M = \mathrm{mono}(J)$. $\quad\square$

Lemma 4.4.4 gives rise to an alternative algorithm, using Gröbner bases in D, for computing the operator $\mathrm{mono}(\,\cdot\,)$. We shall apply this lemma in the situation where $J = \mathrm{in}_w(I_A)$ is an initial ideal of the toric ideal I_A. For special weight vectors w this is not a monomial ideal but it is generated by some monomials ∂^a and some binomials $\partial^b - \partial^c$. Here is a formula for the torus-invariant part of an arbitrary Gröbner deformation of the GKZ hypergeometric system with parametric right hand side vector $s = (s_1, \ldots, s_d)$.

Proposition 4.4.5. *For $w \in \mathbf{R}^n$ let $M = \mathrm{mono}(\mathrm{in}_w(I_A))$. Then*

$$\mathrm{in}_{(-w, w, 0)}(H_A[s]) \cap \mathbf{k}[\theta_1, \ldots, \theta_n, s_1, \ldots, s_d] = \widetilde{M} + \langle A \cdot \theta - s \rangle. \quad (4.35)$$

Proof. We have seen in Section 3.1, specifically in (3.5), that

$$\text{in}_{(-w,w,0)}(H_A[s]) \;=\; D[s] \cdot \text{in}_w(I_A) + D[s]\langle A\theta - s\rangle.$$

The intersection of this ideal with $\mathbf{k}[\theta, s]$ equals $\langle A\theta - s\rangle$ plus the intersection of $D[s] \cdot \text{in}_w(I_A)$ with $\mathbf{k}[\theta, s]$. The latter equals $\widetilde{M}$ by Lemma 4.4.4. $\square$

Example 4.4.6. Let $d = 2, n = 5$, $w = (3, 2, 2, 0, 0)$ and $A = \begin{pmatrix} 1 & 1 & 1 & 1 & 1 \\ 0 & 1 & 2 & 3 & 4 \end{pmatrix}$. The initial ideal of the toric ideal I_A for the weight vector w equals

$$\text{in}_w(I_A) \;=\; \langle \partial_2^3 - \partial_1^2\partial_4,\ \partial_2\partial_5 - \partial_3\partial_4,\ \partial_1\partial_3,\ \partial_1\partial_5,\ \partial_2\partial_3,\ \partial_3^2,\ \partial_3\partial_5 \rangle$$

This is not a monomial ideal, which means that w is not generic for I_A. Using Algorithm 4.4.2 we find that $\text{in}_w(I_A)$ has the maximal monomial subideal

$$M \;=\; \text{mono}(\text{in}_w(I_A)) \;=\; \langle \partial_1\partial_3,\ \partial_1\partial_5,\ \partial_2\partial_3,\ \partial_3^2,\ \partial_3\partial_5,\ \partial_2^2\partial_5,\ \partial_2\partial_5^2 \rangle.$$

The initial ideal of the parametric GKZ system is $\text{in}_{(-w,w,0)}(H_A[s_1, s_2]) =$

$$D[s] \cdot \text{in}_w(I_A) + D[s] \cdot \{\, \theta_1 + \theta_2 + \theta_3 + \theta_4 + \theta_5 - s_1,\ \theta_2 + 2\theta_3 + 3\theta_4 + 4\theta_5 - s_2 \,\}.$$

If we intersect this left $D[s_1, s_2]$-ideal with $\mathbf{k}[\theta_1, \theta_2, \theta_3, \theta_4, \theta_5, s_1, s_2]$, then, according to Proposition 4.4.5, we get the ideal

$$\widetilde{M} \;=\; \langle\, \theta_1\theta_3,\ \theta_1\theta_5,\ \theta_2\theta_3,\ \theta_3(\theta_3 - 1),\ \theta_3\theta_5,\ \theta_2(\theta_2 - 1)\theta_5,\ \theta_2\theta_5(\theta_5 - 1)\,\rangle$$

plus the two linear relations $\theta_1 + \theta_2 + \theta_3 + \theta_4 + \theta_5 - s_1$, $\theta_2 + 2\theta_3 + 3\theta_4 + 4\theta_5 - s_2$.

Our next task is to interpret the monomial ideal $M = \text{mono}(\text{in}_w(I_A))$ in terms of the optimal solutions to the integer programming problem (4.34).

Lemma 4.4.7. *Let $u \in \mathbf{N}^n$ and set $\beta := A \cdot u$. The vector u is an optimal solution to the program (4.34) if and only if the monomial ∂^u is not in M.*

Proof. The vector u is not an optimal solution to (4.34) if and only if there exists $v \in \mathbf{N}^n$ with $A \cdot v = \beta$ and $w \cdot v < w \cdot u$, or, equivalently, if and only if some binomial $\partial^u - \partial^v$ in $\text{in}_w(I_A)$ has w-leading form ∂^u. Since $\text{in}_w(I_A)$ is a binomial ideal, this implies that u is not an optimal solution to (4.34) if and only if ∂^u lies in $\text{in}_w(I_A)$. But $\partial^u \in \text{in}_w(I_A)$ is equivalent to $\partial^u \in M = \text{mono}(\text{in}_w(I_A))$. This proves the contrapositive of our assertion. $\square$

Readers with limited experience in integer programming will find it useful to carefully verify the preceding lemma in examples such as the following.

Example 4.4.6 continued. Here our integer program (4.34) is the following:

Minimize $3u_1 + 2u_2 + 2u_3$ for non-negative integers u_1, u_2, u_3, u_4, u_5
satisfying $u_1 + u_2 + u_3 + u_4 + u_5 = \beta_1$ and $u_2 + 2u_3 + 3u_4 + 4u_5 = \beta_2$.

To determine all possible optimal solutions, for any integral right hand sides β_1, β_2, we examine the ideal of all monomials in $\text{in}_w(I_A)$, that is, $M = \langle \partial_1\partial_3, \partial_1\partial_5, \partial_2\partial_3, \partial_3^2, \partial_3\partial_5, \partial_2^2\partial_5, \partial_2\partial_5^2 \rangle$. We must characterize all monomials not in M. This is best done by computing the standard pairs:

$$\mathcal{S}(M) \quad = \quad \{ (1, \{1, 2, 4\}), \ (1, \{4, 5\}), \ (\partial_3, \{4\}), \ (\partial_2\partial_5, \{4\}) \}.$$

Lemma 4.4.7 now tells us that a given vector $(u_1, u_2, u_3, u_4, u_5)$ in $\mathbf{N}^5$ is an optimal solution if and only if it looks like one of the following four types,

$$(*, *, 0, *, 0), \ (0, 0, 0, *, *), \ (0, 0, 1, *, 0), \ (0, 1, 0, *, 1).$$

Here $*$ stands for an arbitrary non-negative integer. Since w is not generic, there is generally more than one optimal solution $(u_1, u_2, u_3, u_4, u_5)$. For instance, for $\beta_1 = \beta_2 = 6$ the solutions are $(0, 6, 0, 0, 0)$, $(2, 3, 0, 1, 0)$ and $(4, 0, 0, 2, 0)$, and for $\beta_1 = 6, \beta_2 = 17$ they are $(0, 0, 1, 5, 0)$ and $(0, 1, 0, 4, 1)$. We recall from Corollary 3.2.3 that the standard pairs of M give us the prime components of the distraction of M:

$$\widetilde{M} \quad = \quad \langle \theta_3, \theta_5 \rangle \cap \langle \theta_1, \theta_2, \theta_3 \rangle \cap \langle \theta_1, \theta_2, \theta_3 - 1, \theta_5 \rangle \cap \langle \theta_1, \theta_2 - 1, \theta_3, \theta_5 - 1 \rangle.$$

Thus $\widetilde{M}$ has codimension 2, and $\widetilde{M} + \langle A\theta - \beta \rangle$ has codimension 4. We conclude that the Frobenius ideal $\widetilde{M} + \langle A \cdot \theta - s \rangle$ is generally *not holonomic*.

We are now prepared to prove the main theorem in this section.

Proof (of Theorem 4.4.1). Since the coordinates of $s = (s_1, \ldots, s_d)$ are indeterminates, and $\widetilde{M}$ is a radical, the ideal (4.35) is a radical ideal in $\mathbf{k}[\theta_1, \ldots, \theta_n, s_1, \ldots, s_d]$. Its variety is the Zariski closure of all points $(u, A \cdot u) \in \mathbf{N}^n \times \mathbf{Z}^d$, where u is a zero of $\widetilde{M}$. In view of Lemma 4.4.7, these are the points (u, β) where $u \in \mathbf{N}^n$ is an optimal solution of (4.34) for the right hand side β. The image of this set under the linear map $\mathbf{k}^n \times \mathbf{k}^d \to \mathbf{k}^d \times \mathbf{k}, (u, \beta) \mapsto (\beta, w \cdot u)$ is precisely $\text{Val}_{A,w}$. Thus the radical ideal of $\text{Val}_{A,w}$ is gotten from (4.35) by intersection with the subring $\mathbf{k}[s_1, \ldots, s_d, t]$, where $t = \sum_{i=1}^n w_i\theta_i$, This completes the proof.   □

The non-zero polynomials in the ideal

$$I(\text{Val}_{A,w}) \quad = \quad \text{in}_{(-w,w,0)}(H_A[s]) \cap \mathbf{k}[s_1, \ldots, s_d, t]$$

are called *parametric b-functions* with respect to w for the GKZ hypergeometric system defined by A. In Section 5.1 we shall introduce the notion of the *b-function* for an arbitrary holonomic D-ideal. For the GKZ-ideal $H_A(\beta)$ and $w \in \mathbf{R}^n$, the b-function is the generator of the principal ideal

$$\mathrm{in}_{(-w,w)}(H_A(\beta)) \cap \mathbf{k}[t], \qquad \text{with } t = \sum_{i=1}^{n} w_i \theta_i \text{ as before.}$$

The point is that β is a *specific parameter vector* in $\mathbf{k}^d$. The above ideal is principal because all ideals in $\mathbf{k}[t]$ are. The following remark is easy to show.

Remark 4.4.8. If $b(s,t)$ is a parametric b-function and $\beta \in \mathbf{k}^d$ then $b(\beta,t) \in \mathbf{k}[t]$ is either zero or divisible by the b-function. For generic parameters β, the b-function is the greatest common divisor in $\mathbf{k}[t]$ of the polynomials $b(\beta,t)$ where b runs over a generating set of the optimal value ideal $I(\mathrm{Val}_{A,w})$.

Example 4.4.6 continued.   To compute the optimal value ideal $I(\mathrm{Val}_{A,w})$, we add the linear relation $3\theta_1 + 2\theta_2 + 2\theta_3 - t$ to the nine generators of $\widetilde{M} + \langle A \cdot \theta - s \rangle$, and we compute a Gröbner basis for these ten polynomials with respect to an elimination term order of the form $\{\theta_1, \theta_2, \theta_3, \theta_4, \theta_5\} \succ \{s_1, s_2, t\}$. The resulting elimination ideal in $\mathbf{k}[s_1, s_2, t]$ equals

$$
\begin{aligned}
I(\mathrm{Val}_{A,w}) &= \big\langle\, t(t - 3s_1 + s_2)(t-2),\ t(t - 3s_1 + s_2)(3s_1 - s_2 - 1)\,\big\rangle \\
&= \langle t - 3s_1 + s_2 \rangle \cap \langle t \rangle \cap \langle t - 2, 3s_1 - s_2 - 1 \rangle.
\end{aligned}
$$

For instance, the optimal value for $(6,6)$ is 12, and the optimal value for $(6,17)$ is 2, which means that $(6,6,12)$ and $(6,17,2)$ are zeros of $I(\mathrm{Val}_{A,w})$. The polynomials $t(t-3s_1+s_2)(t-2)$ and $t(t-3s_1+s_2)(3s_1-s_2-1)$ are parametric b-functions. If we substitute a generic vector $(\beta_1, \beta_2) \in \mathbf{k}^2$ for (s_1, s_2), then the first polynomial becomes a multiple of the second polynomial, and we conclude that the b-function is the univariate polynomial $t(t - 3\beta_1 + \beta_2)$.

The explicit form of the prime decomposition of the optimal value ideal in the previous example can be generalized as follows. Fix the monomial ideal $M = \mathrm{mono}(\mathrm{in}_w(I_A))$. Suppose that (∂^u, σ) is a standard pair of M. Let $r = r(\sigma) := d - \mathrm{rank}(A_\sigma)$ and choose a $\mathbf{k}$-basis $\ell_\sigma^1, \dots, \ell_\sigma^r$ for the space of row vectors $\ell \in \mathbf{R}^d$ such that $\ell \cdot a_i = 0$ for $i \in \sigma$. Then there exists a row vector v_σ in $\mathbf{R}^d$, unique modulo $\langle \ell_\sigma^1, \dots, \ell_\sigma^r \rangle$, such that

$$v_\sigma \cdot a_i + w_i = 0 \qquad \text{for } i \in \sigma.$$

Formulated equivalently in matrix notation, the row vectors $(v_\sigma, 1)$, $(\ell_\sigma^1, 0)$, $\dots, (\ell_\sigma^r, 0)$ form a basis for the subspace of $\mathbf{R}^{d+1}$ consisting of all vectors whose product on the left with the $(d+1) \times n$-matrix $\binom{A}{w}$ has support disjoint from σ.

Theorem 4.4.9. *The prime decomposition of the optimal value ideal equals*

$$
I(\mathrm{Val}_{A,w}) =
$$
$$
\bigcap_{(\partial^u, \sigma) \in \mathcal{S}(M)} \big\langle\, t - v_\sigma \cdot (Au - s) - w \cdot u,\ \ell_\sigma^{(1)}(Au - s), \dots, \ell_\sigma^{(r(\sigma))}(Au - s) \,\big\rangle.
$$

Proof. The radical ideal $I(\text{Val}_{A,w})$ is computed by eliminating $\theta_1, \ldots, \theta_n$ from the intersection over all $(\partial^u, \sigma) \in \mathcal{S}(M)$ of the prime ideals

$$\langle \theta_i - u_i : i \notin \sigma \rangle + \langle A \cdot \theta - s \rangle + \langle \sum_{i=1}^{n} \theta_i w_i - t \rangle. \tag{4.36}$$

Since all ideals involved are radical and elimination commutes with intersection, it suffices to show that $(4.36) \cap \mathbf{k}[s,t]$ is generated by the $r(\sigma)+1$ linear forms listed above. To see that this is the case, we write (4.36) in matrix form

$$\langle \theta_i - u_i : i \notin \sigma \rangle + \langle \begin{pmatrix} A \\ w \end{pmatrix} \cdot \theta - \begin{pmatrix} s \\ t \end{pmatrix} \rangle$$

Intersection with the polynomial ring $\mathbf{k}[s,t][\theta_i : i \in \sigma]$ yields the ideal

$$\langle \begin{pmatrix} A \\ w \end{pmatrix}_\sigma \cdot \theta_\sigma + \begin{pmatrix} A \\ w \end{pmatrix} \cdot u - \begin{pmatrix} s \\ t \end{pmatrix} \rangle, \tag{4.37}$$

where θ_σ denotes the column vector with coordinates θ_i, $i \in \sigma$. To eliminate θ_σ from (4.37) we multiply the column vector in (4.37) with any row vector $\ell \in \mathbf{R}^{d+1}$ such that $\ell \cdot \begin{pmatrix} A \\ w \end{pmatrix}_\sigma = 0$. The resulting expressions $\ell \cdot \begin{pmatrix} A \\ w \end{pmatrix} \cdot u - \ell \cdot \begin{pmatrix} s \\ t \end{pmatrix}$ generate the elimination ideal $(4.36) \cap \mathbf{k}[s,t] = (4.37) \cap \mathbf{k}[s,t]$. Now take for ℓ the basis vectors $(v_\sigma, 1)$, $(\ell_\sigma^1, 0), \ldots, (\ell_\sigma^r, 0)$ to get the theorem. $\square$

The prime components arising from different standard pairs in Theorem 4.4.9 need not be distinct. For instance, in Example 4.4.6, the two lower-dimensional standard pairs $(\partial_3, \{4\})$ and $(\partial_2 \partial_5, \{4\})$ give rise to the same prime component $\langle t - 2, 3s_1 - s_2 - 1 \rangle$ of $I(\text{Val}_{A,w})$.

We add a few comments on the polyhedral geometry which we get from the varieties of the ideals $\text{in}_w(I_A)$ and $M = \text{mono}(\text{in}_w(I_A))$. The radical of the initial ideal $\text{in}_w(I_A)$ can be interpreted as a regular polyhedral subdivision Δ_w of A; see e.g. [96, Theorem 10.10]. The subdivision Δ_w is the regular triangulation familiar from Section 3.2 if $\text{in}_w(I_A)$ is a monomial ideal, but in general Δ_w is a subdivision of $Q = \text{conv}(A)$ into polytopes not all of which are simplices. Each facet σ in the polyhedral subdivision Δ_w is regarded as a subset of $\{1, 2, \ldots, n\}$, namely, consisting of all points a_i on that facet.

Remark 4.4.10. A subset σ of $\{1, 2, \ldots, n\}$ determines a facet of the polyhedral subdivision Δ_w if and only if $(1, \sigma)$ is a standard pair of $M = \text{mono}(\text{in}_w(I_A))$.

This remark implies that the simplicial complex underlying the radical of M consists of the same faces σ but now regarded as simplices. These simplices are of higher dimension than the corresponding facets of Δ_w when the facets are not simplices. For instance, if Δ_w is a subdivision of an octagon into a triangle, a quadrangle and a pentagon, then the simplicial complex of M consists of a triangle, a tetrahedron, and a 4-simplex, glued according to Δ_w.

4.5 The Exceptional Hyperplane Arrangement

The *exceptional set* of a matrix A is defined to be a set of parameters β at which the holonomic rank of $H_A(\beta)$ is not equal to (hence strictly larger than) vol(A), In Section 4.2 we determined the exceptional set explicitly for $d = 2$, and in Section 4.3 we showed that the exceptional set is empty if I_A is Cohen-Macaulay. For arbitrary A, we have seen that the holonomic rank of $H_A(\beta)$ equals vol(A) for generic parameters β. This implies the following

Remark 4.5.1. The exceptional set lies in a proper algebraic subvariety of $\mathbf{k}^d$.

It is a difficult problem to determine the exceptional set exactly. We believe that examples exist where the exceptional set is not Zariski closed in $\mathbf{k}^d$ but at the moment no such example is known to us. Clearly, much more research needs to be done in understanding the exceptional set for $d \geq 3$.

In this section we introduce a natural arrangement of hyperplanes which contains the exceptional set. For a facet σ of the cone pos(A) = $\mathbf{R}_+A$, denote by F_σ the *primitive integral support function* of σ, namely F_σ is a linear function on $\mathbf{R}A = \mathbf{R}^d$ uniquely determined by the conditions:

1. $F_\sigma(\mathbf{Z}A) = \mathbf{Z}$,
2. $F_\sigma(a_i) \geq 0$ for all $i = 1, \ldots, n$,
3. $F_\sigma(a_i) = 0$ for all $a_i \in \sigma$.

Let $\mathcal{H}_A$ denote the arrangement consisting of the affine hyperplanes

$$\{ u \in \mathbf{k}^d \ : \ F_\sigma(u) = j \},$$

where σ runs over the facets of pos(A) and j is any nonnegative integer. We call $\mathcal{H}_A$ the *exceptional arrangement* of the matrix A. For each facet σ there are infinitely many parallel hyperplanes in $\mathcal{H}_A$. Adolphson [2, Theorem 5.15] showed that the exceptional set is contained in the exceptional arrangement.

Theorem 4.5.2. (Adolphson) *If* $\beta \in \mathbf{k}^d \backslash \mathcal{H}_A$ *then* rank($H_A(\beta)$) = vol(A).

Parameters β outside the exceptional arrangement $\mathcal{H}_A$ were called *semi-nonresonant* by Adolphson. Gel'fand, Kapranov and Zelevinsky [38] had called β *nonresonant* if $F_\sigma(\beta)$ is not an integer for any facet σ of pos(A). An important subset in the difference between semi-nonresonant parameters and nonresonant parameters is the *Euler-Jacobi cone*, which is the interior of $-$pos(A). Thus the following is a special case of Adolphson's theorem.

Corollary 4.5.3. *If* $\beta \in -\text{int}(\text{pos}(A))$ *then* rank($H_A(\beta)$) = vol(A).

We shall present a self-contained proof of Theorem 4.5.2. It uses Gröbner bases and is different from Adolphson's homological approach. It is possible to replace the infinite hyperplane arrangement in Theorem 4.5.2 by a finite arrangement having the same property. One such choice is the locus of parameters β which are not w-flat. This is the subject of Section 4.6. In this section we keep things simple and stay with the infinite arrangement $\mathcal{H}_A$.

In the proof of Theorem 4.5.2 we shall apply the results of Section 4.4 for cost vectors of the form $w = -e_i$, that is, the negative of any unit vector $e_i = (0, \ldots, 0, 1, 0, \ldots, 0)$. The corresponding integer program takes the form

$$\text{Maximize } u_i \text{ subject to } A \cdot u = \beta \text{ and } u \in \mathbf{N}^n. \tag{4.38}$$

The facet linear forms F_σ arise in the prime decomposition of the optimal value ideal $I(\text{Val}_{A,-e_i})$ as follows. The variable ∂_i does not appear in any of the generators of the initial ideal $\text{in}_{-e_i}(I_A)$, and therefore ∂_i does not appear among the generators of the monomial ideal $M = \text{mono}(\text{in}_{-e_i}(I_A))$. Therefore each standard pair of M has the form $(\partial^u, \{i\} \cup \tau)$ for some $\tau \in \{1, \ldots, i-1, i+1, \ldots, n\}$. The sets τ must lie in the boundary of $\text{pos}(A)$:

Lemma 4.5.4. *Let* $M = \text{mono}(\text{in}_{-e_i}(I_A))$. *If* $(\partial^u, \{i\} \cup \tau) \in \mathcal{S}(M)$ *then* τ *lies in a facet* σ *of the cone* $\text{pos}(A)$ *which does not contain the point* a_i.

Proof. Suppose that τ is not in a facet of $\text{pos}(A)$ which does not contain the point a_i. Then there exist integer vectors $v, v' \in \mathbf{N}^n$ such that $Av = Av'$, $\text{supp}(v) \subseteq \tau$ and $i \in \text{supp}(v')$. This means that v is not the optimal solution of the integer program (4.38) since the i-th coordinate of v' is positive while the i-th coordinate of v is zero. By Lemma 4.4.7, this shows that $\partial^v \in M$ and therefore $(\partial^u, \{i\} \cup \tau)$ cannot be a standard pair of M, a contradiction. $\square$

For each standard pair $(\partial^u, \{i\} \cup \tau) \in \mathcal{S}(M)$ we fix a facet σ as in Lemma 4.5.4, and we write $F_\tau := F_\sigma$ for simplicity of notation. Note that $F_\tau(a_j) = 0$ for $j \in \tau$. If we set $w = -e_i$ then this means that

$$F_\tau(a_j) + w_j \cdot F_\tau(a_i) \quad = \quad 0 \quad \text{for all } j \in \tau \cup \{i\}.$$

Dividing this equation by the positive integer $F_\tau(a_i)$, we see that the linear form $\frac{1}{F_\tau(a_i)} \cdot F_\tau$ satisfies all the defining properties of the linear form $v_{\tau \cup \{i\}}$ in Theorem 4.4.9. Thus for $w = -e_i$, Theorem 4.4.9 can be written as follows. Note that the term $w \cdot u$ disappears since the support of u never contains i.

Corollary 4.5.5. *The prime decomposition of the optimal value ideal for the integer program (4.38) has the following form:*

$$I(\text{Val}_{A,-e_i}) = \bigcap_{(\partial^u, \tau \cup \{i\}) \in \mathcal{S}(M)} \left\langle F_\tau(a_i) \cdot t - F_\tau(Au-s), \ \ldots, \ \ell^{(j)}_{\tau \cup \{i\}}(Au-s), \ldots \right\rangle.$$

This implies that the following product is a parametric b-function:

$$\prod_{(\partial^u, \tau \cup \{i\}) \in \mathcal{S}(M)} \left(F_\tau(s) + F_\tau(a_i) \cdot t - F_\tau(Au) \right). \tag{4.39}$$

Our next goal is to give an upper bound for the constant term $F_\tau(Au)$ appearing in the linear factors of this parametric b-function. We first state two general lemmas on monomial ideals.

Lemma 4.5.6. *If M is any monomial ideal in $\mathbf{k}[\partial]$ generated by monomials whose ∂_j-degree is at most D and $(\partial^u, \sigma) \in \mathcal{S}(M)$ then $u_j < D$.*

Proof. This is easy to see for Artinian ideals, where $\sigma = \emptyset$. The general case is reduced to the Artinian case by the construction in Algorithm 3.2.5. $\square$

Let $\mathcal{D}(A)$ denote the largest absolute value of any $d \times d$-subdeterminant of the matrix A.

Lemma 4.5.7. *The ideal $M = \text{mono}(\text{in}_w(I_A))$ is generated by monomials whose ∂_j-th coordinate is at most $(n - d)\mathcal{D}(A)$.*

Proof. A binomial $\partial^u - \partial^v \in I_A$ is called *primitive* if there exists no other binomial $\partial^{u'} - \partial^{v'} \in I_A$ with $\partial^{u'} \mid \partial^u$ and $\partial^{v'} \mid \partial^v$. We claim that each minimal generator of M consists of one of the terms of a primitive binomial. Let ∂^u be a minimal generator of M. Then there exists $v \in \mathbf{N}^n$ with $Au = Av$ and $w \cdot u > w \cdot v$. Suppose that $\partial^u - \partial^v$ is not primitive. Then there exist $u' \neq u$, $v' \neq v \in \mathbf{N}^n$ with $\partial^{u'} \mid \partial^u$, $\partial^{v'} \mid \partial^v$, and $Au' = Av'$. If $w \cdot u' > w \cdot v'$, then $\partial^{u'} \in M$. If $w \cdot u' \leq w \cdot v'$, then $\partial^{u-u'} \in M$. Both cases contradict ∂^u being a minimal generator of M. Hence we completed the proof of the claim. The conclusion now follows from [96, Theorem 4.7]. $\square$

We now apply these two lemmas to the case $w = -e_i$:

Theorem 4.5.8. *If $(\partial^u, \tau \cup \{i\})$ is a standard pair of $M = \text{mono}(\text{in}_{-e_i}(I_A))$ then $F_\tau(Au)$ is an integer between 0 and $n(n - d)\mathcal{D}(A)^2$.*

Proof. Clearly $F_\tau(Au)$ is a nonnegative integer. Each coordinate of u satisfies $u_j < (n - d)\mathcal{D}(A)$ by Lemmas 4.5.6 and 4.5.7. We find

$$F_\tau(Au) \quad = \quad \sum_{j=1}^n u_j \cdot F_\tau(a_j) \quad \leq \quad (n - d)\mathcal{D}(A) \cdot \sum_{j=1}^n F_\tau(a_j).$$

We have $F_\tau(a_j) \leq \mathcal{D}(A)$ since the point a_i has lattice distance at most $\mathcal{D}(A)$ from any facet of $\text{pos}(A)$. This gives the desired upper bound

$$F_\tau(Au) \quad \leq \quad (n - d)\mathcal{D}(A) \cdot \sum_{i=1}^n \mathcal{D}(A) \quad = \quad n(n - d)\mathcal{D}(A)^2.$$

$\square$

The bound in this theorem (and hence in Theorem 4.5.2) is certainly not best possible, and it would be interesting to find an optimal bound. For our purposes, the following conclusion will turn out to be sufficient.

Corollary 4.5.9. *The following product of linear forms is a parametric b-function, that is, it lies in the optimal value ideal* $I(\mathrm{Val}_{A,-e_i})$ *for (4.38):*

$$\prod_{\substack{\sigma:\,\text{facet}\\a_i\notin\sigma}}\prod_{j=0}^{n(n-d)\mathcal{D}(A)^2}\big(F_\sigma(s)+F_\sigma(a_i)\cdot t-j\big)\tag{4.40}$$

We now shift gears in order to connect all this to the rank of the hypergeometric GKZ ideal. Consider the following ideal in $\mathbf{k}[s]=\mathbf{k}[s_1,\ldots,s_d]$:

$$B_{A,i}\ :=\ (H_A[s]+D[s]\partial_i)\cap\mathbf{k}[s]\tag{4.41}$$

The importance of this ideal is described in the next theorem. Put

$$M_A(\beta)\ :=\ D/H_A(\beta).\tag{4.42}$$

Multiplication by ∂_i from the right defines a morphism of left D-modules from $M_A(\beta-a_i)$ to $M_A(\beta)$, which is denoted by ∂_i again. This morphism is referred to as *parameter shift* of hypergeometric functions. It generalizes the classical concept of *contiguity relations*. For additional information see [89].

Theorem 4.5.10. (cf. [88, Corollary 5.4]) *If a parameter β does not lie in the zero set of $B_{A,i}$, then $\partial_i:M_A(\beta-a_i)\longrightarrow M_A(\beta)$ is an isomorphism.*

To prove Theorem 4.5.10, we need the following lemma.

Lemma 4.5.11. *Let $P\in D$, and $c\in\mathbf{k}[\theta_1,\ldots,\theta_n]$. If $P\partial_i\equiv c(\theta_1,\ldots,\theta_n)$ modulo $D\cdot I_A$, then $\partial_iP\equiv c(\theta_1,\ldots,\theta_{i-1},\theta_i+1,\theta_{i+1},\ldots,\theta_n)$ modulo $D\cdot I_A$.*

Proof. Suppose $P\partial_i\equiv c(\theta_1,\ldots,\theta_n)$ modulo $D\cdot I_A$. Multiply both sides by ∂_i from the left. Then the RHS equals $c(\theta_1,\ldots,\theta_{i-1},\theta_i+1,\theta_{i+1},\ldots,\theta_n)\partial_i$. Hence the assertion follows because ∂_i is not a zero-divisor modulo I_A.   □

Proof (of Theorem 4.5.10). Fix $b(s)\in B_{A,i}$ with $b(\beta)\neq0$. Then there exists $P\in D$ such that $P\partial_i\equiv b(s)$ in $M_A[s]=D[s]/H_A[s]$, for if $P(s)\partial_i\equiv b(s)$ in $M_A[s]$ with $P(s)=\sum_\alpha P_\alpha s^\alpha$ $(P_\alpha\in D)$, then we have $P(A\theta+a_i)\partial_i\equiv b(s)$ in $M_A[s]$. By considering a Gröbner basis with respect to any term order satisfying $x_k,\partial_k\prec s_i$, we see $P\partial_i\equiv b(A\theta)$ modulo $D\cdot I_A$. Note that Lemma 4.5.11 implies $\partial_iP\equiv b(A\theta+a_i)$ modulo $D\cdot I_A$.

Next we claim that multiplication by P from the right induces a D-module homomorphism $M_A(\beta)\longrightarrow M_A(\beta-a_i)$. Let $u,v\in\mathbf{N}^n$ satisfy $Au=Av$. Then modulo $D\cdot I_A$,

$$(\partial^u-\partial^v)P\partial_i\equiv(\partial^u-\partial^v)b(A\theta)\equiv b(A\theta+Au)\partial^u-b(A\theta+Av)\partial^v\equiv0.$$

Since ∂_i is not a zero-divisor modulo I_A, we conclude that $(\partial^u-\partial^v)P\equiv0$ modulo $D\cdot I_A$. By the same argument, we also see

$$(A\theta)P \equiv P(A\theta + a_i).$$

Hence we have proved the claim.

If $P\partial_i \equiv b(\beta) \neq 0$ in $M_A(\beta)$, then $\partial_i P \equiv b(\beta - a_i + a_i) \neq 0$ in $M_A(\beta - a_i)$. Therefore the map ∂_i is bijective and an isomorphism of D-modules.   □

The relation to integer programming is given by the following result.

Proposition 4.5.12. *The ideal $B_{A,i}$ in (4.41) is the image of the optimal value ideal $I(\mathrm{Val}_{A,-e_i})$ under the map $\mathbf{k}[s,t] \to \mathbf{k}[s]$, $t \mapsto 0$.*

Proof. The weight vector $w = -e_i$ has the characteristic property

$$\mathrm{in}_{-e_i}(I_A) + \langle \partial_i \rangle \;\; = \;\; I_A + \langle \partial_i \rangle \;\; \subset \;\; \mathbf{k}[\partial].$$

This implies

$$H_A[s] + D[s] \cdot \langle \partial_i \rangle \;\; = \;\; \mathrm{in}_{(e_i, -e_i, 0)}(H_A[s]) + D[s] \cdot \langle \partial_i \rangle.$$

We set $t = -\theta_i = -x_i \partial_i$ and intersect with the subring $\mathbf{k}[s,t]$ to find

$$\mathbf{k}[s,t] \cdot B_{A,i} + \langle t \rangle \;\; = \;\; I(\mathrm{Val}_{A,-e_i}) + \langle t \rangle.$$

The assertion follows from this identity.   □

It follows from Corollary 4.5.9 and Proposition 4.5.12 that the product

$$\prod_{\substack{\sigma\, facet \\ a_i \notin \sigma}} \prod_{j=0}^{n(n-d)\mathcal{D}(A)^2} \big(F_\sigma(s) - j\big) \qquad \text{lies in the ideal } B_{A,i}. \qquad (4.43)$$

Each of the factors in the above product defines one of the hyperplanes of the exceptional hyperplane arrangement $\mathcal{H}_A$ in $\mathbf{k}^d$, which was defined at the beginning of this section. Theorem 4.5.10 implies the following.

Corollary 4.5.13. *The zero set of $B_{A,i}$ lies in the exceptional arrangement $\mathcal{H}_A$. If $\beta \notin \mathcal{H}_A$ then $\partial_i : M_A(\beta - a_i) \longrightarrow M_A(\beta)$ is a D-module isomorphism.*

Proof (of Theorem 4.5.2). Suppose that β does not lie in the exceptional arrangement $\mathcal{H}_A$. Then $\beta - \sum_{i=1}^n \nu_i a_i$ does not lie in $\mathcal{H}_A$, for any $\nu_1, \ldots, \nu_n \in \mathbf{N}$. It follows from Corollary 4.5.13 that the hypergeometric D-modules

$$M_A\Big(\beta - \sum_{i=1}^n \nu_i a_i\Big) \qquad \text{for } \nu_1, \ldots, \nu_n \in \mathbf{N}$$

all have the same rank. Since the set of parameter vectors under consideration is Zariski-dense in $\mathbf{k}^d$, we conclude from Remark 4.5.1 that this common rank equals $\mathrm{vol}(A)$. In particular, $H_A(\beta)$ is holonomic of rank $\mathrm{vol}(A)$.   □

A fundamental open problem is to stratify the parameter space $\mathbf{k}^d$ into isomorphism classes for $H_A(\beta)$ as D-modules or R-modules. A coarse stratification is that by the holonomic rank. Theorem 4.5.10 gives partial information on this. The combinatorial techniques introduced in this book are expected to lay the foundation for algorithms to carry out such stratifications in general. However, these are not enough. We need a method to decide whether $H_A(\beta)$ and $H_A(\beta')$ are isomorphic or not. A general algorithm to decide if two holonomic D-modules are isomorphic is currently being developed by Takayama based on algorithms in Chapter 5

4.6 w-flatness

We fix a generic weight vector $w = (w_1, \ldots, w_n)$ in $\mathbf{R}^n$ so that $M = \mathrm{in}_w(I_A)$ is a monomial ideal in $\mathbf{k}[\partial]$. For any $\beta \in \mathbf{k}^d$, the indicial ideal $\widetilde{\mathrm{in}}_{(-w,w)}(H_A(\beta))$ contains the fake indicial ideal $\widetilde{\mathrm{fin}}_w(H_A(\beta))$. If β is generic in $\mathbf{k}^d$ then

$$\widetilde{\mathrm{fin}}_w(H_A(\beta)) \;\; = \;\; \widetilde{\mathrm{in}}_{(-w,w)}(H_A(\beta)), \qquad (4.44)$$

$$\mathrm{rank}(\widetilde{\mathrm{fin}}_w(H_A(\beta))) \;\; = \;\; \mathrm{rank}(H_A(\beta)) \;\; = \;\; \mathrm{vol}(A). \qquad (4.45)$$

This follows from Corollary 3.1.6 and Corollary 4.1.4 respectively. In this section we will show that (4.45) defines a natural notion of genericity. Note that (4.44) does not imply (4.45). For instance, in Example 4.2.7 the indicial ideal equals the fake indicial ideal but $\mathrm{rank}(H_A(\beta)) = \mathrm{vol}(A) + 2$. The following theorem combines objects introduced in Sections 4.1 and 4.3.

Theorem 4.6.1. *For $\beta \in \mathbf{k}^d$ the following three conditions are equivalent:*
(a) $\mathrm{rank}(\widetilde{\mathrm{fin}}_w(H_A(\beta))) = \mathrm{vol}(A)$;
(b) *The d coordinates of $A\theta - \beta$ form a regular sequence in $\mathbf{k}[\theta]/\widetilde{\mathrm{in}}_w(I_A)$;*
(c) *The simplicial complex $\mathcal{U}_p(M)$ is Cohen-Macaulay of dimension $d-1$, for each fake exponent $p \in V(\widetilde{\mathrm{fin}}_w(H_A(\beta)))$.*

Proof. We first show the equivalence of (a) and (c). The rank of $\widetilde{\mathrm{fin}}_w(H_A(\beta))$ is the sum of the ranks of the primary ideals $Q_{A,p,w}$ where p runs over all fake exponents. Let $m(p) = \#\mathcal{S}_p(M)$ as before, and let $m_{\mathrm{top}}(p)$ be the number of top-dimensional standard pairs in $\mathcal{S}_p(M)$. By Theorem 4.1.13 we have

$$\mathrm{rank}(Q_{A,p,w}) \;\geq\; m_{\mathrm{top}}(p). \qquad (4.46)$$

By summing over all fake exponents p we get the inequality

$$\mathrm{rank}(\widetilde{\mathrm{fin}}_w(H_A(\beta))) = \sum_p \mathrm{rank}(Q_{A,p,w}) \;\geq\; \sum_p m_{\mathrm{top}}(p) = \mathrm{vol}(A). \quad (4.47)$$

This gives a new proof of Theorem 4.1.5 (1). From (4.46) and (4.47) we see that (a) holds if and only if $\mathrm{rank}(Q_{A,p,w}) = m_{\mathrm{top}}(p)$ for each fake exponent

p. But $\mathrm{rank}(Q_{A,p,w}) = m_{\mathrm{top}}(p)$ if and only if $\mathcal{U}_p(M)$ is a $(d-1)$-dimensional Cohen-Macaulay complex. Here the "if"-direction is Lemma 4.1.14, and the "only-if"-direction is as in Corollary 4.1.15. Thus (a) and (c) are equivalent.

The condition (b) can be tested locally, that is, after localization at any maximal ideal of $\mathbf{k}[\theta]/\widetilde{\mathrm{in}}_w(I_A)$ which contains all coordinates of $A\theta - \beta$. Such maximal ideals are precisely those defining the fake exponents p. The coordinates of $A\theta - \beta$ are a regular sequence in the localization of $\mathbf{k}[\theta]/\widetilde{\mathrm{in}}_w(I_A)$ at p if and only if the coordinates of $A\theta$ are a regular sequence in the Stanley-Reisner ring of $\mathcal{U}_p(M)$. This happens if and only if $\mathcal{U}_p(M)$ is a $(d-1)$-dimensional Cohen-Macaulay complex. Hence (b) and (c) are equivalent. $\square$

If the three conditions in Theorem 4.6.1 are satisfied then the parameter vector β is called w-flat. This notion of genericity clearly depends on w.

Remark 4.6.2. If β is w-flat then $\mathrm{rank}(H_A(\beta)) = \mathrm{vol}(A)$.

Proof. Immediate from Theorem 2.5.1 and Theorem 3.5.1. $\square$

The following contrapositive formulation from Remark 4.6.2 is instructive. We do not know at present whether the converse to Proposition 4.6.3 holds for all A and all β.

Proposition 4.6.3. *If* $\mathrm{rank}(H_A(\beta)) > \mathrm{vol}(A)$ *then the parameter vector* $\beta \in \mathbf{k}^d$ *fails to be w-flat for all $w \in \mathbf{R}^n$.*

Our second main theorem in this section states that (4.45) implies (4.44).

Theorem 4.6.4. *If the parameter vector β is w-flat then the indicial ideal* $\mathrm{in}_{(-w,w)}(H_A(\beta))$ *coincides with the fake indicial ideal* $\widetilde{\mathrm{fin}}_w(H_A(\beta))$.

Before embarking towards the proof of Theorem 4.6.4, it is useful to compare w-flatness with the following related combinatorial notions. A standard pair (∂^a, σ) of M is called *top-dimensional* if $|\sigma| = d$ and *embedded* if $|\sigma| < d$. We say that (∂^a, σ) is *associated to* β if $(\partial^a, \sigma) \in \mathcal{S}_\beta(M)$; see (4.15). Equivalently, the ideal on the left hand side of (4.14) is a proper (hence maximal) ideal and defines a unique fake exponent $p = \beta^{(\partial^a, \sigma)} \in \mathbf{k}^n$ as in Theorem 3.2.10. Clearly, every top-dimensional standard pair is associated to β, which implies $\#\mathcal{S}_\beta(M) \geq \mathrm{vol}(A)$. The following two conditions are equivalent:
(d) No embedded standard pair of M is associated to β;
(d) $\#\mathcal{S}_\beta(M) = \mathrm{vol}(A)$.

Consider also the following genericity condition:
(e) The map $\mathcal{T}(M) \to \mathbf{k}^n$, $(\partial^a, \sigma) \mapsto \beta^{(\partial^a, \sigma)}$ is injective.
(e) Distinct top-dimensional standard pairs define distinct fake exponents.

The two conditions (d) and (e) are independent of each other:

Example 4.6.5. (Continuation of Examples 3.1.4, 3.2.6 , 4.1.12)

Let $A = \begin{pmatrix} 1 & 1 & 1 & 1 \\ 0 & 1 & 2 & 3 \end{pmatrix}$, and $w = (1, 3, 0, 0)$. There are three top-dimensional standard pairs: $(1, \{3, 4\})$, $(1, \{1, 3\})$ and $(\partial_4, \{1, 3\})$, and one embedded standard pair $(\partial_2, \{1\})$. In this example, the parameter vector $\beta = (\beta_1, \beta_2)$ is w-flat if and only if (d) holds if and only if $\beta_2 \neq 1$. The condition (e) holds if and only if $\beta_2 \neq 2\beta_1$ and $\beta_2 \neq 2\beta_1 + 1$.

Proposition 4.6.6. *If (d) and (e) hold then β is w-flat, and if β is w-flat then (d) holds, but for both implications the converse does not hold.*

Proof. The condition (d) is equivalent to saying that, for each fake exponent p, the simplicial complex $\mathcal{U}_p(M)$ is pure $(d-1)$-dimensional. The conjunction of (d) and (e) is equivalent to saying that, for each fake exponent p, the simplicial complex $\mathcal{U}_p(M)$ is a $(d-1)$-simplex.

Clearly, a simplex is a Cohen-Macaulay complex and a Cohen-Macaulay complex is pure. Using part (c) of Theorem 4.6.1, this proves the two implications. Example 4.1.6 with $\beta = (8, 131)^T$ satisfies (d) but is not w-flat. The Gauss' hypergeometric function in Example 4.2.6 is w-flat but does not satisfy (e), for any w.   □

Example 4.6.7. Let $H_A(\beta)$ be the maximally degenerate hypergeometric system associated with a reflexive polytope, as in Section 3.6. There is only one fake exponent p and $\beta = Ap$ lies in the core of the unimodular regular triangulation Δ_w. Here β is w-flat since $\mathcal{U}_p(M) = \Delta_w$ is a Cohen-Macaulay complex but (e) is false since all standard pairs give the same exponent.

Remark 4.6.8. The set of parameters β where (d) or (e) fails is a finite union of affine subspaces in $\mathbf{k}^d$. In particular, it is a proper Zariski closed subset of $\mathbf{k}^d$. Each generic w gives such a finite subspace arrangement which contains the exceptional set by Proposition 4.6.3 and Proposition 4.6.6. Taking the common intersection of these arrangements with the exceptional arrangement $\mathcal{H}_A$, we get a substantially stronger version of Adolphson's Theorem 4.5.2.

Example 4.6.9. We shall demonstrate that Remark 4.6.8 provides an effective tool for determining the exceptional set. Consider the "pinched hexagon"
$$A = \begin{pmatrix} 1 & 1 & 1 & 1 & 1 & 1 \\ 0 & 1 & 1 & 0 & -1 & -1 \\ -1 & -1 & 0 & 1 & 1 & 0 \end{pmatrix}. \text{ Fix } w \in \mathbf{R}^6 \text{ to give the initial ideal}$$
$\mathrm{in}_w(I_A)$ in Example 4.2.7. There are six top-dimensional standard pairs,

$$
\begin{array}{ll}
(1, \{1, 2, 6\}) & (*, *, 0, 0, 0, *) \\
(1, \{2, 3, 6\}) & (0, *, *, 0, 0, *) \\
(\partial_4, \{2, 3, 6\}) & (0, *, *, 1, 0, *) \\
(1, \{3, 4, 6\}) & (0, 0, *, *, 0, *) \\
(\partial_5, \{3, 4, 6\}) & (0, 0, *, *, 1, *) \\
(1, \{4, 5, 6\}) & (0, 0, 0, *, *, *),
\end{array}
$$

and two embedded standard pairs

$$(\partial_1\partial_3, \{2,6\}) \quad (1, *, 1, 0, 0, *)$$
$$(\partial_1\partial_5, \{6\}) \quad (1, 0, 0, 0, 1, *).$$

The vector notation for standard pairs, such as $(1,0,0,0,1,*)$ is to indicate all the optimal solutions for the corresponding integer program (4.34).

The violation of the condition (e) can happen for nine possible pairings:

$$
\begin{array}{lcll}
(*,*,0,0,0,*) & = & (0,*,*,0,0,*) & \langle \beta_1 + \beta_2 + 2\beta_3 \rangle \\
(*,*,0,0,0,*) & = & (0,0,*,*,0,*) & \langle \beta_3, \beta_1 + \beta_2 \rangle \\
(*,*,0,0,0,*) & = & (0,0,0,*,*,*) & \langle \beta_3, \beta_1 + \beta_2 \rangle \\
(0,*,*,0,0,*) & = & (0,0,*,*,0,*) & \langle \beta_3 \rangle \\
(0,*,*,0,0,*) & = & (0,0,0,*,*,*) & \langle \beta_3, \beta_1 + \beta_2 \rangle \\
(0,*,*,1,0,*) & = & (0,0,*,*,0,*) & \langle \beta_3 - 1 \rangle \\
(0,*,*,1,0,*) & = & (0,0,0,*,*,*) & \langle \beta_1 + \beta_2 - 1, \beta_3 - 1 \rangle \\
(0,0,*,*,0,*) & = & (0,0,0,*,*,*) & \langle \beta_1 + \beta_2 - \beta_3 \rangle \\
(0,0,*,*,1,*) & = & (0,0,0,*,*,*) & \langle \beta_1 + \beta_2 - \beta_3 + 1 \rangle
\end{array}
$$

The last row of this table means that the two standard pairs $(\partial_5, \{3,4,6\})$ and $(1, \{4,5,6\})$ have the same image under the map $T(M) \to \mathbf{k}^3$, $(\partial^a, \sigma) \mapsto \beta^{(\partial^a, \sigma)}$ if and only if the parameter vector is a zero of the ideal $\langle \beta_1 + \beta_2 - \beta_3 + 1 \rangle$. The intersection of the nine given ideals is the principal ideal

$$\langle\, (\beta_1 + \beta_2 + 2\beta_3) \cdot \beta_3 \cdot (\beta_3 - 1) \cdot (\beta_1 + \beta_2 - \beta_3) \cdot (\beta_1 + \beta_2 - \beta_3 + 1) \,\rangle$$

Thus the locus where (e) fails is a union of five planes in $\mathbf{k}^3$.

The failure of (d) happens on a line and a plane in $\mathbf{k}^3$. The embedded standard pair $(\partial_1\partial_3, \{2,6\})$ is associated to β if and only if $\beta_1 + \beta_2 + 2\beta_3 = 1$, and $(\partial_1\partial_5, \{6\})$ is associated to parameters of the form $\beta = (p+1, -p, 0)$:

$$
\begin{array}{ll}
(1, *, 1, 0, 0, *) & \langle \beta_1 + \beta_2 + 2\beta_3 - 1 \rangle \\
(1, 0, 0, 0, 1, *) & \langle \beta_3, \beta_1 + \beta_2 - 1 \rangle.
\end{array}
$$

In summary, the above tables define an explicit arrangement of six planes and one line in $\mathbf{k}^3$ with the property that $\operatorname{rank}(H_A(\beta)) = \operatorname{vol}(A)$ holds for all β which lie outside that arrangement.

We can now repeat the same calculation for other weight vectors w and take the intersection of the resulting "failure of (d) or (e)" arrangements. For our example of the pinched hexagon, we get twelve more initial ideals by applying the dihedral group generated by the permutations (123456) and $(15)(24)$ to the above standard pairs. Intersecting the resulting arrangements, all that is left is the central line $\{\beta_2 = \beta_3 = 0\}$ plus a few isolated points such as $(2, -1, 1)$. The exceptional set for A is contained in that arrangement. Finally, let us show that the exceptional set is finite. It is enough to prove that the exceptional points on $(\beta_1, 0, 0)$ are finite. If $\beta_1 \notin \mathbf{N}$, then $(\beta_1, 0, 0)$ is semi-nonresonant. If $\beta_1 > n(n-d)\mathcal{D}(A)^2$, then, by the same method as in Section 4.5 for the positive cone $\operatorname{pos}(A)$, we see that parameters $(\beta_1, 0, 0)$ are not in the exceptional set.

We shall prove Theorem 4.6.4 using the characterization of w-flatness via regular sequences (Theorem 4.6.1). To this end we introduce a Koszul complex $K_\bullet^\beta(\mathrm{gr}^{(-w,w)}(D^\pm/D^\pm I_A))$. We work in the localized Weyl algebra

$$D^\pm \quad := \quad \mathbf{k}\langle x_1^{\pm 1}, \ldots, x_n^{\pm 1}, \partial_1, \ldots, \partial_n\rangle.$$

For every element p in $D^\pm$ the initial form $\mathrm{in}_{(-w,w)}(p)$ is defined as before.

The Koszul complex $K_\bullet^\beta(\mathrm{gr}^{(-w,w)}(D^\pm/D^\pm I_A))$ is defined exactly like in Section 4.3:

$$\cdots \xrightarrow{d_2} K_1^\beta(\mathrm{gr}^{(-w,w)}(D^\pm/D^\pm I_A)) \xrightarrow{d_1} K_0^\beta(\mathrm{gr}^{(-w,w)}(D^\pm/D^\pm I_A)) \longrightarrow 0$$

where

$$K_p^\beta(\mathrm{gr}^{(-w,w)}(D^\pm/D^\pm I_A)) \quad = \bigoplus_{1\le i_1 < \cdots < i_p \le d} (D^\pm/D^\pm \mathrm{in}_w(I_A))\, e_{i_1 \cdots i_p},$$

$$\text{and} \qquad d_p(e_{i_1 \cdots i_p}) \quad = \quad \sum_{r=1}^p (-1)^{r-1} \cdot (A\theta - \beta)_{i_r} \cdot e_{i_1 \cdots \hat{i}_r \cdots i_p}.$$

Lemma 4.3.4 guarantees that the differentials d_p are well-defined.

The following lemma justifies the notation $K_\bullet^\beta(\mathrm{gr}^{(-w,w)}(D^\pm/D^\pm I_A))$.

Lemma 4.6.10. *The two relevant initial ideals are preserved under the extension from the Weyl algebra D to the localized Weyl algebra $D^\pm$:*

$$\mathrm{in}_{(-w,w)}(D^\pm I_A) \quad = \quad D^\pm \mathrm{in}_w(I_A).$$

$$\mathrm{in}_{(-w,w)}(D^\pm H_A(\beta)) \quad = \quad D^\pm \mathrm{in}_{(-w,w)}(H_A(\beta)).$$

Proof. This follows from Lemma 2.3.2.   □

The next lemma will be used in the proof of Proposition 4.6.12 below.

Lemma 4.6.11. *Let $I, J \subset \mathbf{k}[\theta]$ be Frobenius ideals. Then $D^\pm I = D^\pm J$ if and only if $I = J$.*

Proof. It suffices to show that $D^\pm I \subset D^\pm J$ implies $I \subset J$. Let $f = f(\theta) \in I$. Since $f \in D^\pm J$ there exists a monomial x^a such that $x^a \cdot f \in DJ$. The D-ideal DJ has a Gröbner basis $\mathcal{G}$ consisting of polynomials in $\mathbf{k}[\theta]$, by Corollary 2.3.8. We know that $x^a \cdot f(\theta)$ reduces to zero modulo $\mathcal{G}$, and therefore $f(\theta)$ reduces to zero modulo $\mathcal{G}$. We conclude that $f(\theta) \in J$ as desired.   □

Proposition 4.6.12. *If the first homology of the Koszul complex vanishes,*

$$H_1(K_\bullet^\beta(\mathrm{gr}^{(-w,w)}(D^\pm/D^\pm I_A))) \quad = \quad 0,$$

then (4.44) holds, that is, the indicial ideal equals the fake indicial ideal.

Proof. We can prove similarly to Theorem 4.3.5 that if the first homology module $H_1(K_\bullet^\beta(\mathrm{gr}^{(-w,w)}(D^\pm/D^\pm I_A)))$ vanishes then $D^\pm \mathrm{in}_{(-w,w)}(H_A(\beta))$ is generated by $\mathrm{in}_w(I_A)$ and $A\theta - \beta$. Hence by Lemma 4.6.11, the indicial ideal $\widetilde{\mathrm{in}}_{(-w,w)}(H_A(\beta))$ coincides with the fake indicial ideal $\widetilde{\mathrm{fin}}_w(H_A(\beta))$.   □

Proof (of Theorem 4.6.4). The left ideal $D^\pm \mathrm{in}_w(I_A)$ is homogeneous with respect to $(-w, w)$. Consider the induced weight decomposition of the quotient $D^\pm/D^\pm \mathrm{in}_w(I_A)$ and of the Koszul complex $K_\bullet^\beta(\mathrm{gr}^{(-w,w)}(D^\pm/D^\pm I_A))$. Since the elements $(A\theta - \beta)_i$ are of weight 0, each differential d_p preserves the $(-w, w)$-weight. Therefore each homology module of the Koszul complex admits a weight space decomposition:

$$H_p(K_\bullet^\beta(\mathrm{gr}^{(-w,w)}(D^\pm/D^\pm I_A))) = \bigoplus_\mu H_p(K_\bullet^\beta(\mathrm{gr}^{(-w,w)}(D^\pm/D^\pm I_A)))_\mu.$$

Here μ runs over the additive subgroup $\mathbf{Z}\{w_1, \ldots, w_n\}$ of $\mathbf{R}$ generated by $w_1, \ldots, w_n$. Since $H_p(K_\bullet^\beta(\mathrm{gr}^{(-w,w)}(D^\pm/D^\pm I_A)))$ is a $\mathbf{k}[x_1^{\pm 1}, \ldots, x_n^{\pm 1}]$-module, all weight spaces $H_p(K_\bullet^\beta(\mathrm{gr}^{(-w,w)}(D^\pm/D^\pm I_A)))_\mu$ are isomorphic as $\mathbf{k}$-vector spaces. In particular, for any weight μ in the group $\mathbf{Z}\{w_1, \ldots, w_n\}$ we have the following isomorphism of $\mathbf{k}$-vector spaces:

$$H_p(K_\bullet^\beta(\mathrm{gr}^{(-w,w)}(D^\pm/D^\pm I_A)))_\mu \simeq H_p(K_\bullet^\beta(\mathrm{gr}^{(-w,w)}(D^\pm/D^\pm I_A)))_0.$$

Since the Gröbner cone containing the vector w is open in $\mathbf{R}^n$, we may assume that the weights $w_1, \ldots, w_n$ are $\mathbf{Q}$-linearly independent, i.e., $\mathbf{Z}\{w_1, \ldots, w_n\} \simeq \mathbf{Z}^n$. This assumption implies that the weight space of $D^\pm$ with weight 0 is exactly the polynomial ring $\mathbf{k}[\theta]$. Therefore the distraction $\widetilde{\mathrm{in}}_w(I_A)$ coincides with the weight space $(D^\pm \mathrm{in}_w(I_A))_0$ of the ideal $D^\pm \mathrm{in}_w(I_A)$. We conclude that the weight space $H_p(K_\bullet^\beta(\mathrm{gr}^{(-w,w)}(D^\pm/D^\pm I_A)))_0$ equals the p-th homology module of the following Koszul complex $K_\bullet^\beta(\mathbf{k}[\theta]/\widetilde{\mathrm{in}}_w(I_A))$ over the commutative ring $\mathbf{k}[\theta]/\widetilde{\mathrm{in}}_w(I_A)$:

$$\cdots \xrightarrow{d_2} K_1^\beta(\mathbf{k}[\theta]/\widetilde{\mathrm{in}}_w(I_A)) \xrightarrow{d_1} K_0^\beta(\mathbf{k}[\theta]/\widetilde{\mathrm{in}}_w(I_A)) \longrightarrow 0$$

where $\qquad K_p^\beta(\mathbf{k}[\theta]/\widetilde{\mathrm{in}}_w(I_A)) = \bigoplus_{1 \le i_1 < \cdots < i_p \le d} (\mathbf{k}[\theta]/\widetilde{\mathrm{in}}_w(I_A)) e_{i_1 \cdots i_p},$

and each differential d_p is defined as before. In summary, we have that $H_p(K_\bullet^\beta(\mathrm{gr}^{(-w,w)}(D^\pm/D^\pm I_A)))$ vanishes if and only if $H_p(K_\bullet^\beta(\mathbf{k}[\theta]/\widetilde{\mathrm{in}}_w(I_A))) \simeq (H_p(K_\bullet^\beta(\mathrm{gr}^{(-w,w)}(D^\pm/D^\pm I_A))))_0$ vanishes.

Recall our assumption that $(A\theta - \beta)_1, \ldots, (A\theta - \beta)_d$ form a regular sequence on $\mathbf{k}[\theta]/\widetilde{\mathrm{in}}_w(I_A)$. This implies $H_1(K_\bullet^\beta(\mathbf{k}[\theta]/\widetilde{\mathrm{in}}_w(I_A))) = 0$ by Theorem 4.3.1. We conclude that $H_1(K_\bullet^\beta(\mathrm{gr}^{(-w,w)}(D^\pm/D^\pm I_A))) = 0$. Proposition 4.6.12 implies that the fake indicial ideal equals the indicial ideal.   □

5. Integration of D-modules

Hypergeometric functions arise naturally from integrals of the form

$$\Phi(x) \quad = \quad \int_C f(x,t)^\alpha \, t_1^{\gamma_1} \cdots t_m^{\gamma_m} dt_1 \cdots dt_m, \tag{5.1}$$

where $\alpha, \gamma_i \in \mathbf{C}$, f is a polynomial in m variables $t = (t_1, \ldots, t_m)$ and n parameters $x = (x_1, \ldots, x_n)$, and C is a suitable cycle in $\mathbf{C}^m$. It is our goal to present algorithms for computing asymptotic expansions of the integral $\Phi(x)$. Our strategy is to proceed in three steps: First we compute the D-ideal consisting of all operators which annihilate $f(x,t)^\alpha$; next we find annihilators of $\Phi(x)$ via the machinery of D-module theoretic integration, and finally we apply Chapter 2 to compute Nilsson series expansions of $\Phi(x)$.

The GKZ hypergeometric ideal $H_A(\beta)$ for $A = \begin{pmatrix} 1 & 1 & \cdots & 1 \\ a_1 & a_2 & \cdots & a_n \end{pmatrix}$ and generic β, studied in the previous chapters, is precisely the annihilating ideal of such an integral $\Phi(x)$ when $f(x,t)$ is a *generic sparse polynomial*,

$$f(x,t) \quad = \quad \sum_{i=1}^n x_i \cdot t^{a_i}.$$

In fact, GKZ hypergeometric systems also appear when $f(x,t)^\alpha$ is a twisted product of sparse generic polynomials, as seen for linear forms in Section 1.4.

This chapter is organized to also serve a more general purpose, namely, as an introduction to algorithms in algebraic geometry based on D-modules. We develop the interplay between b-functions and Gröbner bases for holonomic D-modules, we use this to represent powers of polynomials as holonomic functions, and we show how to restrict holonomic functions to subspaces. The GKZ system $H_A(\beta)$ remains our running example for all general concepts and constructions concerning D-modules, much in the same spirit as toric varieties have served as a ubiquitous source of examples in algebraic geometry.

5.1 b-functions for Holonomic D-ideals

In this section, we introduce the b-function of a holonomic D-ideal with respect to a weight vector w. We show how to compute this b-function and we determine it combinatorially for GKZ systems with generic parameters.

Definition 5.1.1. Let I be a holonomic D-ideal and $w \in \mathbf{R}^n$ be a non-zero weight vector. Consider the elimination ideal

$$\mathrm{in}_{(-w,w)}(I) \ \cap \ \mathbf{k}[w_1\theta_1 + \cdots + w_n\theta_n]. \tag{5.2}$$

This is a principal ideal in the univariate polynomial ring $\mathbf{k}[s]$, where $s = \sum_{i=1}^n w_i\theta_i$. The generator $b(s)$ of the principal ideal (5.2) is called the *b-function* of the holonomic D-ideal I with respect to the weight vector w.

Theorem 5.1.2. *If I is a holonomic D-ideal and $w \in \mathbf{R}^n \backslash \{0\}$ then the ideal (5.2) is non-zero and hence the b-function $b(s)$ is not the zero polynomial.*

Several proofs of this theorem are known and can be found in the literature; see e.g. [64]. The idea behind these proofs is to interpret the operator $s = \sum_{i=1}^n w_i\theta_i$ as a left D-endomorphism of the holonomic module $D/\mathrm{in}_{(-w,w)}(I)$; for $p \in \mathrm{in}_{(-w,w)}(I)$, the operator $p \cdot s$ is in $\mathrm{in}_{(-w,w)}(I)$ and hence

$$D/\mathrm{in}_{(-w,w)}(I) \ni \ell \mapsto \ell s \in D/\mathrm{in}_{(-w,w)}(I)$$

is a left D-endomorphism. The theory of D-modules guarantees that

$$\mathrm{Hom}_D\big(D/\mathrm{in}_{(-w,w)}(I), D/\mathrm{in}_{(-w,w)}(I)\big)$$

is a finite-dimensional $\mathbf{k}$-vector space; see e.g. [15, p.20, p.77]. Hence the endomorphism $s = \sum_{i=1}^n w_i\theta_i$ has a well-defined minimal polynomial, which is precisely our b-function $b(s)$.

The b-function is sometimes called the *indicial polynomial* in the literature. We will not use this terminology in this book in order to avoid possible confusion with our earlier discussion of indicial ideals and fake indicial ideals.

To keep things elementary and algorithmic, we will not present the details of the proof sketched above in this book. Instead we restrict our attention to the case when the weight vector w is generic, i.e., w lies in the interior of a full-dimensional cone in the small Gröbner fan of I. This means that $\mathrm{in}_{(-w,w)}(I)$ is torus-fixed. Under this hypothesis, there is a nice proof of Theorem 5.1.2 which is based on elementary combinatorial considerations only. The precise relationship between the b-function for generic w and that for non-generic w is still somewhat mysterious to us and deserves further study.

Theorem 5.1.3. *Let I be a holonomic D-ideal and $w \in \mathbf{R}^n$ a generic weight vector. Then the commutative ideal $\mathrm{in}_{(-w,w)}(I) \cap \mathbf{k}[\theta_1, \ldots, \theta_n]$ is Artinian.*

Proof. The D-ideal $J = \mathrm{in}_{(-w,w)}(I)$ is holonomic and torus fixed. Fix a term order $\prec$ on D and write $\prec'$ for the induced term order on $\mathbf{k}[\theta]$. The reduced Gröbner basis of J can be written as follows:

$$\mathcal{G} \ = \ \big\{ x^{a_1}p_1(\theta)\partial^{b_1}, \ x^{a_2}p_2(\theta)\partial^{b_2}, \ \ldots, x^{a_r}p_r(\theta)\partial^{b_r} \big\} \tag{5.3}$$

If $\text{in}_{\prec'}(p_i) = \theta^{c_i}$, then

$$\text{in}_{\prec}(J) = \langle x^{a_1+c_1}\xi^{b_1+c_1}, x^{a_2+c_2}\xi^{b_2+c_2}, \ldots, x^{a_r+c_r}\xi^{b_r+c_r}\rangle.$$

This is a monomial ideal in $\mathbf{k}[x, \xi]$. It has dimension n since J is holonomic. Our goal is to show that the following ideal is Artinian:

$$\langle\, [\theta]^{a_1} p_1(\theta - b_1)[\theta]_{b_1}, \ [\theta]^{a_2} p_2(\theta - b_2)[\theta]_{b_2}, \ \ldots, \ [\theta]^{a_r} p_r(\theta - b_r)[\theta]_{b_r}\,\rangle. \quad (5.4)$$

This ideal is contained in $J \cap \mathbf{k}[\theta]$. By an argument similar to the proof of Theorem 2.3.4, we see that the two ideals are equal. Note that $[\theta]_{b_i}$ denotes falling factorials as before and $[\theta]^{a_i}$ denotes rising factorials as in

$$[\theta]^{a_i} = \prod_{j=1}^{n}\prod_{l=1}^{a_{ij}}(\theta_j + l).$$

Proceeding by contradiction, let us assume that $J \cap \mathbf{k}[\theta]$ has positive dimension, i.e., $J \cap \mathbf{k}[\theta]$ has infinitely many roots over the algebraic closure of $\mathbf{k}$. After relabeling the elements of $\mathcal{G}$, we see that some ideal of the form

$$L = \langle [\theta]^{a_1}, \ldots, [\theta]^{a_i}, [\theta]_{b_{i+1}}, \ldots, [\theta]_{b_j}, p_{j+1}(\theta - b_{j+1}), \ldots, p_r(\theta - b_r)\rangle$$

is positive-dimensional. Decomposing the falling and rising factorials, and relabeling variables, we see that there exists a positive-dimensional ideal

$$L' = \langle \theta_1+u_1, \ldots, \theta_s+u_s, \theta_{s+1}-v_{s+1}, \ldots, \theta_t-v_t, p_{j+1}(\theta-b_{j+1}), \ldots, p_r(\theta-b_r)\rangle,$$

where $u_i > 0$, $v_i \geq 0$, $\langle [\theta]^{a_1}, \ldots, [\theta]^{a_i}\rangle \subseteq \langle \theta_1 + u_1, \ldots, \theta_s + u_s\rangle$, and $\langle [\theta]_{b_{i+1}}, \ldots, [\theta]_{b_j}\rangle \subseteq \langle \theta_{s+1}-v_{s+1}, \ldots, \theta_t-v_t\rangle$. Since L' is positive-dimensional, the monomial ideal generated by the $\prec'$-leading terms of the given $t + r - j$ generators is also positive-dimensional. After relabeling variables, we may thus assume that these leading terms all lie in $\langle \theta_1, \theta_2, \ldots, \theta_{n-1}\rangle$.

Our construction implies the following three inclusions of ideals in $\mathbf{k}[x, \xi]$:

$$\langle x^{a_1}, \ldots, x^{a_i}\rangle \subseteq \langle x_1, \ldots, x_s\rangle$$
$$\langle \xi^{b_{i+1}}, \ldots, \xi^{b_j}\rangle \subseteq \langle \xi_{s+1}, \ldots, \xi_t\rangle$$
$$\langle (x\xi)^{c_{j+1}}, \ldots, (x\xi)^{c_r}\rangle \subseteq \langle x_1, \ldots, x_s, \xi_{s+1}, \ldots, \xi_t, x_{t+1}\xi_{t+1}, \ldots, x_{n-1}\xi_{n-1}\rangle.$$

We conclude that

$$\text{in}_{\prec}(J) \subseteq \langle x_1, \ldots, x_s, \xi_{s+1}, \ldots, \xi_t, x_{t+1}\xi_{t+1}, \ldots, x_{n-1}\xi_{n-1}\rangle,$$

and hence $\text{in}_{\prec}(J)$ has codimension $\leq n - 1$. Therefore J is not holonomic. This contradiction completes the proof of Theorem 5.1.3. $\qquad\square$

Theorem 5.1.2 for generic weight vectors w follows as a corollary, and we also get a simple algorithm for computing the *b*-function in this case.

Corollary 5.1.4. *For w generic, the b-function $b(s)$ of I is not zero.*

Proof. Let $J = \text{in}_{(-w,w)}(I)$ be the Artinian ideal in the proof above. Then

$$\langle\, b(\sum_{i=1}^{n} w_i \theta_i)\,\rangle \;=\; (J \cap \mathbf{k}[\theta]) \cap \mathbf{k}[w_1\theta_1 + \cdots + w_n\theta_n] \;\neq\; 0.$$

$\square$

The roots of b-functions play an important role in many algorithms for D-modules. In particular, we shall employ them later in this chapter to compute restrictions and integrals of D-modules and to find the annihilating D-ideals of powers of polynomials. Our discussion above gives rise to the following algorithm to compute the b-function of a holonomic D-ideal when w is generic.

Algorithm 5.1.5 (Finding the b-function for generic weights w)
Input: Generators of a holonomic D-ideal I, generic weight vector $w \in \mathbf{R}^n$.
Output: the b-function $b(s)$.

1. Compute a Gröbner basis $\mathcal{G}$ of the D-ideal $J = \text{in}_{(-w,w)}(I)$ as in (5.3).
2. Replace each element $x^{a_i} p_i(\theta) \partial^{b_i}$ in the Gröbner basis $\mathcal{G}$ by the polynomial $[\theta]^{a_i} p_i(\theta - b_i)[\theta]_{b_i}$ to get generators for the ideal $J \cap \mathbf{k}[\theta]$ in (5.4).
3. Add a new variable s and the new relation $s - \sum_{i=1}^{n} w_i \theta_i$ so as to get $r+1$ polynomials in the $(n+1)$-dimensional polynomial ring $\mathbf{k}[\theta_1, \ldots, \theta_n, s]$.
4. Eliminate $\theta_1, \ldots, \theta_n$ from this Artinian ideal. The result equals $b(s)$.

For arbitrary weight vectors w, the b-function can be obtained by the following algorithm, which appeared in [78] and [80, Algorithm 4.6]. However, the algorithm below requires more computer resources than Algorithm 5.1.5.

Algorithm 5.1.6 (Finding the b-function for arbitrary weights w)
Input: Generators of a holonomic D-ideal I, a weight vector $w \in \mathbf{R}^n \backslash \{0\}$.
Output: the b-function $b(s)$.

1. Compute a generating set $\{g_1, g_2, \ldots, g_m\}$ of the initial ideal $\text{in}_{(-w,w)}(I)$. (Use Algorithm 1.2.6.)
2. Put $P = \{i \,|\, w_i \neq 0\}$ and introduce new variables u_i, v_i for each $i \in P$. These variables are central, so we are now working in $D[u_i, v_i; i \in P]$.
3. Let $\tilde{g}_i$ denote the image of g_i under replacing x_i by $u_i x_i$ and ∂_i by $v_i \partial_i$.
4. By Gröbner basis elimination, compute generators of the intersection

$$D[u_i, v_i; i \in P] \cdot \{\tilde{g}_1, \ldots, \tilde{g}_m, u_i v_i - 1; i \in P\} \cap \mathbf{k}\langle x_i, \partial_i; i \in P\rangle. \quad (5.5)$$

Each generator has the form $x^a p(\theta) \partial^b$, in the variables x_i, ∂_i for $i \in P$.
5. Replace each generator $x^a p(\theta) \partial^b$ of the ideal (5.5) by the polynomial $[\theta]^a p(\theta - b)[\theta]_b$ to get a commutative polynomial ideal J in $\mathbf{k}[\theta_i; i \in P]$.
6. Compute the generator b of the principal ideal $J \cap \mathbf{k}[\sum_{i \in P} w_i \theta_i]$.

The last step 6 is done by commutative Gröbner bases. The correctness of the algorithm follows from Theorem 5.1.2 since $b(\sum w_i \theta_i)$ lies in J by construction. Note that the $\mathbf{k}[x_i, \partial_i; i \in P]$-ideal (5.5) need not be holonomic and the commutative ideal J need not be Artinian, unless w is generic.

Example 5.1.7. We compute all possible b-functions for the D-ideal I generated by $3x^2 \partial_y + 2y\partial_x$ and $2x\partial_x + 3y\partial_y + 6$. This D-ideal consists of all operators which annihilate the rational function $1/(x^3 - y^2)$. In Example 2.1.12 we saw that the small Gröbner fan of I consists of the hyperplane $3w_1 - 2w_2$ and its two open half spaces. When $3w_1 - 2w_2 > 0$, the initial ideal is generated by $2x\partial_x + 3y\partial_y + 6, 2y\partial_x, -3y^2\partial_y - 6y$. The b-function is equal to $s + 2w_2$. When $3w_1 - 2w_2 < 0$, the initial ideal is generated by $2x\partial_x + 3y\partial_y + 6, x^2\partial_y, 3xy\partial_y^2 + 5x\partial_y, 9y^2\partial_y^3 + 45y\partial_y^2 + 35\partial_y$. The b-function is equal to $s + 3w_1$. When $3w_1 - 2w_2 = 0$, the initial ideal equals I and the b-function is $s + 3w_1 = s + 2w_2$.

It is instructive to contemplate the meaning of the b-function $b(s)$ for the solutions to a regular holonomic D-ideal I. The following paragraph is an informal discussion. Suppose that $\mathbf{k} = \mathbf{C}$ and $\{f_1(x), \ldots, f_r(x)\}$ is a basis of local holomorphic solutions of I, where each function $f_i(x)$ is written as

$$f_i(x) \quad = \quad \text{in}_w(f_i)(x) + w\text{-higher terms},$$

and $\{\text{in}_w(f_1)(x), \ldots, \text{in}_w(f_r)(x)\}$ is a basis of solutions to $\text{in}_{(-w,w)}(I)$. The operator $\sum_{i=1}^n w_i\theta_i$ essentially commutes with the action of the $(-w, w)$-homogeneous generators of the D-ideal $\text{in}_{(-w,w)}(I)$. Therefore the function

$$\left(\sum_{i=1}^n w_i\theta_i \right) \bullet \text{in}_w(f_j)$$

is also a solution of $\text{in}_{(-w,w)}(I)$, for each j, and can hence be expressed in terms of the solution basis. This gives a complex $r \times r$-matrix A such that

$$\left(\sum_{i=1}^n w_i\theta_i \right) \bullet \begin{pmatrix} \text{in}_{(-w,w)}(f_1) \\ \vdots \\ \text{in}_{(-w,w)}(f_r) \end{pmatrix} \quad = \quad A \cdot \begin{pmatrix} \text{in}_{(-w,w)}(f_1) \\ \vdots \\ \text{in}_{(-w,w)}(f_r) \end{pmatrix}.$$

If $p(s)$ is the minimal polynomial of the matrix A then $p(\sum_{i=1}^n w_i\theta_i)$ annihilates all solutions to $\text{in}_{(-w,w)}(I)$, and therefore the b-function $b(s)$ either equals $p(s)$ or is a multiple of $p(s)$. If λ is a root of $p(s) = 0$, then $p(\sum_{i=1}^n w_i\theta_i) \bullet g = 0$ admits solutions of the form $g = x_1^{a_1} \cdots x_n^{a_n}, \sum w_i a_i = \lambda$. We may think of $b(s)$ as giving information about the "monodromy in direction w" of the holonomic system I.

We next present a technical lemma on b-functions which will be crucial for the algorithms in the next section. Let I be a holonomic D-ideal, let

$w \in \mathbf{R}^n$ be weight vector such that $w_i > 0$ for $i = 1, \ldots, m$ and $w_j = 0$ for $j = m+1, \ldots, n$, and let s_0 be the largest root of the b-function $b(s)$ of I with respect to w. We shall use s_0 as a degree bound concerning the $\mathbf{k}$-vector space $I + x_1 D + \cdots + x_m D$. Note that this is not a D-ideal, since the right D-ideal generated by $x_1, \ldots, x_m$ is added to the left D-ideal I.

Lemma 5.1.8. *If $\beta \in \mathbf{N}^n$ with $w \cdot \beta > s_0$ then there exists an element f in the $\mathbf{k}$-vector space $I + x_1 D + \cdots + x_m D$ which satisfies $\mathrm{in}_{(-w,w)}(f) = \partial^\beta$.*

Proof. Abbreviate $s_1 := w \cdot \beta > s_0$. Then $b(s_1)$ is a non-zero number. By the definition of the b-function, there exists an element $g \in I$ such that

$$b\left(\sum_{i=1}^m w_i \theta_i\right) \;=\; \mathrm{in}_{(-w,w)}(g).$$

If we multiply this identity by ∂^β on the left, then we get

$$\mathrm{in}_{(-w,w)}(\partial^\beta \cdot g) \;=\; \partial^\beta \cdot \mathrm{in}_{(-w,w)}(g)$$
$$= \quad \partial^\beta \cdot b\left(\sum_{i=1}^m w_i \theta_i\right) \;=\; b\left(s_1 + \sum_{i=1}^m w_i \theta_i\right) \cdot \partial^\beta$$
$$= \quad b(s_1) \cdot \partial^\beta + \sum_{i=1}^m x_i u_i \qquad \text{for some } u_1, \ldots, u_m \in D.$$

In fact, the u_i are $(-w, w)$-homogeneous operators. Hence the element

$$f \quad := \quad \frac{1}{b(s_1)} \cdot \left(\partial^\beta g - \sum_{i=1}^m x_i u_i\right)$$

has the desired properties $f \in I + x_1 D + \cdots + x_m D$ and $\mathrm{in}_{(-w,w)}(f) = \partial^\beta$. $\square$

To draw a connection to the previous two chapters, let us calculate the b-function $b(s)$ for the GKZ-hypergeometric system $H_A(\beta)$ for generic parameters $\beta \in \mathbf{k}^d$ and generic weights $w \in \mathbf{R}^n$. Fix $\beta \in \mathbf{k}^d$ and $w \in \mathbf{R}^n$ generic. Then $M = \mathrm{in}_w(I_A)$ is a monomial ideal in $\mathbf{k}[\partial_1, \ldots, \partial_n]$. We recall from Section 3.2 that $H_A(\beta)$ is holonomic of rank $\mathrm{vol}(A)$ and that the exponents of $H_A(\beta)$ in direction w are the $\mathrm{vol}(A)$ many vectors $\beta^{(\partial^\beta, \sigma)} \in \mathbf{k}^n$ where (∂^β, σ) runs over the set $\mathcal{T}(M)$ of top-dimensional standard pairs.

Proposition 5.1.9. *Let $\beta \in \mathbf{k}^d$ and $w \in \mathbf{R}^n$ both be generic. Then the b-function of the GKZ-hypergeometric ideal $H_A(\beta)$ with respect to w equals*

$$b(s) \quad = \quad \prod_{(\partial^\beta, \sigma) \in \mathcal{T}(M)} \left(s - w \cdot \beta^{(\partial^\beta, \sigma)}\right).$$

Proof. Since β is generic, Theorem 3.1.3 and Corollary 3.1.6 imply

$$\mathrm{in}_{(-w,w)}\big(H_A(\beta)\big) \cap \mathbf{k}[\theta_1, \ldots, \theta_n] \quad = \quad \widetilde{\mathrm{in}}_{(-w,w)} H_A(\beta).$$

In view of Algorithm 5.1.5, we can compute the b-function from the indicial ideal $\widetilde{\text{in}}_{(-w,w)}H_A(\beta)$ by elimination as follows:

$$\langle\, b(\sum_{i=1}^{n} w_i\theta_i)\,\rangle \;\;=\;\; \widetilde{\text{in}}_{(-w,w)}H_A(\beta) \,\cap\, \mathbf{k}[\sum_{i=1}^{n} w_i\theta_i]. \qquad (5.6)$$

By Theorem 3.2.10, the indicial ideal $\widetilde{\text{in}}_{(-w,w)}H_A(\beta)$ equals the vanishing radical ideal of the points $\beta^{(\partial^\beta,\sigma)}$ where (∂^β,σ) runs over $\mathcal{T}(M)$. The linear functional defined by w takes on $\text{vol}(A)$ many distinct values on the points $\beta^{(\partial^\beta,\sigma)}$, since it is generic. Therefore the elimination ideal (5.6) is generated by the square-free polynomial whose roots are these values.   □

It would be very interesting to better understand the behavior of the b-function of the hypergeometric ideal $H_A(\beta)$ for special values of β and w.

5.2 Computing Restrictions

In this section we examine the operation of restricting a holonomic function f on affine n-space to an m-dimensional coordinate subspace. More precisely, fix $m \in \{1,\ldots,n\}$, set $D' = \mathbf{k}\langle x_{m+1},\ldots,x_n,\partial_{m+1},\ldots,\partial_n\rangle$, and suppose we are given a holonomic D-ideal I which annihilates a certain function

$$f(x_1,\ldots,x_m,x_{m+1},\ldots,x_n).$$

Then our problem is to compute a holonomic D'-ideal J which annihilates

$$f(0,\ldots,0,x_{m+1},\ldots,x_n).$$

For example, consider the polynomial function $f(x_1,x_2) = (x_1+x_2)^3$, which is represented by the GKZ-hypergeometric ideal for $A = (\,1\ 1\,)$ and $\beta = 3$,

$$I = D \cdot \{\theta_1 + \theta_2 - 3,\, \partial_1 - \partial_2\}.$$

Then the D'-ideal $J = D' \cdot \{\theta_2 - 3,\, \partial_2^4\}$ represents the restricted function $f(0,x_2)$. In this section we explain how to calculate J directly from I.

In order to do the restriction algorithm right, at this point, it is necessary to switch from D-ideals to D-*submodules*. Everything we said about the Gröbner basis theory of D-ideals holds almost verbatim for left D-submodules M of the free left D-module D^r. This enables us to work with finitely generated D-modules of the form D^r/M rather than just cyclic D-modules D/I as before. A detailed introduction to Gröbner bases for modules over a commutative polynomial ring can be found in [1, Chapter 3] and in [28, Chapter 5]. To show how Gröbner bases work for us in this more general setting, we next define the characteristic variety of an arbitrary finitely generated D-module.

Fix a positive integer r and let D^r denote the free left D-module with basis $e_1, \ldots, e_r$. Every element of D^r can be written uniquely as a $\mathbf{k}$-linear combination of *monomials* $x^a \partial^b e_i$ where $a, b \in \mathbf{N}^n$. For the weight vector (u, v) of the Weyl algebra, the (u, v)-*weight* of $x^a \partial^b e_i$ is $a \cdot u + b \cdot v$. Let $f \in D^r$ and write $f = \sum c_{a,b,i} \cdot x^a \partial^b e_i$. The *initial form* $\mathrm{in}_{(u,v)}(f)$ is the subsum of all non-zero summands $c_{a,b,i} \cdot x^a \partial^b e_i$ which have maximum (u, v)-weight. The initial form is to be regarded as an element of the free module $\left(\mathrm{gr}_{(u,v)}(D) \right)^r$. Let M be a D-submodule of D^r. It is finitely generated. The $\mathrm{gr}_{(u,v)}(D)$-submodule generated by the initial forms $\mathrm{in}_{(u,v)}(f)$ for $f \in M$ is denoted by $\mathrm{in}_{(u,v)}(M)$ and is called the *initial module* of M for (u, v).

The initial module is computed by refining the partial order in D^r defined by (u, v) to a total order. See Theorem 1.1.6 for an analogous algorithm for an ideal I. Consider $\mathrm{in}_{(0,e)}(M)$. It is a $\mathbf{k}[x, \xi]$-submodule of $\left(\mathrm{gr}_{(0,e)}(D) \right)^r = \mathbf{k}[x, \xi]^r$. Let $\mathcal{O}$ be the sheaf of regular functions on (x, ξ)-space $\mathbf{k}^{2n}$.

Definition 5.2.1. We call the set

$$\{ (p, q) \in \mathbf{k}^{2n} \mid \mathcal{O} \cdot \mathrm{in}_{(0,e)}(M) \neq \mathcal{O}^r \ \text{ at } \ (x, \xi) = (p, q) \} \qquad (5.7)$$

the *characteristic variety* of D^r/M and denote it by $\mathrm{ch}(D^r/M)$ or $\mathrm{ch}(M)$.

For $r = 1$ this definition agrees with that given in the beginning of Section 1.4. The characteristic variety always has a natural determinantal presentation. The following proposition is a standard fact of commutative algebra.

Proposition 5.2.2. *Assume that* $\mathrm{in}_{(0,e)}(M)$ *is generated by*

$$h_1 = \sum_{i=1}^{r} h_{1i}(x, \xi) e_i, \quad \ldots, \quad h_s = \sum_{i=1}^{r} h_{si}(x, \xi) e_i \in \mathbf{k}[x, \xi]^r.$$

The characteristic variety $\mathrm{ch}(M)$ *equals the common zero set in* $\mathbf{k}^{2n}$ *of the* $r \times r$-*minors of the* $s \times r$-*matrix* (h_{ij}).

It was shown by Oaku [76] (see also Theorem 1.4.1) that this notion of characteristic variety agrees with the standard definition for $\mathcal{D}$-modules. The fundamental theorem of algebraic analysis (Theorem 1.4.6) also holds for the more general characteristic variety $\mathrm{ch}(M)$. A D-module D^r/M is called *holonomic* when the characteristic variety is of dimension n. The number

$$\dim_{\mathbf{k}(x)} \left[\left(\mathbf{k}(x)[\xi] \right)^r / \mathbf{k}(x)[\xi] \, \mathrm{in}_{(0,e)}(M) \right]$$

is the *holonomic rank* of D^r/M. For $\mathbf{k} = \mathbf{C}$ it agrees with the dimension of the space of classical solutions at a generic point x, where M is regarded as a system of linear differential equations for a vector-valued function. We say that D^r/M is *regular holonomic* if it satisfies the $\mathcal{D}$-module theoretic definition in Section 2.4. Theorem 2.5.1 remains valid for D-submodules M.

We are now prepared to define the restriction operation at the level of D-modules and to present an algorithm for computing it. The main technique of the algorithm is the use of the integral roots of b-functions. This idea was first proposed by Oaku in [78] and it was further developed in [80] and [81].

Inside affine n-space $X = \mathbf{k}^n$ we fix the subspace $Y = \mathbf{k}^{n-m}$ of points whose first m coordinates are zero. Let $D = D_X$ be the Weyl algebra of differential operators on X and $D' = D_Y$ the Weyl algebra of differential operators on Y. We define the following quotient of $\mathbf{k}$-vector spaces

$$D_{Y \to X} \quad := \quad D/(x_1 D + \cdots + x_m D).$$

Notice that $D_{Y \to X}$ is a left D'-module and a right D-module. But it is not a left D-module since $x_1 D + \cdots + x_m D$ is not a left D-ideal but a right D-ideal.

Consider any left D-ideal I. We can form the tensor product over the Weyl algebra D of the right D-module $D_{Y \to X}$ with the left D-module D/I.

$$N \quad := \quad D_{Y \to X} \otimes_D D/I \quad \simeq \quad D/(I + x_1 D + \cdots + x_m D). \qquad (5.8)$$

Such tensor products of bimodules are explained e.g. in [26, Chapter 12]. Namely, the left factor $D_{Y \to X}$ is a D'-D-bimodule and the right factor D/I is a D-$\mathbf{k}$-bimodule, hence their tensor product N is a D'-$\mathbf{k}$-bimodule. In particular, N has the structure of a left D'-module. This left D'-module N is called the *restriction module* of D/I with respect to the variables $x_1, \ldots, x_m$. In the literature on D-modules, it is customary to call N simply the *restriction* of D/I to the linear subspace $\{x_1 = \cdots = x_m = 0\}$.

The restriction N can also be expressed as a tensor product of modules over the commutative polynomial ring $\mathbf{k}[x] = \mathbf{k}[x_1, \ldots, x_n]$ as follows:

$$N \quad = \quad D_{Y \to X} \otimes_D D/I \quad = \quad (\mathbf{k}[x]/\langle x_1, \ldots, x_m \rangle) \otimes_{\mathbf{k}[x]} D/I. \qquad (5.9)$$

The formula (5.9) is derived by replacing $D_{Y \to X}$ with the right hand side of

$$D_{Y \to X} \quad = \quad (\mathbf{k}[x]/\langle x_1, \ldots, x_m \rangle) \otimes_{\mathbf{k}[x]} D_X.$$

This states that $D_{Y \to X}$ is simply the restriction of the (infinitely generated) $\mathbf{k}[x]$-module $D = D_X$ to Y in the usual sense of algebraic geometry. The formula (5.9) states that N is the restriction of the $\mathbf{k}[x]$-module D/I to Y in the usual sense of algebraic geometry. Thus our definition of restriction as a D-module agrees with the usual definition of restriction for $\mathbf{k}[x]$-modules.

The left D'-module structure on the right hand side of (5.9) is given by

$$\partial_i : (p \otimes q) \quad \mapsto \quad (\partial_i \bullet p) \otimes q + p \otimes \partial_i q$$
$$x_i : (p \otimes q) \quad \mapsto \quad x_i \cdot p \otimes q$$

for $m + 1 \le i \le n$, $q \in D/I$ and $p \in \mathbf{k}[x]/\langle x_1, \ldots, x_m \rangle = \mathbf{k}[x_{m+1}, \ldots, x_n]$. See [26, §15.1] for further algebraic information. Here is a simple example.

Example 5.2.3. Let $n = 2$ and $I = D\{\partial_1^2, \partial_2^2\}$ which is holonomic of rank 4. Let $m = 1$. The restriction module of D/I with respect to x_1 equals

$$N \;=\; D/(D\{\partial_1^2, \partial_2^2\} + x_1 D) \;=\; \mathbf{k}[x_2] \otimes_{\mathbf{k}[x_1, x_2]} D/D\{\partial_1^2, \partial_2^2\}.$$

Every element of N can be written uniquely as

$$a_0(x_2) + a_1(x_2)\partial_2 + \big(b_0(x_2) + b_1(x_2)\partial_2\big) \cdot \partial_1$$

where a_0, a_1, b_0, b_1 are univariate polynomials. We must regard N as a left module over $D' = \mathbf{k}\langle x_2, \partial_2\rangle$, and as such it is not cyclic. In fact, we have

$$N \;\simeq_{D'}\; (D')^2 \Big/ \mathrm{image} \begin{pmatrix} \partial_2^2 & 0 \\ 0 & \partial_2^2 \end{pmatrix}.$$

Thus N is a holonomic D'-module of rank 4. What does it mean in terms of differential equations ? The D'-module N represents the system of equations

$$\frac{\partial^2}{\partial x_2^2} g(x_2) \;=\; \frac{\partial^2}{\partial x_2^2} h(x_2) \;=\; 0$$

for a vector-valued function $\begin{pmatrix} g(x_2) \\ h(x_2) \end{pmatrix}$. The 4-dimensional solution space to this system consists of the vectors $\begin{pmatrix} f(0, x_2) \\ \frac{\partial f}{\partial x_1}(0, x_2) \end{pmatrix}$ where $f(x_1, x_2)$ runs over the 4-dimensional solution space of the original system I. We leave it to the reader to work out the case $m = 0$ where the restriction module is nothing but a 4-dimensional $\mathbf{k}$-vector space. This space encodes the germs of polynomial functions of the form $\alpha + \beta x_1 + \gamma x_2 + \delta x_1 x_2$ at the origin.

Returning to our general discussion, we are now prepared to address the question raised in the beginning of this section. Consider the cyclic D'-submodule generated by the element $1 \otimes 1$ in the restriction module (5.9). The element $1 \otimes 1$ in (5.9) corresponds to the element 1 in the representation (5.8), so this cyclic D'-module equals the quotient of D' by the left D'-ideal

$$(I + x_1 D + \cdots + x_m D) \cap D'. \tag{5.10}$$

We call the D'-ideal (5.10) the *restriction ideal* of I with respect to the variables $x_1, \ldots, x_m$.

Proposition 5.2.4. *Let I be a holonomic D-ideal and let $f(x_1, \ldots, x_n)$ be any function which is annihilated by I. Then the restricted function $f(0, \ldots, 0, x_{m+1}, \ldots, x_n)$ is annihilated by the restriction ideal (5.10).*

Proof. An operator $p \in D'$ which lies in (5.10) can be written as

$$p(x, \partial) \;=\; \sum_{i=1}^{m} x_i q_i(x, \partial) + r(x, \partial) \qquad \text{where} \quad r \in I.$$

Since $r \bullet f = 0$, we have

$$p \bullet f(x_1, \ldots, x_n) \quad = \quad \sum_{i=1}^{m} x_i \cdot (q_i \bullet f).$$

Replace $x_1, \ldots, x_m$ by zero in this identity. This leaves the operator p unchanged, and we see that $p \bullet f(0, \ldots, 0, x_{m+1}, \ldots, x_n) = 0$ as desired. ☐

The class of (regular) holonomic D-modules is stable under restrictions.

Theorem 5.2.5. *Let I be a holonomic D-ideal. Then the restriction module (5.8) is a finitely presented holonomic D'-module D'^r/M, and the restriction ideal (5.10) is a holonomic D'-ideal. Moreover, if I is regular holonomic then the restriction module and the restriction ideal are regular holonomic.*

Proofs of the first statement can be found in the text books [15, Chapter 1], [17], [26, Chapter 18], [54, §3.2]. A proof of the second assertion on regularity can be found in [17], [54, p.99]. Our aim in what follows is to give an algorithm to compute, from the given generators of I, an explicit presentation D'^r/M of the restriction module.

We fix a non-negative integer weight vector $w \in \mathbf{Z}^n$ such that

$$w_j > 0 \quad \text{for} \quad j = 1, \ldots, m \quad \text{and} \quad w_j = 0 \quad \text{for} \quad j = m+1, \ldots, n. \quad (5.11)$$

The weight vector w defines a filtration of the Weyl algebra D by D'-modules. Indeed, for any real number k, the following subset of D is a left D'-submodule:

$$F_k \quad := \quad \Big\{ \sum c_{\alpha, \beta} x^\alpha \partial^\beta \mid (\beta - \alpha) \cdot w \le k \Big\}. \quad (5.12)$$

The module F_k is not finitely generated over D'. However, if we take the quotient of F_k by the left D'-submodule

$$\sum_{j=1}^{m} x_i \cdot F_{k+w_i} \quad = \quad \Big(\sum_{j=1}^{m} x_i \cdot D \Big) \cap F_k,$$

then this quotient is a finitely generated D'-module. More precisely, let $\mathcal{B}_d$ denote the set of all monomials $\partial_1^{i_1} \partial_2^{i_2} \cdots \partial_m^{i_m}$ with the property $i_1 w_1 + i_2 w_2 + \cdots + i_m w_m \le d$. This set is finite since $w_j > 0$ for $j = 1, \ldots, m$, and we have

$$F_d / (x_1 F_{d+w_1} + \cdots + x_m F_{d+w_m}) \quad = \quad D' \cdot \mathcal{B}_d.$$

Our computations will take place in this free D'-module with basis $\mathcal{B}_d$.

Let $\{g_1, \ldots, g_r\}$ be a Gröbner basis of the holonomic D-ideal I with respect to the weight vector $(-w, w)$, and let $m_i := \mathrm{ord}_{(-w,w)}(g_i)$ denote the maximal $(-w, w)$-degree of any monomial appearing in g_i. Consider the b-function $b(s)$ of I with respect to w, and let s_0 be any integer which is larger than or equal to the maximal integer root of $b(s) = 0$.

Theorem 5.2.6. *The inclusion $F_{s_0} \subset D$ defines the following isomorphism of left D'-modules:*

$$F_{s_0} \Big/ \left(\sum_{i=1}^{r} F_{s_0-m_i} g_i + \sum_{j=1}^{m} x_j F_{s_0+w_j} \right) \quad \simeq_{D'} \quad D/(I + x_1 D + \cdots + x_m D).$$

Proof. The inclusion $F_{s_0} \subset D$ gives a well-defined map of left D'-modules from the left hand side to the right hand side, since the parenthesized submodule on the left hand side is clearly contained in $I + x_1 D + \cdots + x_m D$.

This map is surjective by Lemma 5.1.8. It implies that every monomial $x^\alpha \partial^\beta$ can be rewritten modulo $I + x_1 D + \cdots + x_m D$ as a k-linear combination of monomials of $(-w, w)$-degree less than or equal to s_0. Hence each element in $D/(I + x_1 D + \cdots + x_m D)$ can be represented by an operator in F_{s_0}.

It remains to be shown that the map in question is injective. Suppose that p is an element of F_{s_0} which lies in $I + x_1 D + \cdots + x_m D$. We must prove that p lies in $\sum_{i=1}^{r} F_{s_0-m_i} g_i + \sum_{j=1}^{m} x_j F_{s_0+w_j}$. Write

$$p \; = \; q + \sum_{j=1}^{m} x_j u_j \qquad \text{with } q \in I \tag{5.13}$$

in such a way that $s_1 := \max_{j=1,\dots,m} \mathrm{ord}_{(-w,w)}(x_j u_j)$ is minimal over all possible expressions of this form. Suppose that $s_1 > s_0$. Then, we have

$$\sum_{j=1}^{m} x_j u_j \; = \; 0 \quad \text{in} \quad G_{s_1} := F_{s_1} \big/ \left((I \cap F_{s_1}) + F_{s_1-1} \right).$$

We will prove in Proposition 5.2.16 that the Koszul complex on the D'-module G_{s_1} defined by the sequence $x_1, \dots, x_m$ is exact. Here the hypothesis $s_1 > s_0$ is essential. Therefore each multiplier u_i can be rewritten as follows:

$$u_i \; = \; \sum_{j<i} (-x_j) u_{ji} + \sum_{k>i} x_k u_{ik} + h_i + v_i$$

where $h_i \in I$ and $v_i \in F_{s_1+w_i-1}$. If we replace u_i by this expression in $p = q + \sum_{j=1}^{m} x_j u_j$ then the terms involving u_{ji} and u_{ik} cancel and we find

$$p \; = \; \tilde{q} + \sum_{j=1}^{m} x_j v_j$$

where $\tilde{q}$ is a new element in I. We have shown that $\sum_{j=1}^{m} x_j v_j \in F_{s_1-1}$, which is a contradiction to the minimality in our choice of s_1.

We have shown that each element $p \in (I + x_1 D + \cdots + x_m D) \cap F_{s_0}$ permits a representation (5.13) where $q \in I \cap F_{s_0}$ and $u_j \in F_{s_0+w_j}$ for $j = 1, \dots, m$. Since $\{g_1, \dots, g_s\}$ is a Gröbner basis for I with respect to $(-w, w)$, we have

$$q = \sum_{i=1}^{r} c_i \cdot g_i \quad \text{with} \quad \mathrm{ord}_{(-w,w)}(c_i) \leq s_0 - m_i$$

by Theorem 1.2.10 on standard representations. We conclude that p lies in $\sum_{i=1}^{r} F_{s_0-m_i} \cdot g_i + \sum_{j=1}^{m} x_j F_{s_0+w_j}$ as desired. □

Corollary 5.2.7. *If the b-function of I with respect to w has no non-negative integer root, then the restriction module $D/(I + x_1 D + \cdots + x_m D)$ is zero.*

Proof. The D'-isomorphism in Theorem 5.2.6 holds for $s_0 = -1$. However, the D'-module on the left hand side is zero since every monomial $x^\alpha \partial^\beta$ of negative $(-w, w)$-weight must contain one of the variables $x_1, \ldots, x_m$, i.e.,

$$F_{-1} = \sum_{j=1}^{m} x_j \cdot F_0.$$

This implies that $D/(I + x_1 D + \cdots + x_m D) = 0$. □

Theorem 5.2.6 gives rise to the following algorithm for computing the restriction module $(D')^r / M$ of the holonomic D-ideal I with respect to the variables $x_1, \ldots, x_m$.

Algorithm 5.2.8 (Computing the restriction module) [78], [80], [81].
Input: Generators of a holonomic D-ideal I
Output: A presentation $(D')^r / M$ of the restriction module (5.8)

1. Choose a weight vector $(-w, w)$ satisfying (5.11) and compute a Gröbner basis of I with respect to $(-w, w)$. Let $\{g_1, \ldots, g_p\}$ be the Gröbner basis.
2. Compute the b-function $b(s)$ of I with respect to $(-w, w)$.
3. Let s_0 be the maximal integral root of $b(s) = 0$. If $s_0 < 0$ or there is no integral root at all, then output "the restriction module is zero".
4. Let $(D')^r$ be the free left D'-module with basis $\mathcal{B}_{s_0}$.
5. For each g_i let $m_i := \mathrm{ord}_{(-w,w)}(g_i)$ and form the finite set $\mathcal{B}_{s_0-m_i}$.
6. For each i and each $\partial^\beta \in \mathcal{B}_{s_0-m_i}$ compute the expansion of $\partial^\beta \cdot g_i$ as a $\mathbf{k}$-linear combination of monomials $x^u \partial^v$, and substitute $x_1 = \cdots = x_m = 0$ into this expansion. The result is an element of $D' \cdot \mathcal{B}_{s_0} = (D')^r$.
7. Let M be the D'-submodule of $(D')^r$ generated by all these elements.
8. Return the restriction module $(D')^r / M$.

Example 5.2.9. Consider the simple case $m = n = 1$ and $I = D \cdot \{x\partial - \alpha\}$. We will compute the restriction of D/I to $x = 0$. Here $w = 1$ and $D' = \mathbf{k}$. The b-function equals $b(s) = s - \alpha$. Hence, if $\alpha \notin \mathbf{N}$ then the restriction is 0 by Corollary 5.2.7. This makes sense analytically, as the differential equation $x\partial - \alpha$ has a holomorphic solution in a neighborhood of $x = 0$ only if $\alpha \in \mathbf{N}$.

If α is a non-negative integer then we can apply Algorithm 5.2.8 with $s_0 = \alpha$. The ideal generator $g_1 := x\partial - \alpha$ alone is already a Gröbner basis of

I with respect to $(-1, 1)$. In step 4 we have $\mathcal{B}_{s_0} = \{1, \partial, \partial^2, \ldots, \partial^\alpha\}$ so that $(D')^r = \mathbf{k} \cdot \mathcal{B}_{s_0} = \mathbf{k}^{\alpha+1}$. In step 5 we note $m_1 = 0$. In step 6 we compute

$$\partial^i \cdot g_1 \quad = \quad x\partial^{i+1} + (i - \alpha)\partial^i,$$

and in Step 7 we conclude that M is a $\mathbf{k}$-linear subspace of codimension 1:

$$M \quad = \quad \mathbf{k} \cdot \{(i - \alpha)\partial^i \; : \; i = 0, 1, \ldots, \alpha\} \quad = \quad \mathbf{k} \cdot \{1, \partial^2, \ldots, \partial^{\alpha-1}\} \subset \mathbf{k} \cdot \mathcal{B}_{s_0}.$$

Therefore the restriction is the one-dimensional vector space $\mathbf{k}\partial^\alpha \simeq \mathbf{k}$.

Example 5.2.10. Put $I = D \cdot \{x_1, \partial_2 - 1\}$. The restriction of this D-ideal was presented in Example 2.4.7, but we did not give a proof that the restriction is 0. Let us compute the restriction of I to $x_1 = 0$ by our algorithm. Put $(-w, w) = (-1, 0, 1, 0)$. The Gröbner basis is $\{x_1, \partial_2 - 1\}$. The b-function is $s + 1$. Hence, the restriction module is 0.

Remark 5.2.11. The set $\mathcal{B}_{s_0}$ may be made smaller by analyzing integral roots of b-functions. A method to bound its size from below is given in [78], [80].

The output of Algorithm 5.2.8 can always be interpreted as a system of linear differential equations for a vector-valued function $v = v(x_{m+1}, \ldots, x_n)$ of $n - m$ variables. The indeterminate vector v has r coordinates which are indexed by the elements of $\mathcal{B}_{s_0}$, say, $v = (v^{(\alpha)})_{\partial^\alpha \in \mathcal{B}_{s_0}}$ where $v^{(\alpha)} = v^{(\alpha)}(x_{m+1}, \ldots, x_n)$ is an indeterminate function of $n - m$ variables.

The operator constructed from $\partial^\beta \cdot g_i$ by normal expansion and setting $x_1 = \cdots = x_m = 0$ in step 6 of Algorithm 5.2.8 takes the form

$$\sum_{0 \le \alpha \cdot w \le s_0} p_{\beta i \alpha} \cdot \partial^\alpha, \qquad p_{\beta i \alpha} \in D'.$$

These are the generators of M. They represent the system of differential equations

$$\sum_{0 \le \alpha \cdot w \le s_0} p_{\beta i \alpha} \bullet v^{(\alpha)}(x_{m+1}, \ldots, x_n) = 0, \quad (\partial^\beta \in \mathcal{B}_{s_0 - m_i}, \; i = 1, \ldots, p). \quad (5.14)$$

Suppose that $f = f(x_1, \ldots, x_n)$ is a holomorphic solution to I whose iterated derivatives with respect to $x_1, \ldots, x_m$ define holomorphic functions on $\{x_1 = \cdots = x_m = 0\}$. Then the vector $v = (v^{(\alpha)})_{\partial^\alpha \in \mathcal{B}_{s_0}}$ with coordinates

$$v^{(\alpha)}(x_{m+1}, \ldots, x_n) \quad = \quad \frac{\partial^{|\alpha|} f}{\partial x_1^{\alpha_1} \cdots \partial x_m^{\alpha_m}}(0, \ldots, 0, x_{m+1}, \ldots, x_n) \quad (5.15)$$

is a solution to the system of differential equations (5.14).

The first coordinate of the vector v described in (5.15) is the restriction

$$v^{(0)}(x_{m+1}, \ldots, x_n) \quad = \quad f(0, \ldots, 0, x_{m+1}, \ldots, x_n) \quad (5.16)$$

We can compute the holonomic annihilating D-ideal (5.10), which represents the function (5.16), from the restriction module M as follows.

Algorithm 5.2.12 (Computing the restriction ideal)
Input: Generators of a holonomic D-ideal I.
Output: A Gröbner basis of the restriction ideal $(I + x_1 D + \cdots + x_m D) \cap D'$.

1. Compute generators for the restriction module M using Algorithm 5.2.8.
2. Identify D' with the submodule of the free module $(D')^r$ which is spanned by the first vector ∂^0 in the basis $\mathcal{B}_{s_0}$ of $(D')^r$.
3. Choose a *POT term order* (see [1, Def. 3.5.3], [28, §5.2]) on the free module $(D')^r$ for which the first position (indexed by $\partial^0 \in \mathcal{B}_{s_0}$) is the cheapest.
4. Compute a Gröbner basis G of the module M with respect to the POT term order $\prec$.
5. Output the set $G \cap D'$. It consists of those vectors in G for which only the first coordinate is non-zero. This set is a Gröbner basis for the D'-ideal $M \cap D' = (I + x_1 D + \cdots + x_m D) \cap D'$.

The correctness of this algorithm follows from the elimination property of the POT term order ("position over term"), which guarantees that if G is a Gröbner basis for M then $G \cap D'$ is a Gröbner basis for $M \cap D'$.

Remark 5.2.13. The restriction of D/I to any smooth algebraic variety Y can be defined in a similar way and there are algorithms to compute it via local cohomology modules. These topics are out of the scope of this book. See [78], [80, Algorithm 7.3, Proposition 7.4] and [101] for these algorithms.

We note that Algorithm 5.2.8 is of interest even in the extreme case $m = n$ when Y is a point, that is, $Y = \{x_1 = \cdots = x_n = 0\}$. In that case, $D' = \mathbf{k}$ is just the ground field, and the restriction module

$$D_{Y \to X} \otimes_D D/I \quad = \quad D/(I + x_1 D + \cdots + x_n D).$$

is a finite-dimensional $\mathbf{k}$-vector space. For $\mathbf{k} = \mathbf{C}$, this vector space represents the formal power series solutions to I in a neighborhood of the origin.

Proposition 5.2.14. (see, e.g., [65, p.428]) *For $m = n$ there is a $\mathbf{k}$-linear isomorphism*

$$D_{Y \to X} \otimes_D D/I \quad \simeq_{\mathbf{k}} \quad \{ f \in \mathbf{k}[[x_1, \ldots, x_n]] \mid I \bullet f = 0 \}.$$

Proof (sketch). The power series solutions to I are encoded by the constant terms of their various derivatives, that is, we can represent $f \in \mathbf{k}[[x_1, \ldots, x_n]]$ by a sufficiently long vector $v = (v^{(\alpha)})_{\alpha \in \mathbf{N}^n}$ with coordinates

$$v^{(\alpha)} \quad = \quad \frac{\partial^{|\alpha|} f}{\partial x_1^{\alpha_1} \cdots \partial x_n^{\alpha_n}}(0, \ldots, 0)$$

The differential equations (5.14) encoding the restriction module $D_{Y \to X} \otimes_D D/I = (D')^r/M$ are simply linear equations with coefficients $p_{\beta i \alpha}$ in $\mathbf{k}$:

$$\sum_{0 \leq \alpha \cdot w \leq s_0} p_{\beta i \alpha} \cdot v^{(\alpha)} = 0, \qquad (\partial^\beta \in \mathcal{B}_{s_0 - m_i}, \, i = 1, \dots, p).$$

This system of k-linear equations is precisely the system of recurrence relations for the coefficients of all the power series solutions to I. $\square$

If $f \in k[[x_1, \dots, x_n]]$ is a solution to I and $w \geq 0$ then $\mathrm{in}_w(f)$ is a series in $k[[x_1, \dots, x_n]]$ which is a solution to the initial D-ideal $\mathrm{in}_{(-w,w)}(I)$. This was shown in much greater generality in Theorem 2.5.5. Proposition 5.2.14 and Algorithm 5.2.8 therefore imply the following two inequalities.

Corollary 5.2.15. *For any non-negative weight vector $w \geq 0$, we have*

$$\dim_k D_{Y \to X} \otimes_D D/I \quad \leq \quad \dim_k D_{Y \to X} \otimes_D D/\mathrm{in}_{(-w,w)}(I)$$

$$\leq \quad \#\{\beta \in \mathbf{N}^n \mid \sum_{i=1}^n w_i \beta_i \leq s_0\}.$$

Here, s_0 is the maximal integral root of the b-function of I with respect to w.

Both of these inequalities can be strict. The first inequality is strict for $n = 1$, $w = 1$ and $I = D \cdot \{\theta(\theta - 3) - x(\theta + 1)(\theta + 1)\}$. The dimension of formal power series solutions is 1, but $\mathrm{in}_{(-1,1)}(I) = D \cdot \{\theta(\theta - 3)\}$ has two solutions 1 and x^3. The second inequality is strict for $n = 2$, $w = (1, 1)$ and $I = D \cdot \{\partial_1^2, \partial_2^2\}$, namely, the three numbers in question are 4, 4 and 6. It is an interesting problem to study the case of equality in Corollary 5.2.15.

We shall prove the exactness of the Koszul complex on G_k by the sequence $x_1, \dots, x_m$ which was used in the proof of Theorem 5.2.6. The proof of Proposition 5.2.16 is somewhat technical and has been postponed to this place so as to not interrupt our discussion of the restriction algorithm. Take a weight vector w satisfying (5.11). Let $b(s)$ be the b-function of the holonomic ideal I with respect to the weight vector w. Put $G_k = F_k/((F_k \cap I) + F_{k-1})$.

Proposition 5.2.16 ([80, Proposition 5.2]). *If $b(k) \neq 0$, then the Koszul complex on G_k defined by the sequence $x_1, \dots, x_m$ is exact.*

To prove this we use the idea presented in Lemma 5.1.8 together with a result in homological algebra. Suppose that we are given the following double complex of abelian groups K^i, L^j and morphisms d_K^i, d_L^j, u^i:

$$
\begin{array}{ccccccccc}
0 = K^0 \longrightarrow & K^1 & \xrightarrow{d_K^1} & K^2 & \xrightarrow{d_K^2} & \cdots & \xrightarrow{d_K^{p-1}} & K^p & \xrightarrow{d_K^p} & K^{p+1} = 0 \\
& \downarrow u^1 & & \downarrow u^2 & & & & \downarrow u^p & & \\
0 = L^0 \longrightarrow & L^1 & \xrightarrow{d_L^1} & L^2 & \xrightarrow{d_L^2} & \cdots & \xrightarrow{d_L^{p-1}} & L^p & \xrightarrow{d_L^p} & L^{p+1} = 0
\end{array}
\tag{5.17}
$$

By the standard transformation from the double complex to the total complex, we obtain the following complex:

$$\cdots \longrightarrow K^k \oplus L^{k-1} \longrightarrow \qquad K^{k+1} \oplus L^k \qquad \longrightarrow \cdots$$

$$\Cup \qquad\qquad\qquad \Cup \qquad\qquad\qquad\qquad (5.18)$$

$$(f,g) \qquad \longmapsto \quad (-d_K^k(f), u^k(f) + d_L^{k-1}(g))$$

It is called the *mapping cone* of $(K^{\cdot}, d_K^{\cdot}) \xrightarrow{u^{\cdot}} (L^{\cdot}, d_L^{\cdot})$.

Example 5.2.17. Consider the double complex

$$
\begin{array}{ccc}
0 \longrightarrow G_{k+w_1+w_2} & \xrightarrow{x_1} & G_{k+w_2} \longrightarrow 0 \\
\downarrow x_2 & & \downarrow x_2 \\
0 \longrightarrow \quad G_{k+w_1} & \xrightarrow{x_1} & G_k \quad \longrightarrow 0,
\end{array}
$$

then the mapping cone is

$$
\begin{array}{ccc}
(g,h) & \longmapsto & x_1 g + x_2 h \\
\Cap & & \Cap \\
0 \longrightarrow G_{k+w_1+w_2} \longrightarrow G_{k+w_1} \oplus G_{k+w_2} \longrightarrow & G_k & \longrightarrow 0. \\
\Cup & & \\
f \qquad \longmapsto \qquad (x_2 f, -x_1 f)
\end{array}
$$

It is the Koszul complex $K(G_k; x_1, x_2)$ on G_k defined by the sequence x_1 and x_2. Each row of the double complex is the Koszul complex $K(G_{k+w_2}; x_1)$ and $K(G_k; x_1)$ respectively.

In general, the Koszul complex $K(G_k; x_1, \ldots, x_m)$ is the mapping cone of $K(G_{k+w_m}; x_1, \ldots, x_{m-1}) \xrightarrow{x_m} K(G_k; x_1, \ldots, x_{m-1})$.

The following fact is standard in homological algebra:

Theorem 5.2.18. *If* $\mathrm{Ker}\, u^{\cdot}$ *and* $\mathrm{Coker}\, u^{\cdot}$ *are quasi-isomorphic to zero, then the mapping cone is also quasi-isomorphic to zero. In other words, if the complexes*

$$0 \to \mathrm{Ker}\, u^1 \to \mathrm{Ker}\, u^2 \to \cdots \to \mathrm{Ker}\, u^p \to 0$$

and

$$0 \to \mathrm{Coker}\, u^1 \to \mathrm{Coker}\, u^2 \to \cdots \to \mathrm{Coker}\, u^p \to 0$$

are exact, then the mapping cone is also exact.

We now prove that the Koszul complex $K(G_k; x_1, \ldots, x_m)$ is exact. From the definition of the b-function, we have $v \cdot b \left(\sum_{i=1}^m w_i \theta_i \right) \in F_{k-1}$ modulo I for $v \in F_k \backslash F_{k-1}$, i.e., $v \cdot b \left(\sum_{i=1}^m w_i \theta_i \right) = 0$ in G_k. By expressing v in the form

$$v = \sum_{(\beta - \alpha) \cdot w = k} a_{\alpha\beta} x^\alpha \partial^\beta,$$

we can verify that

$$b \left(\sum_{i=1}^m w_i \theta_i + k \right) v = 0 \quad \text{in } G_k. \qquad (5.19)$$

To prove the proposition by induction, we forget that G_k is the quotient $F_k/((F_k \cap I) + F_{k-1})$ and focus only on the property (5.19). We will prove a slightly more general version of Proposition 5.2.16.

Theorem 5.2.19. *Let $\{G_k\}_{k \in \mathbf{Z}}$ be a set of $\mathrm{gr}_{(-w,w)}(D)$-modules satisfying $F_{k'} G_k \subseteq G_{k'+k}$. Suppose that there exists a polynomial $b(s)$ independent of k such that*

$$b\left(\sum_{i=1}^{m} w_i \theta_i + k\right) v = 0 \quad \text{in } G_k \qquad \text{for all } k \in \mathbf{Z}.$$

If $b(k) \neq 0$, then the Koszul complex $K(G_k; x_1, \ldots, x_m)$ is exact.

Proof. The proof is by induction on m. For $m = 1$ the Koszul complex is

$$G_{k+w_1} \xrightarrow{x_1} G_k.$$

Let us prove that the map x_1 is surjective. For $v \in G_k$, we have $b(w_1 \theta_1 + k)v = 0$ in G_k. Then, we have

$$b(k)v + x_1(\cdots)v = 0 \quad \text{in } G_k.$$

Since $b(k) \neq 0$, we conclude $v \in \mathrm{Im}\, x_1$.

Next, we prove that it is injective. Take $v \in G_{k+w_1}$ such that $x_1 v = 0$. From the assumption of the theorem, we have $b(w_1 \theta_1 + k + w_1)v = 0$. Since

$$b(w_1 \theta_1 + k + w_1)v = b(w_1 \partial_1 x_1 + k)v = (\cdots)x_1 v + b(k)v = b(k)v = 0,$$

we conclude $v = 0$. Our theorem is proved for $m = 1$.

Let us now proceed with the induction step. We consider the complex

$$0 \longrightarrow G_{i+w_m} \xrightarrow{x_m} G_i \longrightarrow 0.$$

We put

$$G_i' = \mathrm{Ker}\, x_m, \quad G_i'' = \mathrm{Coker}\, x_m = G_i / x_m \cdot G_{i+w_m}.$$

It will follow from Theorem 5.2.18 that the Koszul complex $K(G_k; x_1, \ldots, x_m)$ is exact, if we show that the smaller Koszul complexes $K(G_k'; x_1, \ldots, x_{m-1})$ and $K(G_k''; x_1, \ldots, x_{m-1})$ are exact. Let us show that

$$b\left(\sum_{i=1}^{m-1} w_i \theta_i + k\right) v' = 0 \quad \text{in } G_k' \text{ for any } v' \in G_k'$$

$$\text{and} \quad b\left(\sum_{i=1}^{m-1} w_i \theta_i + k\right) v'' = 0 \quad \text{in } G_k'' \text{ for any } v'' \in G_k''.$$

If we show these, then the smaller Koszul complexes become exact by the induction hypothesis and we are done.

For the first equality take $v' \in G_{k+w_m}$ such that $x_m v' = 0$. Then, we have

$$b \left(\sum_{i=1}^{m-1} w_i \theta_i + w_m \theta_m + k + w_m \right) v'$$

$$= b \left(\sum_{i=1}^{m-1} w_i \theta_i + w_m \partial_m x_m + k \right) v'$$

$$= b \left(\sum_{i=1}^{m-1} w_i \theta_i + k \right) v' \quad (x_m v' = 0 \text{ in } G_k)$$

$$= 0 \quad \text{in } G_{k+w_m}.$$

Hence, we have $b \left(\sum_{i=1}^{m-1} w_i \theta_i + k \right) v' = 0$ in G'_k for any $v' \in G'_k$. Let us derive the second equality. Take $v'' \in G_k$. Then, we have

$$b \left(\sum_{i=1}^{m-1} w_i \theta_i + w_m \theta_m + k \right) v''$$

$$= b \left(\sum_{i=1}^{m-1} w_i \theta_i + k \right) v'' + x_m (\cdots) v'' = 0 \quad \text{in } G_k.$$

Therefore, we have $b \left(\sum_{i=1}^{m-1} w_i \theta_i + k \right) v'' = 0$ in G''_k for any $v'' \in G''_k$.   □

We close this section with a bibliographic note. Let (D^{b_i}, φ_i) be a free resolution of D/I. The i-th cohomology group of the complex $(D_{Y \to X} \otimes_D D^{b_i}, \text{id} \otimes \varphi_i)$ is called the i-th restriction. An algorithm to compute the i-th restriction is given in [78], [80], [82]. An algorithm to compute the i-th restriction of a given complex $M^{\cdot}$ is given by Walther [102]. In this section, we only introduced an algorithm to get the 0-th restriction.

5.3 Powers of Polynomials

In this section we introduce Oaku's algorithm [77, Theorem 19, 24] for representing powers of polynomials as holonomic functions and finding the b-function of a polynomial. Let f be an element in $\mathbf{k}[x_1, \ldots, x_n]$ and $\alpha \in \mathbf{k}$. We are interested in the annihilating ideal

$$\text{Ann}(f^\alpha) = \{ p \in D : p \bullet f^\alpha = 0 \}.$$

Here differential operators act in the usual way on powers, for instance,

$$\partial_i \bullet f^\alpha = \alpha \cdot f^{\alpha-1} \cdot \frac{\partial f}{\partial x_i}.$$

An example of such an annihilating ideal was discussed in Example 5.1.7:

$$\text{Ann}(f^{-1}) \quad = \quad D \cdot \{3x^2\partial_y + 2y\partial_x, \; 2x\partial_x + 3y\partial_y + 6\} \qquad \text{for} \quad f = x^3 - y^2.$$

We wish to emphasize that in general the exponent α will not be an integer.

The general theory of $\mathcal{D}$-modules provides the following regularity result due to Kashiwara and Kawai. For the proof we refer to the paper [60].

Theorem 5.3.1. $\text{Ann}(f^\alpha)$ *is a regular holonomic D-ideal for any* $\alpha \in \mathbf{k}$.

We start out with the following construction due to Malgrange [67]. Let t be a new variable and consider the $(n+1)$-dimensional Weyl algebra $D\langle t, \partial_t \rangle$. The polynomial $f \in \mathbf{k}[x_1, \ldots, x_n]$ defines the following left ideal in $D\langle t, \partial_t \rangle$:

$$I_f \quad := \quad D\langle t, \partial_t \rangle \cdot \{t - f(x), \; \partial_1 + \frac{\partial f}{\partial x_1}\partial_t, \; \ldots, \; \partial_n + \frac{\partial f}{\partial x_n}\partial_t\}. \qquad (5.20)$$

The ideal I_f annihilates the delta-function $\delta(t - f(x))$.

Proposition 5.3.2. *The ideal* I_f *is maximal, i.e,* $D\langle t, \partial_t \rangle / I_f$ *is a simple* $D\langle t, \partial_t \rangle$-module.

Proof. Choose a term order $\prec$ on $D\langle t, \partial_t \rangle$ such that the given $n+1$ generators of I_f have the leading terms $t, \partial_1, \ldots, \partial_n$. Using Buchberger's S-pair criterion, one checks that the given $n+1$ generators form a Gröbner basis for I_f. (This implies in particular that I_f is a proper ideal). The (commutative) initial ideal $\text{in}_{\prec}(I_f) = \langle t, \xi_1, \ldots, \xi_n \rangle$ is maximal with the property of being middle-dimensional. The Fundamental Theorem of Algebraic Analysis now implies that I_f is a maximal left ideal in $D\langle t, \partial_t \rangle$. □

Let s be a new indeterminate and let E_s be the difference operator with respect to s. The ring $D\langle s, E_s \rangle$ of difference-differential operators is defined over D by the commutation relation $E_s s = (s+1)E_s$. Consider the function f^s and the $\mathbf{k}[x, s]$-module $\mathbf{k}[s, x, f^{-1}, f^s]$. It admits a natural structure of a left $D\langle s, E_s \rangle$-module which is given by the following actions:

$$
\begin{aligned}
x_i \bullet g(s,x)f^{s+j} \quad &= \quad x_i g(s,x)f^{s+j}, \\
\partial_i \bullet g(s,x)f^{s+j} \quad &= \quad \frac{\partial g}{\partial x_i}f^{s+j} + (s+j)g\frac{\partial f}{\partial x_i}f^{s+j-1}, \\
s \bullet g(s,x)f^{s+j} \quad &= \quad sg(s,x)f^{s+j}, \\
E_s \bullet g(s,x)f^{s+j} \quad &= \quad g(s+1,x)f^{s+1+j}.
\end{aligned}
$$

Next consider the subring $D\langle t\partial_t, t \rangle$ of the $(n+1)$-dimensional Weyl algebra $D\langle \partial_t, t \rangle$. The generators of this subring satisfy the commutation relations

$$(t\partial_t) \cdot t \quad = \quad t \cdot (t\partial_t + 1).$$

Therefore the *Mellin transform*

$$s + 1 \mapsto -t\partial_t, \quad E_s \mapsto t \qquad (5.21)$$

induces an algebra isomorphism between $D\langle s, E_s \rangle$ and $D\langle t\partial_t, t \rangle$. Thus $D\langle t\partial_t, t \rangle$ acts on $\mathbf{k}[s, x, f^{-1}, f^s]$ via the isomorphism given by the Mellin transform. The following action of the Weyl algebra $D\langle t, \partial_t \rangle$ on $\mathbf{k}[s, x, f^{-1}, f^s]$ is the natural extension of the action by its subring $D\langle t\partial_t, t \rangle \simeq D\langle s, E_s \rangle$:

$$
\begin{aligned}
x_i \bullet g(s, x) f^{s+j} &= x_i g(s, x) f^{s+j}, \\
\partial_i \bullet g(s, x) f^{s+j} &= \frac{\partial g}{\partial x_i} f^{s+j} + (s + j) g \frac{\partial f}{\partial x_i} f^{s+j-1}, \\
t \bullet g(s, x) f^{s+j} &= g(s + 1, x) f^{s+j+1}, \\
\partial_t \bullet g(s, x) f^{s+j} &= -s g(s - 1, x) f^{s-1+j}.
\end{aligned}
$$

To see that this is indeed an action, one must check that $g(s, x) f^{s+i}$ is fixed by $\partial_t t - t\partial_t$. Furthermore, consider what happens for $j = 0$ and $g = 1$:

$$
\begin{aligned}
x_i \bullet f^s &= x_i f^s, \\
\partial_i \bullet f^s &= s \frac{\partial f}{\partial x_i} f^{s-1}, \\
t \bullet f^s &= f^{s+1}, \\
\partial_t \bullet f^s &= -s f^{s-1}.
\end{aligned}
$$

From this restricted action we derive the following lemma:

Lemma 5.3.3. *The left ideal I_f in $D\langle t, \partial_t \rangle$ equals the annihilator of f^s.*

Proof. The $n + 1$ generators of I_f annihilate f^s under the above action. Since I_f is a maximal ideal (Proposition 5.3.2), it equals the annihilator of f^s. □

We are interested in the annihilator of f^s under the action by $D[s]$. The Mellin transform takes s to $-t\partial_t - 1$ and thereby identifies $D[s]$ with the subring $D[t\partial_t]$ of the Weyl algebra $D\langle t, \partial_t \rangle$. This proves the following theorem.

Theorem 5.3.4. *The ideal of operators in $D[s]$ which annihilate f^s equals the image of the intersection $I_f \cap D[t\partial_t]$ under the substitution $t\partial_t \mapsto -s - 1$.*

Our proof of Theorem 5.3.4 was purely algebraic and did not involve any analysis. However, to better understand the remarkable properties of the Mellin transform (5.21), it is useful to take a peek at its analytic counterpart.

Remark 5.3.5. The Mellin transform is the following integral transform

$$F(t) \longmapsto G(s) = \int_0^\infty F(t) \cdot t^s dt. \qquad (5.22)$$

If the function F is rapidly decreasing in the variable t and the exponent s ranges over a suitable open subset of the complex plane, then integration by parts leads to the following identities:

$$sE_s^{-1} \bullet G(s) = -\int_0^\infty (\partial_t \bullet F(t))t^s dt,$$

$$E_s \bullet G(s) = \int_0^\infty t \cdot F(t)t^s dt.$$

As a consequence of these identities, if F is annihilated by a linear differential operator $q(t, \partial_t)$ then G is annihilated by the difference operator $q(E_s, -sE_s^{-1})$, and conversely. The map which takes difference operators annihilating $G(s)$ to differential operators annihilating $F(t)$ is the algebraic Mellin transform (5.21). To interpret Theorem 5.3.4 analytically, we note that the analytic Mellin transform (5.22) maps the delta-function $F(t) = \delta(t - f(x))$, which is annihilated by I_f, to the classical function $G(s) = f(x)^s$, whose annihilator we are interested in.

An immediate consequence of Theorem 5.3.4 is the following algebraic algorithm for computing a generating set of the left $D[s]$-ideal $\mathrm{Ann}(f^s)$.

Algorithm 5.3.6 (Computing $\mathrm{Ann}(f^s)$)
Input: A polynomial f in $\mathbf{k}[x_1, \ldots, x_n]$.
Output: Generators of the left $D[s]$-ideal $\mathrm{Ann}(f^s)$.

1. Introduce two new central indeterminates u and v, and fix a term order on $D\langle t, \partial_t\rangle[u, v]$ which eliminates u and v.
2. Using Gröbner bases for the above order, compute the intersection ideal

$$D\langle t, \partial_t\rangle[u, v] \cdot \{tu - f, \frac{\partial f}{\partial x_i}v\partial_t + \partial_i \ (i = 1, \ldots, n), uv - 1\} \cap D\langle t, \partial_t\rangle. \quad (5.23)$$

   Each generator has the form $t^a \cdot p(t\partial_t, x, \partial_x) \cdot \partial_t^b$ for some $a, b \in \mathbf{N}$ and some operator $p \in D[t\partial_t]$.
3. Replace each $t^a \cdot p(t\partial_t, x, \partial_x) \cdot \partial_t^b$ by $[t\partial_t]^a \cdot p(t\partial_t - b, x, \partial_x) \cdot [t\partial_t]_b \in D[t\partial_t]$. The resulting set of operators generates $I_f \cap D[t\partial_t]$.
4. Replace $t\partial_t$ by $-s - 1$ in each generator. Output the result.

Remark 5.3.7. Algorithm 5.3.6 also works when the polynomial f contains parameters. In this case, the field $\mathbf{k}$ is the rational function field in these parameters over the base field, say, $\mathbf{Q}$ or $\mathbf{C}$.

Example 5.3.8. The simplest example for Algorithm 5.3.6 is $n = 1$ and $f = x$. Here the input is the left ideal in $\mathbf{k}\langle x, \partial_x, t, \partial_t\rangle[u, v]$ generated by $tu - x$, $v\partial_t + \partial_x$ and $uv - 1$. Eliminating u and v from these three expressions, we find that (5.23) is the principal left in $\mathbf{k}\langle x, \partial_x, t, \partial_t\rangle$ generated by $x\partial_x + t\partial_t + 1$. Replacing $t\partial_t$ by $-s - 1$, we conclude $\mathrm{Ann}(x^s) = \mathbf{k}\langle x, \partial_x\rangle[s] \cdot \{x\partial_x - s\}$.

Example 5.3.9. Let $n = 5$ and fix the polynomial $f = x_1 + x_2 z + x_3 z^2 + x_4 z^3$. This is the general univariate cubic, regarded as a polynomial in five variables x_1, x_2, x_3, x_4, z. The output of Algorithm 5.3.6 for this polynomial equals

$$
\begin{aligned}
\operatorname{Ann}(f^s) \;=\; D[s] \cdot \big\{ & z\partial_1 - \partial_2,\ z\partial_2 - \partial_3,\ z\partial_3 - \partial_4, \\
& \partial_2^2 - \partial_1\partial_3,\ \partial_3^2 - \partial_2\partial_4,\ \partial_2\partial_3 - \partial_1\partial_4, \\
& x_2\partial_1 + 2x_3\partial_2 + 3x_4\partial_3 - \partial_z,\ x_2\partial_2 + 2x_3\partial_3 + 3x_4\partial_4 - z\partial_z, \\
& x_1\partial_1 - x_3\partial_3 - 2x_4\partial_4 + z\partial_z - s,\ 3x_4 z\partial_4 - z^2\partial_z + x_2\partial_3 + 2x_3\partial_4 \big\}.
\end{aligned}
$$

Our real problem is to compute the annihilator of f^α in D for some specific number $\alpha \in \mathbf{k}$. Clearly, if we replace the indeterminate s by α in all generators of $\operatorname{Ann}(f^s) \subset D[s]$, then we obtain operators in D which annihilate f^α. However, these elements may not suffice to generate $\operatorname{Ann}(f^\alpha)$ for special choices of α. This can be seen already for Example 5.3.8, where $n = 1$ and $f = x$. Replacing s by α in $\operatorname{Ann}(x^s)$ gives the D-ideal generated by $x\partial_x - \alpha$. This is the annihilator of x^α if and only if $\alpha \notin \mathbf{N}$. If α is a non-negative integer, then $\operatorname{Ann}(x^\alpha) = D \cdot \{\partial_x^{\alpha+1}, x\partial_x - \alpha\}$. To understand this phenomenon, we need to introduce the global b-function of a polynomial.

Definition 5.3.10. Let $w = (1, 0, \ldots, 0) \in \mathbf{R}^{n+1}$ be the weight vector which gives weight 1 to ∂_t and weight 0 to each ∂_i. Let $B(s)$ be the b-function for the $D\langle t, \partial_t\rangle$-ideal I_f with respect to w, as defined in Section 5.1. We call $b(s) := B(-s - 1)$ the *global b-function* of the polynomial $f \in \mathbf{k}[x_1, \ldots, x_n]$. If there is no risk of confusion, we call it the b-function of the polynomial f.

Lemma 5.3.11. *The global b-function $b(s)$ of f is the unique (up to scaling) polynomial of minimal degree in $\mathbf{k}[s]$ which satisfies an identity*

$$
L \bullet f^{s+1} \;=\; b(s) \cdot f^s \qquad \text{for some } L \in D[s]. \tag{5.24}
$$

Proof. Suppose (5.24) holds. Then $Lf - b(s)$ is an element of $\operatorname{Ann}(f^s)$, and hence its Mellin transform $Lf - b(-t\partial_t - 1)$ is an element of I_f. Also $L(t - f)$ lies in I_f and hence so does $Lf - b(-t\partial_t - 1) + L(t - f) = -b(-t\partial_t - 1) + Lt$. We conclude that $b(-t\partial_t - 1) \in \operatorname{in}_{(-w,w)}(I_f) \cap \mathbf{k}[\theta_t]$, which means that $b(s)$ is a multiple of the global b-function.

Conversely, if $b(s)$ is a multiple of the global b-function then $b(-t\partial_t - 1) \in \operatorname{in}_{(-w,w)}(I_f) \cap \mathbf{k}[\theta_t]$. There exists $R \in D\langle t, t\partial_t\rangle$ such that $b(-t\partial_t - 1) + Rt \in I_f$. Modulo $D \cdot \{t - f\}$, replace t by f in Rt and apply the inverse of the Mellin transform (5.21) to get an operator of the form $b(s) - Lf$ in $\operatorname{Ann}(f^s)$. This proves the assertion. $\qquad\square$

Remark 5.3.12. To compute the global b-function of a polynomial f we simply apply Algorithm 5.1.6 to I_f. There are several other methods for computing b-functions of special classes of polynomials (see, e.g., the references of [77]), but this algorithm works for any polynomial f. It would be an interesting project to improve its performance and implement a nice system in which several methods for computing b-functions are installed.

It is known that $s + 1$ divides the global b-function, so the number α_0 in the following theorem is always a well-defined negative integer.

Theorem 5.3.13. *Let α_0 be the minimal integral root of the global b-function $b(s)$ of a polynomial $f \in \mathbf{k}[x]$ and suppose that $\alpha \notin \alpha_0 + 1 + \mathbf{N}$. Then $\mathrm{Ann}(f^\alpha) \subset D$ is obtained from $\mathrm{Ann}(f^s) \subset D[s]$ by replacing s with α.*

Proof. Let $P \in D$ be any operator which annihilates f^α. If we apply that operator to f^s, with s an indeterminate, then the result equals

$$P \bullet f^s \;=\; \sum_{i=0}^{r} (s - \alpha) \cdot u_i(x, s) \cdot f^{s-i},$$

where r is the order of P and each u_i is a polynomial in x and in s. Using the identity (5.24) we can write

$$f^{s-i} \;=\; \frac{L^{(i)}}{b(s-i)b(s-i+1)\cdots b(s-1)} \bullet f^s \qquad \text{for some } L^{(i)} \in D[s].$$

The following operator in $D[s]$ annihilates f^s by construction:

$$b(s-i)b(s-i+1)\cdots b(s-1) \cdot P \;-\; \sum_{i=0}^{r} (s-\alpha) \cdot u_i(x,s) \cdot L^{(i)}. \qquad (5.25)$$

If we substitute α for s into (5.25), then the right hand sum vanishes but $b(\alpha - i)b(\alpha - i + 1)\cdots b(\alpha - 1)$ is a non-zero element of $\mathbf{k}$, since $\alpha \notin \alpha_0 + 1 + \mathbf{N}$. Dividing (5.25) by that scalar, we obtain an element in $\mathrm{Ann}(f^s)$ which specializes to P under the substitution $s \mapsto \alpha$. $\qquad \square$

Theorem 5.3.13 explains what we had observed for $n = 1$ and $f = x$, since $b(s) = s + 1$ and $\alpha_0 = -1$ in this case. Here is another example which shows that we might have $\alpha_0 = -1$ even if the global b-function is complicated.

Example 5.3.14. Let $n = 2$ and $f = x^3 - y^2$. We apply Algorithm 5.3.6 to compute the annihilator of f^s. Here the input for step 2 equals

$$ut - x^3 + y^2, \; 3vx^2\partial_t + \partial_x, \; -2vy\partial_t + \partial_y, \; uv - 1.$$

Eliminating u and v and replacing $t\partial_t$ by $-s - 1$, we obtain

$$3x^2\partial_y + 2y\partial_x, \quad 2x\partial_x + 3y\partial_y - 6s. \qquad (5.26)$$

These are the generators of the annihilating ideal $\mathrm{Ann}(f^s)$. Using the algorithm given in Section 5.1, we find that the global b-function for $f = x^3 - y^2$ equals $b(s) = (6s + 5)(s + 1)(6s + 7)$. Thus $\alpha_0 = -1$, and the annihilator of f^α is gotten from (5.26) by setting $s = \alpha$ unless α is a non-negative integer.

To deal with the exceptional case $\alpha \in \alpha_0 + 1 + \mathbf{N}$ left open by Theorem 5.3.13, we need to introduce one more algorithmic tool, namely, the computation of *syzygies*. A syzygy is a left linear relation among a collection of operators in the Weyl algebra. Fix r operators $p_1, \ldots, p_r \in D$. Then the following subset of the free module D^r is a left D-submodule:

$$\mathrm{Syz}(p_1, \ldots, p_r) \;=\; \{\, (c_1, \ldots, c_r) \in D^r : c_1 p_1 + c_2 p_2 + \cdots + c_r p_r = 0 \,\}.$$

We compute generators for this D-submodule by adapting the well-known algorithm for commutative polynomial rings (see [1, §3.7], [12, §6.1], [28, §5.3]) to the Weyl algebra D. The details of the D-version are described in [80]. The following algorithm is due to [81]. Steps 1 and 2 are combined in practice, but we separate them here for the sake of exposition.

Algorithm 5.3.15 (Computing $\mathrm{Ann}(f^\alpha)$)
Input: A polynomial f in $\mathbf{k}[x_1, \ldots, x_n]$ and a scalar $\alpha \in \mathbf{k}$.
Output: Generators of the left D-ideal $\mathrm{Ann}(f^\alpha)$.

1. Compute a generating set $\{g_1(s), \ldots, g_r(s)\}$ of $\mathrm{Ann}(f^s) \subset D[s]$, using Algorithm 5.3.6.
2. Compute the global b-function $b(s)$ of the polynomial f, using Algorithm 5.1.6 applied to the $D\langle t, \partial_t\rangle$-ideal I_f in (5.20).
3. Let α_0 be the smallest integer root of $b(s)$ and set $d = \alpha - \alpha_0$.
4. Output $\{g_1(\alpha), \ldots, g_r(\alpha)\}$. If d is not a positive integer then STOP.
5. If d is a positive integer, then compute generators for the syzygy module

$$\mathrm{Syz}\big(f^d, g_1(\alpha_0), \ldots, g_r(\alpha_0)\big) \;\subset\; D^{r+1}.$$

6. For each generator $(c_0, c_1, \ldots, c_r)$ of this syzygy module, output the first coordinate c_0, which is an element in the Weyl algebra D.

The correctness of this algorithm follows from the observation that an operator $c_0 \in D$ annihilates $f^\alpha = f^d \cdot f^{\alpha_0}$ if and only if the operator $c_0 f^d$ annihilates f^{α_0}. In light of Theorem 5.3.13 applied to $\alpha = \alpha_0$, the operator $c_0 f^d$ annihilates f^{α_0} if and only if there exist $c_1, \ldots, c_r \in D$ such that

$$c_0 f^d + c_1 g_1(\alpha_0) + \cdots + c_r g_r(\alpha_0) \;=\; 0.$$

Example 5.3.16. If $n = 1$, $f = x$ and α a non-negative integer, then $\alpha_0 = -1$ and $d = \alpha + 1$. Algorithm 5.3.15 outputs $x\partial_x - \alpha$ in step 4 and it outputs $\partial^{\alpha+1}$ in step 6. The second generator is found in step 5 by means of the syzygy

$$c_0 x^d + c_1(x\partial + 1) = 0 \qquad \text{where} \quad c_0 = \partial^d, \;\; c_1 = \prod_{j=2}^{d}(x\partial + j).$$

Remark 5.3.17. Computing the annihilator of a product of powers of polynomials $\prod_{j=1}^{\ell} f_j^{\alpha_j}$ is analogous to the method in this section. See [79].

The global b-function is the Weyl algebra analogue to the local b-function of a hypersurface singularity $\{f = 0\}$ where f is a polynomial localized at the origin. The global b-function is always a multiple of the local b-function, and the two may be equal. The local b-function of a function germ is invariant under analytic changes of coordinates. As a consequence of this invariance, the b-function provides an invariant in singularity theory. One remarkable property of the local b-function is that all roots are negative rational numbers. This result is due to Kashiwara [58]. Malgrange proved that the local b-function is a generating function of the eigenvalues of a monodromy matrix for an isolated singularity [67]. Many authors call $b(s)$ the *Bernstein-Sato polynomial* of f. We close this section with a discussion of a non-trivial example of such a singularity, namely, the cone over an elliptic curve.

Example 5.3.18. Let $n = 3$ and $f = x_1^3 + x_2^3 + x_3^3$. This is a first example such that $\alpha_0 \neq -1$. First we apply Algorithm 5.3.6 to find

$$\text{Ann}(f^s) = D[s]\{x_1\partial_1 + x_2\partial_2 + x_3\partial_3 - 3s, x_1^2\partial_2 - x_2^2\partial_1, x_1^2\partial_3 - x_3^2\partial_1, x_2^2\partial_3 - x_3^2\partial_2\}.$$

We next use Algorithm 5.1.6 to see that the global b-function of f equals

$$(s + 1)(s + 2)(3s + 4)(3s + 5).$$

Here the global b-function coincides with the local one at $x_1 = x_2 = x_3 = 0$. One can use Oaku's algorithm to stratify $\mathbf{C}^n$ so that the local b-functions are the same on each stratum and to compute the local b-functions [78].

We now compute $\text{Ann}(f^\alpha)$ for the special value $\alpha = -1$. It can be done by computing the syzygies among f and the annihilating ideal of f^{-2}. It is generated by $\text{Ann}(f^s)|_{s=-1}$ and

$$x_2 x_3 \partial_1^2 \partial_2 \partial_3 + x_1 x_3 \partial_2^3 \partial_3 + x_1 x_2 \partial_2 \partial_3^3 + x_2 \partial_1^2 \partial_2 + x_1 \partial_2^3 + x_3 \partial_1^2 \partial_3 + x_1 \partial_3^3 + \partial_1^2.$$

To compute the annihilating ideals for all the special values, we can use the following method. Put $I(\alpha) = \text{Ann}(f^\alpha)$. From the definition of the b-function, there exists an operator $L(s)$ such that $Lf^{s+1} = b(s)f^s$. If $b(\alpha) \neq 0$, then the following gives a left D-isomorphism between Df^α and $Df^{\alpha+1}$:

$$D/I(\alpha) \ni p \mapsto p\frac{L}{b(\alpha)} \in D/I(\alpha + 1),$$

$$D/I(\alpha + 1) \ni q \mapsto qf \in D/I(\alpha).$$

Therefore, $I(\alpha + 1)$ is generated by $\frac{L}{b(\alpha)} \cdot I(\alpha)$. In our example, if we compute generators of $I(0)$, $I(-1/3)$, and $I(-2/3)$ by Algorithm 5.3.15, then the annihilating ideals for all other special values can be expressed in terms of L.

We shall revisit this example in Section 5.5 on integration of D-modules. In particular, we shall compute the third de Rham cohomology group of the complement $\mathbf{C}^3\backslash\{f = 0\}$ of our Fermat surface using only Gröbner bases.

5.4 Hypergeometric Integrals

In this section we introduce hypergeometric integrals, and we show that "generic" hypergeometric integrals are solutions of GKZ systems. Definite integrals are regarded as pairings of homology groups and cohomology groups. This point of view, which is the *de Rham theory*, is important both in theory and in algorithms. An introduction to the de Rham theory is found in the book by Bott and Tu [18]. In what follows we review the relevant basics.

Let T be an m-dimensional complex manifold. It is regarded as a $2m$-dimensional real smooth manifold. Let σ_i be a smooth map from the p-dimensional simplex to T. A finite sum $C = \sum c_i \sigma_i$, $c_i \in \mathbf{C}$ is called a *p-chain*. This p-chain is called a *p-cycle* if $\partial C = \sum c_i \partial \sigma_i = 0$. Here, ∂ is the boundary operator. The homology group $H_p(T, \mathbf{C})$ is the vector space of p-cycles modulo the image of $(p+1)$-chains by the boundary operator. The *de Rham cohomology group* $H^p(T, \mathbf{C})$ is the following quotient space:

$$\frac{\mathrm{Ker}\,(d : \{\text{smooth } p\text{-forms on } T\} \to \{\text{smooth } (p+1)\text{-forms on } T\})}{\mathrm{Im}\,(d : \{\text{smooth } (p-1)\text{-forms on } T\} \to \{\text{smooth } p\text{-forms on } T\})}$$

$$= \frac{\{\ \mathbf{C}\text{-valued smooth } p\text{-forms } \omega \text{ such that } d\omega = 0\}}{d\{\ \mathbf{C}\text{-valued smooth } (p-1)\text{-forms}\}}.$$

An element of the numerator is called a *p-cocycle*.

The de Rham cohomology group $H^p(T, \mathbf{C})$ is a finite-dimensional complex vector space, and there is an isomorphism

$$H_p(T, \mathbf{C}) \quad \simeq \quad H^p(T, \mathbf{C})$$

by Poincaré-Lefschetz duality (see, e.g., [18, p.87, p.183]). Stokes' Theorem states that, for any smooth $(p+1)$-chain C and any smooth p-form ω on T,

$$\int_{\partial C} \omega = \int_C d\omega.$$

For a p-cycle C and a p-cocycle ω, we define the integral

$$\langle C, \omega \rangle := \int_C \omega.$$

The value of this integral is a complex number which depends only on the homology class of C and the cohomology class of ω, by virtue of Stokes' Theorem. Therefore the above integral represents a $\mathbf{C}$-bilinear form

$$\langle\ ,\ \rangle : H_p(T, \mathbf{C}) \times H^p(T, \mathbf{C}) \longrightarrow \mathbf{C}.$$

Example 5.4.1. This is an example from elementary complex analysis. Let $T = \mathbf{C}^* = \mathbf{C}\backslash\{0\}$, regarded as a 2-dimensional smooth real manifold with coordinates (x, y). Let C be a smooth closed path in T and put

$$\omega = \frac{dz}{z} = \frac{dx + \sqrt{-1}dy}{x + \sqrt{-1}y}, \qquad (z = x + \sqrt{-1}y)$$
$$= \frac{xdx + ydy}{x^2 + y^2} + \sqrt{-1}\frac{xdy - ydx}{x^2 + y^2}.$$

Then, $\int_C \omega = 2\pi\sqrt{-1}q$ where q is the *winding number* of the path C around the origin. The path C is an element of $H_1(T, \mathbf{Z})$ and ω is an element of $H^1(T, \mathbf{C})$. The value $2\pi q$ depends only on the homology class of C and the cohomology class of ω. By allowing complex linear combinations of paths to play the role of C, we extend the above integral to a pairing between $H_1(T, \mathbf{C})$ and $H^1(T, \mathbf{C})$. Thus, the definite integral $\int_C \omega$ defining the winding number should be regarded as pairings between 1-cycles and 1-cocycles in $T = \mathbf{C}^*$.

We call the integral of the product of powers of polynomials with respect to a specified set of variables a *hypergeometric integral*. The integral of the product of the exponential function with a rational function argument and powers of polynomials with respect to a specified set of variables will be called a *confluent hypergeometric integral*. These will appear in Conjecture 5.4.4.

Let us first discuss generic hypergeometric integrals. Let $A = (a_{ij})$ be an integer $d \times n$-matrix of rank d which satisfies the homogeneity condition (3.3). After a unimodular row transformation, which does not change the associated GKZ system $H_A(\beta)$, we may assume that the last row of A is constant:

$$a_{d1} = a_{d2} = \cdots = a_{dn} =: p. \tag{5.27}$$

From now on we always assume (5.27) for some integer $p > 0$, so as to avoid unnecessary complications in manipulating the integrals below.

The j-th column vector of A is denoted by $a_j \in \mathbf{Z}^d$. We consider the generic sparse polynomial with support $A = (a_1, \ldots, a_n)$, that is,

$$f(x, t) = \sum_{i=1}^{n} x_i t^{a_i} = \sum_{i=1}^{n} x_i t_1^{a_{1i}} \cdots t_d^{a_{di}}.$$

The *GKZ hypergeometric integral* is

$$\Phi(\gamma; x) := \int_C f(x, t)^\alpha t^\gamma dt_1 \cdots dt_m$$

where $\gamma_1, \ldots, \gamma_m$ are complex numbers and we put $m = d - 1$, $p\alpha = -\gamma_d$. The integrand does not depend on t_d by (5.27) and the condition $p\alpha = -\gamma_d$.

What is the cycle C in the integral? We assume that α and $\gamma_1, \ldots, \gamma_m$ are negative integers to explain the meaning of C within the de Rham theory. The case of complex α, γ_i requires the *twisted de Rham theory* and will be discussed in the end of this section. Fix a generic $x \in \mathbf{C}^n$ and abbreviate

$$T_x := \{ t \in (\mathbf{C}^*)^m \mid f(t, x) \neq 0 \}.$$

Let C be a m-cycle in $H_m(T_x, \mathbf{C})$. The differential form $f(x,t)^\alpha t^\gamma dt$ is a closed smooth m-form on T_x. Since the support of C is compact, the cycle C belongs to $H_m(T_{x'}, \mathbf{C})$ when x' is sufficiently close to x. Therefore, the integral $\Phi(\gamma; x)$ is a holomorphic function with respect to x and the value depends only on the homology class of C and the cohomology class of the integrand. Thus, the hypergeometric integral is a pairing between homology groups and cohomology groups whose value is a function in $x_1, \ldots, x_n$.

Our hypergeometric integral satisfies the GKZ hypergeometric system.

Theorem 5.4.2. *Assume (5.27), let $\alpha, \gamma_1, \ldots, \gamma_{d-1}$ be integers, set $\gamma_d := -p\alpha$, $m = d-1$, and $\beta = (-\gamma_1 - 1, \ldots, -\gamma_{d-1} - 1, -\gamma_d)^T$. Then, the hypergeometric ideal $H_A(\beta)$ annihilates the integral*

$$\Phi(\gamma; x) \;=\; \int_C \left(\sum_{i=1}^n x_i t^{a_i} \right)^\alpha t^\gamma dt_1 \cdots dt_m \qquad (5.28)$$

Proof. We first claim that if $Au = Av$, $u, v \in \mathbf{N}^n$, then

$$(\partial^u - \partial^v) \bullet \left(\sum_{i=1}^n x_i t^{a_i} \right)^\alpha = 0.$$

In fact, $\quad \partial^u \bullet \left(\displaystyle\sum_{i=1}^n x_i t^{a_i} \right)^\alpha = [\alpha]_{\sum u_i} \cdot \prod_{i=1}^n t^{u_i a_i} \cdot \left(\sum_{i=1}^n x_i t^{a_i} \right)^{\alpha - \sum u_i}$

$$= [\alpha]_{\sum u_i} \cdot t^{Au} \cdot \left(\sum_{i=1}^n x_i t^{a_i} \right)^{\alpha - \sum u_i}.$$

The condition $a_{di} = p$ implies $\sum u_i = \sum v_i$. Hence, $\partial^u \bullet (\sum_{i=1}^n x_i t^{a_i})^\alpha = \partial^v \bullet (\sum_{i=1}^n x_i t^{a_i})^\alpha$. Since $\partial^u - \partial^v$ commutes with the integral sign, we have $(\partial^u - \partial^v) \bullet \Phi(\gamma; x) = 0$.

We next prove that each row of $A\theta - \beta$ annihilates the function $\Phi(\gamma; x)$. By putting $\omega = fg\, dt_1 \wedge \cdots \wedge dt_{k-1} \wedge dt_{k+1} \wedge \cdots \wedge dt_m$ in Stokes' Theorem,

$$\int_C \frac{\partial f}{\partial t_k} g\, dt_1 \cdots dt_m + \int_C f \frac{\partial g}{\partial t_k} dt_1 \cdots dt_m \;=\; 0$$

holds for any m-cycle and smooth functions f and g. Assume $k \leq m = d-1$. Apply the operator $\sum_{j=1}^n a_{kj}\theta_j$ to the function $\Phi(\gamma; x)$. The result equals

$$\int_C \left(\sum_{j=1}^n a_{kj}\theta_j \right) \bullet \left(\sum_{i=1}^n x_i t^{a_i} \right)^\alpha t^\gamma\, dt$$

$$= \int_C \alpha \left(\sum_{j=1}^n a_{kj} x_j t^{a_j} \right) \left(\sum_{i=1}^n x_i t^{a_i} \right)^{\alpha-1} t^\gamma\, dt$$

$$= \int_C t_k \frac{\partial \left(\sum_{i=1}^n x_i t^{a_i}\right)^\alpha}{\partial t_k} t^\gamma\, dt$$

$$= -\int_C \left(\sum_{i=1}^n x_i t^{a_i}\right)^\alpha (\gamma_k + 1) t^\gamma\, dt \qquad \text{(by Stokes' Theorem)}$$

$$= -(\gamma_k + 1)\Phi(\gamma; x).$$

In case of $k = d$, we have

$$\frac{\partial}{\partial t_k}\left(\left(\sum_{i=1}^n x_i t^{a_i}\right)^\alpha t^\gamma\right) = 0.$$

This implies that the image of $\Phi(\gamma; x)$ under the operator $\sum_{j=1}^n a_{dj}\theta_j$ equals

$$\int_C t_k \frac{\partial \left(\sum_{i=1}^n x_i t^{a_i}\right)^\alpha}{\partial t_k} t^\gamma dt \;\; = \;\; -\int_C \gamma_k \left(\sum_{i=1}^n x_i t^{a_i}\right)^\alpha t^\gamma dt \;\; = \;\; -\gamma_k \Phi(\gamma; x).$$

$\square$

Example 5.4.3. Consider the hypergeometric integral

$$\Phi(x) \;\; = \;\; \int_C \frac{1}{x_1 t_3^{-1} + x_2 t_1 t_2^{-1} t_3^{-1} + x_3 t_3 + x_4 t_2 t_3 + x_5 t_1^{-1} + x_6} \frac{dt_1\, dt_2\, dt_3}{t_1 t_2 t_3}.$$

To apply Theorem 5.4.2, we first read off the matrix from the denominator:

$$A = \begin{pmatrix} 0 & 1 & 0 & 0 & -1 & 0 \\ 0 & -1 & 0 & 1 & 0 & 0 \\ -1 & -1 & 1 & 1 & 0 & 0 \\ 1 & 1 & 1 & 1 & 1 & 1 \end{pmatrix}.$$

The GKZ system $H_A(\beta)$ which annihilates $\Phi(x)$ consists of the operators

$$\partial_1\partial_3 - \partial_6^2, \;\; \partial_2\partial_4\partial_5 - \partial_1\partial_3\partial_6, \;\; x_2\partial_2 - x_5\partial_5, \;\; -x_2\partial_2 + x_4\partial_4,$$
$$-x_1\partial_1 - x_2\partial_2 + x_3\partial_3 + x_4\partial_4, x_1\partial_1 + x_2\partial_2 + x_3\partial_3 + x_4\partial_4 + x_5\partial_5 + x_6\partial_6 + 1.$$

What is the holonomic rank of this system ? How does it relate to $\mathrm{conv}(A)$?

For A that does not satisfy the homogeneity condition (3.3) we propose an integral representation whose integrand contains the exponential function.

Conjecture 5.4.4. For "suitable" d-cycles C and generic γ, the integrals

$$\Phi(\gamma; x) \;\; = \;\; \int_C \exp\left(\sum_{i=1}^n x_i t^{a_i}\right) t^\gamma dt_1 \cdots dt_d$$

span the classical solution space of the GKZ system $H_A(\beta)$ with $\beta = (-\gamma_1 - 1, \ldots, -\gamma_d - 1)^T$.

Example 5.4.5. Put $A = (1,3)$, $\beta = (-1)$. The GKZ system is generated by

$$\theta_1 + 3\theta_2 + 1, \ \partial_1^3 - \partial_2 \tag{5.29}$$

and the holonomic rank is 3. Define three half lines in the complex plane,

$$C_1 = [0, +\infty), \ C_2 = [0, +\infty\omega), \ C_3 = [0, +\infty\omega^2),$$

where $\omega^2 + \omega + 1 = 0$. Assume $\operatorname{Re} x_2 < 0$. The integrand $e^{x_1 t + x_2 t^3}$ is rapidly decreasing when t goes to the infinity along the half lines C_i. The integrals

$$\int_{C_i^{-1} C_{i+1}} e^{x_1 t + x_2 t^3} dt, \quad \operatorname{Re} x_2 < 0, \ (i = 1, 2)$$

give two linearly independent solutions to (5.29). The two solutions do not span the solution space of (5.29) but they span the solution space of

$$\theta_1 + 3\theta_2 + 1, \ \partial_1^3 - \partial_2, \ 3x_2\partial_1^2 + x_1.$$

When $x_2 = -1/3$, $x_1 = x$, the integral is nothing but the *Airy integral*. Similarly, the 1×2-matrix $A = (-1, 1)$ corresponds to the *Bessel function*.

Hypergeometric integrals arising in practice often involve products of polynomials. For instance, in Section 1.5 we examined products of linear forms. Such integrals can be described combinatorially as follows. Let

$$A_1 = (a_1, \ldots, a_{n_1}), \ \ldots, \ A_k = (a_{n_{k-1}+1}, \ldots, a_{n_k}), \quad a_i \in \mathbf{Z}^m.$$

To each matrix A_j we associate a generic polynomial with that support:

$$f_j(x, t) = \sum_{i=n_{j-1}+1}^{n_j} x_i t^{a_i}.$$

For negative integers $\alpha_1, \ldots, \alpha_k$ and $\gamma_1, \ldots, \gamma_m$, we consider the integral

$$\Phi(\alpha, \gamma; x) = \int_C \prod_{j=1}^k f_j(x, t)^{\alpha_j} t^\gamma dt_1 \cdots dt_m.$$

Here, C is a cycle in $H_m((\mathbf{C}^*)^m \setminus \{\prod f_j = 0\}, \mathbf{C})$. The function $\Phi(\alpha, \gamma; x)$ is annihilated by the hypergeometric ideal $H_A(\beta)$ where

$$A = \begin{pmatrix} 1 & \cdots & 1 & 0 & \cdots & 0 & & 0 & \cdots & 0 \\ 0 & \cdots & 0 & 1 & \cdots & 1 & & 0 & \cdots & 0 \\ 0 & \cdots & 0 & 0 & \cdots & 0 & & 0 & \cdots & 0 \\ \cdot & \cdots & \cdot & \cdot & \cdots & \cdot & & \cdot & \cdots & \cdot \\ \cdot & \cdots & \cdot & \cdot & \cdots & \cdot & & \cdot & \cdots & \cdot \\ \cdot & \cdots & \cdot & \cdot & \cdots & \cdot & & \cdot & \cdots & \cdot \\ 0 & \cdots & 0 & 0 & \cdots & 0 & & 1 & \cdots & 1 \\ a_1 & \cdots & a_{n_1} & a_{n_1+1} & \cdots & a_{n_2} & & a_{n_{k-1}+1} & \cdots & a_{n_k} \end{pmatrix}$$

and $\beta = (\alpha_1, \ldots, \alpha_k, -\gamma_1 - 1, \ldots, -\gamma_m - 1)^T$. The proof is analogous to those of Theorem 5.4.2 and Theorem 1.5.2 in Chapter 1.

Example 5.4.6. Recall the classical univariate hypergeometric series

$$_pF_{p-1}(c_1, \ldots, c_p; d_2, \ldots, d_p; z) \quad = \quad \sum_{k=0}^{\infty} \frac{(c_1)_k (c_2)_k \cdots (c_p)_k}{(1)_k (d_2)_k \cdots (d_p)_k} z^k.$$

Here, $c_1, \ldots, c_p, d_2, \ldots, d_p$ are complex constants with $d_i \notin \{-1, -2, -3, \ldots\}$. We shall present a matrix A and an integral representation for the classical hypergeometric function $_pF_{p-1}(z)$. To avoid messy expressions with many suffixes, the result is stated only for the typical case $p = 4$. Put

$$A \quad = \quad \begin{pmatrix} 1 & 1 & 0 & 0 & 0 & 0 & 0 & 0 \\ 0 & 0 & 1 & 1 & 0 & 0 & 0 & 0 \\ 0 & 0 & 0 & 0 & 1 & 1 & 0 & 0 \\ 0 & 0 & 0 & 0 & 0 & 0 & 1 & 1 \\ 1 & 0 & 0 & 1 & 0 & 0 & 0 & 0 \\ 0 & 0 & 1 & 0 & 0 & 1 & 0 & 0 \\ 0 & 0 & 0 & 0 & 1 & 0 & 0 & 1 \end{pmatrix}$$

and
$$\beta = (d_1 - c_1 - 1, d_2 - c_2 - 1, d_3 - c_3 - 1, d_4 - c_4 - 1,$$
$$-(c_2 - d_1) - 1, -(c_3 - d_2) - 1, -(c_4 - d_3) - 1)^T, \quad d_1 = 1.$$

The toric ideal I_A is the principal ideal generated by

$$\partial_1 \partial_3 \partial_5 \partial_7 - \partial_2 \partial_4 \partial_6 \partial_8,$$

This shows that the GKZ system $H_A(\beta)$ is holonomic of rank 4. For generic parameters c_i, d_j, we can construct the four canonical series solutions arising from the underlined leading monomial as follows. Solve the indicial equations

$$s_1 s_3 s_5 s_7 = 0, \quad As = \beta$$

to get the four exponents. One of the exponents is

$$s = (0, -c_1, d_2 - 1, -c_2, d_3 - 1, -c_3, d_4 - 1, -c_4),$$

giving the starting monomial

$$x^s \quad = \quad \frac{x_3^{d_2 - 1} x_5^{d_3 - 1} x_7^{d_4 - 1}}{x_2^{c_1} x_4^{c_2} x_6^{c_3} x_8^{c_4}}.$$

The resulting series is

$$x^s \cdot {}_4F_3\left(c_1, c_2, c_3, c_4; d_2, d_3, d_4; \frac{x_1 x_3 x_5 x_7}{x_2 x_4 x_6 x_8}\right). \tag{5.30}$$

The GKZ hypergeometric integral corresponding to A equals

$$\int_C t_1^{c_2-d_1} t_2^{c_3-d_2} t_3^{c_4-d_3} \cdot (x_1 t_1 - x_2 t_0)^{d_1-c_1-1} (x_3 t_2 - x_4 t_1)^{d_2-c_2-1}$$

$$\cdot (x_5 t_3 - x_6 t_2)^{d_3-c_3-1} (x_7 t_4 - x_8 t_3)^{d_4-c_4-1} dt_1 dt_2 dt_3.$$

Here, we put $t_0 = t_4 = 1$ and $d_1 = 1$. This integral equals a constant multiple of (5.30) for a suitable C. Such integrals for general p are discussed in [85].

Example 5.4.7. (*Selberg type integral of type $m = 0$, $N = 4$ in Kaneko [57]*) The hypergeometric integral

$$\int_C \left(x_1 + x_2 \frac{t_1}{t_2} \right)^{\alpha_1} \left(x_3 + x_4 \frac{t_1}{t_3} \right)^{\alpha_2} \left(x_5 + x_6 \frac{t_1}{t_4} \right)^{\alpha_3}$$

$$\cdot \left(x_7 + x_8 \frac{t_2}{t_3} \right)^{\alpha_4} \left(x_9 + x_{10} \frac{t_2}{t_4} \right)^{\alpha_5} \left(x_{11} + x_{12} \frac{t_3}{t_4} \right)^{\alpha_6} t_1^{\gamma_1} \cdots t_4^{\gamma_4} dt_1 \cdots dt_4$$

is a solution of $H_A(\alpha_1, \ldots, \alpha_6, -\gamma_1 - 1, \ldots, -\gamma_4 - 1)$ for

$$A \;=\; \begin{pmatrix} 1 & 1 & 0 & 0 & 0 & 0 & 0 & 0 & 0 & 0 & 0 & 0 \\ 0 & 0 & 1 & 1 & 0 & 0 & 0 & 0 & 0 & 0 & 0 & 0 \\ 0 & 0 & 0 & 0 & 1 & 1 & 0 & 0 & 0 & 0 & 0 & 0 \\ 0 & 0 & 0 & 0 & 0 & 0 & 1 & 1 & 0 & 0 & 0 & 0 \\ 0 & 0 & 0 & 0 & 0 & 0 & 0 & 0 & 1 & 1 & 0 & 0 \\ 0 & 0 & 0 & 0 & 0 & 0 & 0 & 0 & 0 & 0 & 1 & 1 \\ 0 & 1 & 0 & 1 & 0 & 1 & 0 & 0 & 0 & 0 & 0 & 0 \\ 0 & -1 & 0 & 0 & 0 & 0 & 0 & 1 & 0 & 1 & 0 & 0 \\ 0 & 0 & 0 & -1 & 0 & 0 & 0 & -1 & 0 & 0 & 0 & 1 \\ 0 & 0 & 0 & 0 & 0 & -1 & 0 & 0 & 0 & -1 & 0 & -1 \end{pmatrix}.$$

In the remainder of this section we explain the meaning of the hypergeometric integral (5.28) when the parameters α, γ_i are not integers but complex numbers. Fix $\alpha = -\gamma_d/p, \gamma_1, \ldots, \gamma_d \in \mathbf{C}$. For each $x \in \mathbf{C}^n$, put

$$T_x \;=\; \{ t \in \mathbf{C}^m \mid f^\alpha t^\gamma = f^\alpha t_1^{\gamma_1} \cdots t_d^{\gamma_d} \text{ is holomorphic at } t \},$$

and let $\mathcal{L}_{x,\gamma}$ be the locally constant sheaf over $\mathbf{C}$ defined by $f^\alpha t^\gamma$. The precise definition of this sheaf is as follows. Define a rational 1-form ω on T_x by

$$\omega \;=\; \frac{d_t(f^\alpha t^\gamma)}{f^\alpha t^\gamma}.$$

Here, d_t denotes the exterior differential with respect to $t_1, \ldots, t_m$. Let U be an open set in T_x as $2m$-dimensional smooth manifold. The operator $\nabla = d_t - \omega \wedge \cdot$ defines a map from holomorphic functions on U to the space of holomorphic 1-forms on U. When U is simply connected, the kernel $\operatorname{Ker} \nabla(U)$ is a one-dimensional $\mathbf{C}$-vector space spanned by a branch of the multi-valued function $f^\alpha t^\gamma$. The collection of $\operatorname{Ker} \nabla(U)$ together with restriction maps as functions is a presheaf on T_x. The sheaf associated to the presheaf is $\mathcal{L}_{x,\gamma}$.

Fix a generic point $x \in \mathbb{C}^n$. A *chain* C is a finite formal sum of singular simplices σ with coefficients in the space $\Gamma(\sigma, \mathcal{L}_{x,\gamma})$ of global sections:

$$C = \sum_{\text{finite}} c_\sigma \sigma, \quad c_\sigma \in \Gamma(\sigma, \mathcal{L}_{x,\gamma}).$$

The boundary map ∂_ω on chains is defined by

$$\partial_\omega C = \sum c_{\sigma|\partial\sigma} \cdot \partial\sigma.$$

The homology group with respect to these chains and the boundary operator ∂_ω is called the *twisted* homology group and it is denoted by $H_m(T_x, \mathcal{L}_{x,\gamma})$. The twisted (de Rham) cohomology group $H^i(T_x, \mathcal{L}_{x,\gamma})$ is

$$\frac{\text{Ker} \left(\nabla : \{\text{rational } i\text{-forms on } T_x\} \rightarrow \{\text{rational } (i+1)\text{-forms on } T_x\} \right)}{\text{Im} \left(\nabla : \{\text{rational } (i-1)\text{-forms on } T_x\} \rightarrow \{\text{rational } i\text{-forms on } T_x\} \right)}.$$

The comparison theorem of Deligne [30, pp.98–99] states that the de Rham cohomology group defined above agrees with the sheaf-theoretic cohomology group of $\mathcal{L}_{x,\gamma}$ as a sheaf on the $2m$-dimensional smooth manifold T_x.

It is not an easy task to explicitly construct (twisted) cycles in a higher-dimensional space. However, these twisted cycles are well studied in dimension $m = 1$; see, for instance, [103, Chapter 4]. The book of Aomoto and Kita [4] gives an introduction to twisted homology groups, twisted cohomology groups and hypergeometric functions associated to products of linear forms. For a twisted cycle $C = \sum c_\sigma \sigma \in H_m(T_x, \mathcal{L}_{x,\gamma})$, consider the integral

$$\Phi(\gamma; x_1, \ldots, x_n) = \langle C, dt_1 \cdots dt_m \rangle := \sum \int_\sigma c_\sigma dt_1 \cdots dt_m, \qquad (5.31)$$

which is often denoted by the following intuitive expression

$$\int_C f^\alpha t^\gamma dt_1 \cdots dt_m. \qquad (5.32)$$

Let x' be a sufficiently close point to x. All σ will lie in $T_{x'}$. Hence, there is a unique cycle $C' = \sum c'_\sigma \sigma$ such that $\Phi(x') = \langle C', dt_1 \cdots dt_m \rangle$ is obtained by analytic continuation of $\Phi(x)$. Thus (5.32) is a holomorphic function in x.

The following theorem is proved similarly to Theorem 5.4.2 via an analogue of Stokes' Theorem for the twisted homology group $H_m(T_x, \mathcal{L}_{x,\gamma})$ and the twisted cohomology group $H^m(T_x, \mathcal{L}_{x,-\gamma})$ [4, p.30]. Note that $dt_1 \cdots dt_m$ in the integral (5.32) should be regarded as an element of $H^m(T_x, \mathcal{L}_{x,-\gamma})$.

Theorem 5.4.8. [39] *The integral $\Phi(\gamma; x)$ satisfies the GKZ system $H_A(\beta)$, with A and β as in Theorem 5.4.2.*

The converse of Theorem 5.4.8 is not true; GKZ hypergeometric integrals do not always span the solution space of the GKZ system. For example, take A as in Example 2.6.4 and $\beta = (1, 0, 0)^T$. Here the Laurent polynomial f is

$$f(x,t) \quad = \quad x_1 t_1 + x_2 t_1 t_2 + x_3 t_2 + x_4 t_1^{-1} t_2^{-1} + x_5. \tag{5.33}$$

We have $T_x = (\mathbf{C}^*)^2$. Therefore, $H_2(T_x, \mathbf{C}) \simeq \mathbf{C}$. The integral

$$\int_{|t_1|=1} \int_{|t_2|=1} f(x,t) \frac{dt_1 dt_2}{t_1 t_2} \quad = \quad x_5$$

gives a solution of the GKZ system in Example 2.6.4, but there are no homology cycles to express the other three solutions to $H_A(\beta)$.

However, if the parameter vector β is generic then the hypergeometric integrals span the solution space of the GKZ system. The following result is due to Gel'fand, Kapranov and Zelevinsky.

Theorem 5.4.9. [39, Theorem 2.10, 2.11] *If β is generic then the vector space*

$$\{ \langle C, dt_1 \cdots dt_m \rangle \,|\, C \in H_m(T_x, \mathcal{L}_{x,\gamma}) \}$$

agrees with the solution space of the GKZ system $H_A(\beta)$. Moreover, the monodromy representation is irreducible. This implies that the left ideal $R \cdot H_A(\beta)$ is a maximal ideal in $R = \mathbf{C}(x_1, \ldots, x_n)\langle \partial_1, \ldots, \partial_n \rangle$.

It would be desirable to find an elementary proof for the last assertion.

5.5 Computing Integrals

An integral $\int_C f(x,t)^\alpha t^\gamma dt$ satisfies a GKZ system if f is generic. This was shown in the Section 5.4. In this section we allow non-generic polynomials f, and we show that any hypergeometric integral is annihilated by a holonomic D-ideal. An algorithm is given to compute that holonomic D-ideal.

Let $f_1, \ldots, f_p$ be polynomials in $\mathbf{k}[t_1, \ldots, t_m, x_1, \ldots, x_n]$, where $\mathbf{k}$ is a computable subfield in $\mathbf{C}$. We consider the hypergeometric integral

$$F(\alpha; x) \quad = \quad \int_C \prod_{i=1}^{p} f_i(x,t)^{\alpha_i} dt_1 \cdots dt_m,$$

where $\alpha_i \in \mathbf{k}$ and C is an m-cycle in a suitable sense. This integral can be regarded as a pairing of a (twisted) homology group and a (twisted) cohomology group in a similar way to GKZ hypergeometric integrals. The function $F(\alpha; x)$ depends on the homology class of C. Put

$$D \quad = \quad \mathbf{k}\langle t_1, \ldots, t_m, x_1, \ldots, x_n, \partial_{t_1}, \ldots, \partial_{t_m}, \partial_{x_1}, \ldots, \partial_{x_n} \rangle.$$

Theorem 5.5.1. *Let I be a D-ideal that annihilates the function $f_\alpha(x,t) = \prod_{i=1}^{p} f_i(x,t)^{\alpha_i}$. Then, the following left ideal annihilates the function $F(\alpha; x)$:*

$$J \quad = \quad (I + \partial_{t_1} D + \cdots + \partial_{t_m} D) \cap \mathbf{k}\langle x_1, \ldots, x_n, \partial_{x_1}, \ldots, \partial_{x_n} \rangle \tag{5.34}$$

Proof. For simplicity, we assume that the α_i are integers. In case of complex α_i, the proof is similar but uses Stokes' Theorem for twisted homology and cohomology groups. We prove our claim by a direct calculation. For

$$\ell = \ell_0 + \partial_{t_1}\ell_1 + \cdots + \partial_{t_m}\ell_m \in J \qquad \text{with} \quad \ell_0 \in I, \ell_1, \ldots, \ell_m \in D, \quad (5.35)$$

we have

$$\begin{aligned}
\ell \bullet F(\alpha; x) &= \int_C \ell \bullet f_\alpha(x, t)dt \\
&= \int_C \ell_0 \bullet f_\alpha dt + \sum_{i=1}^m \int_C \partial_{t_i}\ell_i \bullet f_\alpha dt \\
&= \sum_{i=1}^m \int_C \frac{\partial(\ell_i \bullet f_\alpha)}{\partial t_i} dt \\
&= 0 \qquad \text{(by Stokes' Theorem and the assumption } \partial C = 0.\text{)}
\end{aligned}$$

We call the left ideal J in $\mathbf{k}\langle x, \partial_x \rangle$ the *integral ideal* of I with respect to t. □

Example 5.5.2. Let us find a linear ordinary differential equation for

$$F(x) = \int_C \frac{dt}{t^2 - x}.$$

The annihilating ideal of $1/(t^2 - x)$ is $I = D \cdot \{2t\partial_x + \partial_t, \theta_t + 2\theta_x + 2\}$. Replacing θ_t by $\partial_t t - 1$ in the second generator, we find that $2\theta_x + 1$ lies in the integral ideal $(I + \partial_t D) \cap \mathbf{k}\langle x, \partial_x \rangle$. Using Algorithm 5.5.4 below we can check that this $\mathbf{k}\langle x, \partial_x \rangle$-ideal is in fact generated by $2\theta_x + 1$. We conclude that the function $F(x)$ satisfies the first-order differential equation

$$2x \cdot \frac{\partial F}{\partial x} = -F.$$

A decomposition of type (5.35) already appeared in the proof of Theorem 1.3.5, but the integration method used there was rather heuristic. In what follows we present an algorithm to compute the integral ideal. We first switch from D-ideals to D-modules and define integral modules. Considering modules is not only necessary to compute an ideal annihilating a definite integral with parameters, but also essential to obtain bases of de Rham cohomology groups. We shall see this at the end of this section.

Let I be a D-ideal. In Section 5.2 we introduced the restriction module of the D-module D/I and we gave an algorithm to compute the restriction module for holonomic I. The integral of the D-module D/I is defined in a similar way. We retain the notation of Section 5.2 for describing our integration algorithm; we put $X = \{(x_1, \ldots, x_n)\} = \mathbf{k}^n$, $Y = \{(x_{m+1}, \ldots, x_n)\} = \mathbf{k}^{n-m}$, $(m \leq n)$. Define the left D_Y and right D_X bimodule $D_{Y \leftarrow X}$ by

$$D_{Y \leftarrow X} := D/(\partial_1 D + \cdots + \partial_m D), \quad D = D_X.$$

The tensor product

$$D_{Y \leftarrow X} \otimes_D D/I \quad \simeq \quad D/(I + \partial_1 D + \cdots + \partial_m D) \qquad (5.36)$$

is called the *integral module* of D/I with respect to $x_1, \ldots, x_m$. Or, we simply call it the *integral* of D/I. It is a left D_Y-module. The *integral ideal* is

$$J \;=\; (I + \partial_1 D + \cdots + \partial_m D) \cap D', \qquad D' = D_Y.$$

The class of (regular) holonomic D-modules is stable under integration.

Theorem 5.5.3.

(1) *If D/I is holonomic, then the integral module (5.36) is a holonomic D'-module, and it can be expressed as $(D')^r/M$ for a suitable left submodule M of D'^r. In particular, when $n = m$, the integral of a holonomic D-module D/I is a finite-dimensional $\mathbf{k}$-vector space.*

(2) *If D/I is regular holonomic, then the integral module (5.36) is a regular holonomic D'-module.*

Proofs of the first statement can be found in the text books [15, Chapter 1], [17], [26, Chapter 18]. A proof of the second assertion on regularity can be found in [17, p.308], [54, p.99]. The proof requires the formalism of derived categories, but the proof is eventually reduced to a question on the growth of functions defined by integrals. This kind of result was first established by Nilsson [73] in the 1960's – that is why we are using the term "Nilsson ring".

An immediate consequence of Theorem 5.5.3 is that if I is (regular) holonomic, then the integral ideal J is also (regular) holonomic. This is because $D'/J = D'/(I + \partial_1 D + \cdots + \partial_m D) \cap D'$ is a submodule of the integral $D/(I + \partial_1 D + \cdots + \partial_m D)$ as a D'-module, and any submodule of a (regular) holonomic D'-module is also (regular) holonomic (see Theorem 2.4.8).

We derive an algorithm for computing the integral of D/I by reducing it to the restriction algorithm in Section 5.2. Consider the algebra endomorphism $\mathcal{F}$ from D to D defined by $x_i \mapsto -\partial_i$ and $\partial_i \mapsto x_i$. This acts on normally ordered elements as follows:

$$\mathcal{F} : \quad \sum a_{\alpha\beta} x^\alpha \partial^\beta$$
$$\mapsto \sum a_{\alpha\beta} (-\partial_1)^{\alpha_1} \cdots (-\partial_m)^{\alpha_m} x_{m+1}^{\alpha_{m+1}} \cdots x_n^{\alpha_n} x_1^{\beta_1} \cdots x_m^{\beta_m} \partial_{m+1}^{\beta_{m+1}} \cdots \partial_n^{\beta_n}.$$

The image $\mathcal{F}[\ell]$ of an operator $\ell \in D$ is called the *Fourier transform* of ℓ with respect to $x_1, \ldots, x_m$. For an element $p = (p_1, \ldots, p_r) \in D^r$, we define $\mathcal{F}[p] = (\mathcal{F}[p_1], \ldots, \mathcal{F}[p_r])$ and $\mathcal{F}[D^r/M]$ means $D^r/\mathcal{F}[M]$. By virtue of the Fourier transform, the integral (5.36) of D/I is equal to

$$\mathcal{F}^{-1}[D/(\mathcal{F}[I] + x_1 D + \cdots + x_m D)].$$

The integral ideal of I is

$$(\mathcal{F}[I] + x_1 D + \cdots + x_m D) \cap D'.$$

Thus, we obtain the following algorithm.

Algorithm 5.5.4 (Computing the integral module and the integral ideal)
Input: Generators $g_1, \ldots, g_k$ of a holonomic D-ideal I and a weight vector $w \in \mathbf{N}^n$ satisfying (5.11).
Output: Generators for a submodule M of $(D')^r$ such that $(D')^r / M$ equals the integral module (5.36), and generators for the integral ideal J in (5.34).

1. Apply Algorithm 5.2.8 to $\{\mathcal{F}[g_1], \ldots, \mathcal{F}[g_k]\}$ with weights w.

2. Let $\mathcal{B}$ and $\left\{\sum_{\partial^\beta \in \mathcal{B}} h_{i\beta} \partial^\beta \mid i = 1, \ldots, \ell\right\}$ be the output of the restriction algorithm. Here, $h_{i\beta} \in D'$ and $\mathcal{B}$ is a finite set of monomials in $\mathbf{k}[\partial_1, \ldots, \partial_m]$. Output $\hat{\mathcal{B}} := \{x^\beta \mid \partial^\beta \in \mathcal{B}\}$, $r = \#(\hat{\mathcal{B}})$, and $M = D' \cdot \left\{\sum_{x^\beta \in \hat{\mathcal{B}}} h_{i\beta}(-1)^{|\beta|} x^\beta \mid i = 1, \ldots, \ell\right\}$. The integral module of D/I is

$$\left(\sum_{x^\beta \in \hat{\mathcal{B}}} D' x^\beta\right) \Bigg/ D' \cdot \left\{\sum_{x^\beta \in \hat{\mathcal{B}}} h_{i\beta}(-1)^{|\beta|} x^\beta \,\middle|\, i = 1, \ldots, \ell\right\}.$$

3. The integral ideal J consists of all elements in M that do not contain any of the variables $x_1, \ldots, x_m$. Using a POT term order for Gröbner bases of submodules in $(D')^r$, eliminate the coordinates x^β with $\beta \neq 0$ from the module M, to get a generating set for J. Output that generating set.

We now return to the discussion of hypergeometric integrals; the variables t_i will become the variables of integration and x_i the parameters, again. We shall explain the meaning of the integral module on the level of functions. Let I be the annihilating ideal of $f_\alpha(x, t)$ and $D_{T \times X}$ and D_X be the Weyl algebras on the (t, x)-space and the x-space respectively. We apply Algorithm 5.5.4 to get the integral $D_{T \times X}/I$ with respect to the variables $t_1, \ldots, t_m$. It outputs a finite set of monomials $\hat{\mathcal{B}}$ in t and relations $\sum_\beta h_{i\beta} t^\beta$, $h_{i\beta} \in D_X$, $(i = 1, \ldots, \ell)$. The integral of $D_{T \times X}/I$ is isomorphic to $\left(\sum_{t^\beta \in \hat{\mathcal{B}}} D_X t^\beta\right)/M$ where M is the D_X-submodule generated by $\{\sum_{t^\beta \in \hat{\mathcal{B}}} h_{i\beta} t^\beta \mid i = 1, \ldots, \ell\}$.

Fix an m-cycle C. For $t^\beta \in \hat{\mathcal{B}}$, we define the function

$$F^\beta(\alpha; x) \;=\; \int_C t^\beta f_\alpha(x, t) dt. \tag{5.37}$$

The integral module gives a (regular holonomic) system of linear differential equations in x which is satisfied by the vector of functions $\left(F^\beta\right)_{t^\beta \in \hat{\mathcal{B}}}$.

Theorem 5.5.5. *For any* $h = \sum_{t^\beta \in \hat{\mathcal{B}}} h_\beta t^\beta \in M$, $h_\beta \in D_X$, *we have*

$$\sum_{t^\beta \in \hat{\mathcal{B}}} h_\beta \bullet F^\beta \;=\; 0.$$

Proof. By definition of the module M, there exist $p_1, \ldots, p_m \in D_{T \times X}$ with

$$\sum_{t^\beta \in \hat{\mathcal{B}}} h_\beta \cdot t^\beta - \sum_{i=1}^m \partial_{t_i} \cdot p_i \in I.$$

From this decomposition, we can prove the theorem by a direct calculation:

$$\sum_\beta h_\beta \bullet \int_C t^\beta f_\alpha dt = \int_C \sum_\beta h_\beta \bullet t^\beta f_\alpha dt = \sum_{i=1}^m \int_C \partial_{t_i} \bullet (p_i f_\alpha) dt = 0.$$

$\square$

Example 5.5.6 (Continuation of Example 5.5.2). A Gröbner basis of the Fourier transform of the annihilating ideal of $1/(t^2 - x)$ with the weight

vector $(-w, w) = \begin{pmatrix} t & x & \partial_t & \partial_x \\ -1, & 0, & 1, & 0 \end{pmatrix}$ is generated by

$$-2\partial_t \partial_x + t, \quad -\theta_t + 2\theta_x + 1, \quad 4x\partial_x^2 + 6\partial_x - t^2.$$

The b-function with respect to w is $s(s-1)$. Hence, $\mathcal{B} = \{1, \partial_t\}$. The relations are the right hand sides of the following operators:

$$-2\partial_t \partial_x + t \to -2\partial_x \partial_t, \quad -\theta_t + 2\theta_x + 1 \to 2\theta_x + 1,$$
$$\partial_t(-\theta_t + 2\theta_x + 1) = -t\partial_t^2 + 2\theta_x \partial_t \to 2\theta_x \partial_t,$$
$$4x\partial_x^2 + 6\partial_x - t^2 \to 4x\partial_x^2 + 6\partial_x,$$
$$\partial_t(4x\partial_x^2 + 6\partial_x - t^2) \to (4x\partial_x^2 + 6\partial_x)\partial_t.$$

Hence, the integral module is $\mathbf{k}\langle x, \partial_x \rangle^2 / M$ where M is generated by $(2\theta_x + 1, 0)$ and $(0, \partial_x)$. The classical solution space is spanned by $(F(x), 0)$ and $(0, \int_C t dt / (t^2 - x))$.

Example 5.5.7. (Automatic derivation of a GKZ hypergeometric system) Consider the generic sparse univariate polynomial

$$f = (x_1 + x_2 t + x_3 t^2 + x_4 t^3) \cdot t^2.$$

The annihilating ideal $I(s)$ of f^s is generated by

$$t\partial_1 - \partial_2, \ t\partial_2 - \partial_3, \ t\partial_3 - \partial_4, \ \partial_2^2 - \partial_1\partial_3, \ \partial_3^2 - \partial_2\partial_4, \ \partial_2\partial_3 - \partial_1\partial_4,$$
$$t\partial_t - 2x_1\partial_1 - 3x_2\partial_2 - 4x_3\partial_3 - 5x_4\partial_4, \ x_1\partial_1 + x_2\partial_2 + x_3\partial_3 + x_4\partial_4 - s.$$

The b-function of f is $(s+1)^2(2s+1)$. Let us specialize s to a generic value, for example, $-1/5$. We integrate $I(-1/5)$ with respect to t, using the weights $w = (1, 0, 0, 0, 0)$. Here ∂_t has weight 1. The b-function of the Fourier transform of $I(-1/5)$ is s^2, and the integral ideal coincides with the GKZ system for

$$A = \begin{pmatrix} 1 & 1 & 1 & 1 \\ 0 & 1 & 2 & 3 \end{pmatrix}, \quad \beta = \left(-\frac{1}{5}, -\frac{3}{5} \right)^T.$$

Example 5.5.8. The conclusion of Example 5.5.7 is often false for special parameters β. For instance, the integral ideal of the annihilating ideal of

$$\frac{1}{(x_1 + x_2 t + x_3 t^2 + x_4 t^3)t}$$

has holonomic rank 3, but it strictly contains the GKZ hypergeometric ideal for A as above and $\beta = (-1, 0)^T$. Both ideals are holonomic of rank 3.

The integral ideal of the annihilating ideal of

$$\frac{1}{\sqrt{(x_1 + x_2 t + x_3 t^2 + x_4 t^3)t}}$$

is of holonomic rank 2. It is generated by the GKZ hypergeometric ideal for A and $\beta = (-1/2, -1/2)^T$ together with the operator $3x_1 \partial_2 + 2x_2 \partial_3 + x_3 \partial_4$.

Example 5.5.9. (The integral module of the Fermat cubic)
Fix the polynomial

$$f(x, t_1, t_2) \quad = \quad x^3 + t_1^3 + t_2^3.$$

The b-function of f is $(s+1)(s+2)(3s+4)(3s+5)$, and $\mathrm{Ann}\,(f^{-2})$ equals

$$D_{T \times X} \cdot \left\{ x \partial_x + t_1 \partial_{t_1} + t_2 \partial_{t_2} + 6,\ t_2^2 \partial_{t_1} - t_1^2 \partial_{t_2},\ t_2^2 \partial_x - x^2 \partial_{t_2},\ t_1^2 \partial_x - x^2 \partial_{t_1} \right\}.$$

We apply the integration algorithm to $D_{T \times X}/\mathrm{Ann}\,(f^{-2})$ with the weight

$$\begin{array}{ccc} \partial_{t_1} & \partial_{t_2} & \partial_x \end{array}$$
vector $w = \begin{pmatrix} 1 & 1 & 0 \end{pmatrix}$. The b-function is $s(s-1)(s-2)(s-4)$. Hence, $\hat{B}$ is the set of monomials in t_1 and t_2 having degree at most 4. In step 2 of Algorithm 5.5.4 we see $r = 10$. The integral module of $D_{T \times X}/\mathrm{Ann}\,(f^{-2})$ with respect to t_1, t_2 is $(D')^{10}/M = (D_X \cdot \hat{B})/M$ where M is generated by

$$t_1^4 - 2t_1 t_2^3,\ 2t_1^3 t_2 - t_2^4,\ t_1^2 t_2^2,\ t_1^3 - t_2^3,\ t_1^2 t_2,\ t_1 t_2^2,$$
$$t_1^2,\ t_2^2,\ t_1 t_2^3 \partial_x - t_1 x^2,\ t_2^4 \partial_x - 2t_2 x^2,\ t_1 t_2 x \partial_x + 2t_1 t_2,$$
$$t_2 x \partial_x + 3t_2,\ t_1 x \partial_x + 3t_1,\ x \partial_x + 4,\ t_2^3 + x^3,\ t_2 x^3,\ t_1 x^3.$$

The holonomic rank is 4. This rank equals the dimension of the cohomology group $H^2(\mathbf{C}^2 \backslash V(t_1^3 + t_2^3 + x^3), \mathbf{C})$ for generic x. See Algorithm 5.5.12 below.

Historically speaking, one of the important motivations behind the development of D-module theory was Deligne's Comparison Theorem for de Rham cohomology groups. The *Riemann-Hilbert correspondence* of Kashiwara and Mebkhout is a far-reaching generalization of the Comparison Theorem. In what follows we present an algorithm, resulting from this correspondence, for calculating the middle-dimensional cohomology of the complement of a given hypersurface. Its main step is the computation of an integral module.

Let $f(t)$ be a non-zero polynomial in $t = (t_1, \ldots, t_m)$ with coefficients in a computable subfield of $\mathbf{C}$, for example, $\mathbf{Q}$. Put $\mathbf{k} = \mathbf{C}$ and $n = 0$, so that $D = \mathbf{C}\langle t_1, \ldots, t_m, \partial_{t_1}, \ldots, \partial_{t_m}\rangle$. Our aim is to compute the m-th cohomology group of $T = \{t \in \mathbf{C}^m \mid f(t) \neq 0\}$, by means of computing the integral module of the annihilating ideal of $f(t)^\alpha$. We assume throughout that

$$\alpha \notin \{ \text{ the roots of the } b\text{-function of } f\} + \{1, 2, 3, \ldots\}. \tag{5.38}$$

It is allowed for α to be a sufficiently small negative integer. Put

$$\omega_i = \frac{\alpha \frac{\partial f}{\partial t_i}}{f} \in \mathbf{C}\left[t, \frac{1}{f}\right] \quad \text{and} \quad \nabla = d_t + \sum_{i=1}^m \omega_i dt_i.$$

Then, $f^{-\alpha}$ belongs to the kernel of ∇. We abbreviate $\mathcal{L}_{-\alpha} = \operatorname{Ker} \nabla$.

Remark 5.5.10. If α is an integer, then the locally constant sheaf $\mathcal{L}_{-\alpha}$ is isomorphic to the constant sheaf $\mathbf{C}$ on T.

Consider the D-module $M = \mathbf{C}\left[t, \frac{1}{f}\right] f^\alpha$ where the action of ∂_{t_i} on M is defined by the product rule from calculus:

$$\partial_{t_i} \bullet (h(t)f^{k+\alpha}) = \left(\frac{\partial h}{\partial t_i} f^k + (k+\alpha)h \frac{\partial f/\partial t_i}{f} f^k\right) f^\alpha, \quad k \in \mathbf{Z}$$

$$t_i \bullet (h(t)f^{k+\alpha}) = t_i h(t) f^{k+\alpha}.$$

Each element of M can be uniquely written as $h(t)f^{k+\alpha}$ where k is an integer and $h(t)$ is a polynomial in t not divisible by f. An algorithm to find a D-ideal I such that $D/I \simeq M$ is given in Algorithm 5.3.6 and Theorem 5.3.13. We can take I such that the isomorphism is given by $1 \mapsto f^\alpha$.

Theorem 5.5.11. *The m-th de Rham cohomology group with respect to ∇,*

$$H^m(T, \mathcal{L}_{-\alpha}) \simeq \frac{\mathbf{C}[t, 1/f]dt_1 \cdots dt_m}{\left\{\sum_{i=1}^m \left(\frac{\partial g_i}{\partial t_i} + \omega_i g_i\right) dt \,\middle|\, (g_1, \ldots, g_m) \in \mathbf{C}\left[t, \frac{1}{f}\right]^m\right\}},$$

is isomorphic to the integral of M,

$$M/(\partial_{t_1} M + \cdots + \partial_{t_m} M) \simeq D/(I + \partial_{t_1} D + \cdots + \partial_{t_m} D).$$

The isomorphism is given by the $\mathbf{C}$-linear map

$$\rho : h(t)f^k dt_1 \cdots dt_m \mapsto h(t)f^k f^\alpha.$$

Proof. We have

$$\rho\left(\nabla \sum g_i(-1)^{i+1} dt_1 \wedge \cdots \wedge dt_{i-1} \wedge dt_{i+1} \wedge \cdots \wedge dt_m\right) \subseteq \partial_{t_1} M + \cdots + \partial_{t_m} M$$

from the definition of the action of ∂_{t_i}, i.e., in the D-module M we have

$$\left(\sum_{i=1}^{m} \frac{\partial g_i}{\partial t_i} + \omega_i g_i \right) f^\alpha = \sum_{i=1}^{m} \partial_{t_i} \cdot (g_i f^\alpha).$$

This shows that ρ is a well-defined **C**-linear map. The map ρ is obviously surjective. It remains to show that the kernel of ρ is equal to 0. Suppose

$$\sum_i \partial_{t_i} M \ni \sum_i \partial_{t_i} (h_i f^{k_i + \alpha}) = \sum_i \left(\frac{\partial h_i}{\partial t_i} f + (k_i + \alpha) h_i \frac{\partial f}{\partial t_i} \right) f^{k_i - 1} f^\alpha.$$

It follows from the uniqueness of the expression of the elements of M that the preimage under ρ of the element above is

$$\left(\sum_i \left(\frac{\partial h_i}{\partial t_i} f + (k_i + \alpha) h_i \frac{\partial f}{\partial t_i} \right) f^{k_i - 1} \right) dt$$

$$= \left(\sum_i \left(\frac{\partial (h_i f^{k_i})}{\partial t_i} + \omega_i h_i f^{k_i} \right) \right) dt$$

$\in$ the denominator of the de Rham cohomology group.

This completes the proof of our claim that ρ is a **C**-isomorphism. $\square$

The following two algorithms are immediate consequences of this theorem.

Algorithm 5.5.12 (Dimension of the middle-dimensional cohomology)
Choose α satisfying the condition (5.38). Then,

$$\dim_\mathbf{C} H^m(T, \mathcal{L}_{-\alpha}) = \dim_\mathbf{C} D/(\partial_{t_1} D + \cdots + \partial_{t_m} D) \otimes_D D/(\mathrm{Ann}\, f^s)_{|s \to \alpha}.$$

The right hand side can be evaluated by Algorithm 5.5.4.

This algorithm appeared in [79]. In that paper it has also been extended to an algorithm for evaluating the dimensions of all other cohomology groups. This is done by constructing a free resolution of $D/(\mathrm{Ann}\,(f^s)_{|s \to \alpha})$ over D. We note that there are several other methods for computing integrals and cohomology groups. However, as far as we know, they require some assumptions on the polynomial f. The algorithm above needs no assumption on f, but it is a future problem to improve it from the efficiency point of view.

Algorithm 5.5.13 (A basis for the m-th de Rham cohomology group)
Let $D/I \simeq \mathbf{C}[t, 1/f] f^\alpha$ and assume that the isomorphism is given by $1 \mapsto f^\alpha$. Apply the integration algorithm to $I \subset \mathbf{C}\langle t_1, \ldots, t_m, \partial_{t_1}, \ldots, \partial_{t_m} \rangle$ with respect to $t_1, \ldots, t_m$. Let $\hat{\mathcal{B}} = \{t^\beta\}$ and $L := \{\sum_{t^\beta \in \hat{\mathcal{B}}} h_{i\beta} t^\beta \mid i = 1, \ldots, \ell\}$ be the output. Find a **C**-basis $\hat{\mathcal{B}}' \subseteq \hat{\mathcal{B}}$ for the vector space $\mathbf{C} \cdot \hat{\mathcal{B}}/\mathbf{C} \cdot L$. Then,

$$\{t^\beta dt \mid t^\beta \in \hat{\mathcal{B}}'\}$$

is a basis of the m-th de Rham cohomology group.

Remark 5.5.14. When α is a negative integer, we have $\mathcal{L}_{-\alpha} \simeq \mathbf{C}$ and hence $H^m(T, \mathcal{L}_{-\alpha}) \simeq H^m(T, \mathbf{C})$. By tracing this isomorphism, we see that

$$\{t^\beta f^\alpha dt \mid t^\beta \in \hat{\mathcal{B}}'\}$$

is a basis of $H^m(T, \mathbf{C})$.

Example 5.5.15. Let us find a basis of the cohomology group $H^1(\mathbf{C} \backslash V(f), \mathbf{C})$ for $f = t(1-t)$. The b-function of f is $s+1$. The annihilating ideal of $1/f$ is generated by $t(1-t)\partial_t - 2t + 1$. We apply the integration algorithm with $w = 1$. The b-function is $s(s-1)$. Hence, $\hat{\mathcal{B}} = \{1, t\}$. There is no relation between 1 and t, and we conclude that $\{\frac{dt}{f}, \frac{t dt}{f}\}$ is a basis for $H^1(\mathbf{C} \backslash V(f), \mathbf{C})$.

Example 5.5.16. This example appears in Dimca's text book on hypersurface singularities [31, Example 2.18 (ii), p.194]. We apply the integration algorithm to $f = x_1^3 + x_2^3 + x_3^3$. The annihilating ideal $I(s)$ of f^s is generated by

$$x_1\partial_1 + x_2\partial_2 + x_3\partial_3 - 3s, \quad x_1^2\partial_2 - x_2^2\partial_1, \quad x_1^2\partial_3 - x_3^2\partial_1, \quad x_2^2\partial_3 - x_3^2\partial_2.$$

The b-function of the polynomial f is

$$(s+1)(s+2)(3s+4)(3s+5).$$

So, we apply the algorithm for integration for $I(-2)$. If we take the weight

$$w = \begin{pmatrix} \partial_1 & \partial_2 & \partial_3 \\ 1 & 2 & 3 \end{pmatrix}$$

for the integration, the integral roots of the b-function become $\{3, 4, 5, 6, 7, 8, 12\}$. The set $\hat{\mathcal{B}}$ consists of 102 monomials $\{x^\beta \mid 0 \le \beta_1 + 2\beta_2 + 3\beta_3 \le 12\}$, but there are 100 linear relations. By eliminating dependent elements in $\hat{\mathcal{B}}$ by 100 relations, we see that

$$H^3(\mathbf{C}^3 \backslash V(f), \mathbf{C}) \simeq \mathbf{C}\frac{x_3^3 dx}{f^2} + \mathbf{C}\frac{x_1 x_2 x_3 dx}{f^2}$$

and $\dim_{\mathbf{C}} H^3(\mathbf{C}^3 \backslash V(f), \mathbf{C}) = \dim_{\mathbf{C}} H_3(\mathbf{C}^3 \backslash V(f), \mathbf{C}) = 2$.

5.6 Asymptotic Expansions of Hypergeometric Integrals

In the first five sections of Chapter 5 we saw how Gröbner bases in the Weyl algebra are used to compute hypergeometric integrals such as

$$\Phi_C(x) = \int_C f(x, t)^\alpha t_1^{\gamma_1} \cdots t_m^{\gamma_m} dt_1 \cdots dt_m.$$

The output of that computation is the D-ideal I of linear partial differential operators with polynomial coefficients which annihilate $\Phi_C(x)$. In this final section we go one step further. We observe that the series expansion methods developed in Chapter 2 can be applied to the D-ideal $I = \mathrm{Ann}(\Phi_C)$. This is possible because I is regular holonomic by Theorem 5.3.1 and Theorem 5.5.3.

Let $w \in \mathbf{R}^n$ be a generic weight vector. Let $\{\, s_1(x),\, s_2(x),\, \ldots,\, s_r(x)\,\}$ be the set of canonical series solutions to I with respect to $(-w, w)$. These series were introduced in Section 2.5, and an algorithm for computing them was given in Section 2.6. The series s_i are elements of the Nilsson ring $N_w(I)$. They have a common domain of convergence $\mathcal{U}$ in $\mathbf{C}^n$, and they form a basis for the holomorphic solutions to I on $\mathcal{U}$. Recall that the domain $\mathcal{U}$ is determined from the open polyhedral cone σ containing w in the small Gröbner fan of I.

Suppose we are given an explicit (twisted) cycle C. Then the integral $\Phi_C(x)$ represents a holomorphic function on $\mathcal{U}$ which is annihilated by I. Therefore, there exist complex numbers $c_1, c_2, \ldots, c_r$ such that

$$\Phi_C(x) \;=\; c_1 s_1(x) + c_2 s_2(x) + \cdots + c_r s_r(x). \qquad (5.39)$$

This is an identity of holomorphic functions on $\mathcal{U}$. Yet, the right hand side of the identity (5.39) is also an element of the Nilsson ring

$$N_w(I) \;=\; \mathbf{C}[[\sigma^* \cap \mathbf{Z}^n]][x^{e^1}, \ldots, x^{e^h}, \log x_1, \ldots, \log x_n].$$

Here $e^1, \ldots, e^h \in \mathbf{C}^n$ are the exponents of I with respect to w, that is, the roots of the indicial ideal $\mathrm{ind}_w(I)$. Note that the Nilsson ring $N_w(I)$ and the indicial ideal $\mathrm{ind}_w(I)$ remain the same if w is replaced by another vector in the interior of the cone σ. This is the meaning of the small Gröbner fan. It follows that the Nilsson ring N_w may be denoted by N_σ.

Definition 5.6.1. The *asymptotic expansion in direction w* of the hypergeometric integral $\Phi_C(x)$ is the Nilsson series on the right hand side of (5.39).

The following important conclusion is now virtually a tautology.

Theorem 5.6.2. *Let σ be a maximal-dimensional cone of the small Gröbner fan of I. The asymptotic expansion of the function $\Phi_C(x)$ is an element of the Nilsson ring $N_\sigma(I)$. The asymptotic expansion is convergent when $(-\log|x_1|, \ldots, -\log|x_n|)$ lies in a translate of the cone σ, i.e.,*

$$(-\log|x_1|, \ldots, -\log|x_n|) \in p + \sigma \quad \text{for a certain fixed vector } p \in \mathbf{R}^n.$$

By combining the algorithms given in Chapters 2 and 5, we can calculate the Nilsson series $s_1(x), \ldots, s_r(x)$ up to any order of accuracy, from the input data $f(x, t)$, α, γ. This calculation is completely independent of the cycle C. The cycle C only enters the picture if we wish to actually determine the constants $c_1, c_2, \ldots, c_r$. If $f(x, t)$ is a generic polynomial, then these algorithms can be enhanced with techniques and results from Chapters 3 and 4. Specifically, we can try to express the output using the explicit series constructed in Sections 3.4 and 3.6. In some cases of interest, we will have enough geometric information about the cycle C to determine the initial series

$\text{in}_w(\Phi(x))$. In other case, only partial information about the initial series will be available. For instance, the geometry might tells us that $\text{in}_w(\Phi(x))$ is a known monomial $x_1^{u_1} \cdots x_n^{u_n}$ multiplied by an unknown cubic polynomial in $\log(x_1), \ldots, \log(x_n)$. Or, imagine a black box for numerical integration which provides numerical approximations to the complex numbers $u_1, \ldots, u_n$. If we are lucky then it might be possible to infer the constants $c_1, \ldots, c_r$ from just looking at $\text{in}_w(s_1), \ldots, \text{in}_w(s_r)$. Here is a simple example where the asymptotic expansion can be determined exactly.

Example 5.6.3. (A residue integral arising from Example 3.1.1)
Let C be a very large circle in the complex plane, and consider the integral

$$\Phi_C(x_1, x_2, x_3) \;\; = \;\; \int_C \frac{dt}{(x_1 t^2 + x_2 t + x_3)^{10} \cdot t^{32}}.$$

By Theorem 5.4.2, this integral satisfies the GKZ system $H_A(\beta)$ for $A = \begin{pmatrix} 2 & 1 & 0 \\ 1 & 1 & 1 \end{pmatrix}$ and $\beta = \begin{pmatrix} 31 \\ -10 \end{pmatrix}$. Fix the weight vector $w = (0, 1, 0)$. The initial ideal $\text{in}_{(-w,w)}(H_A(\beta))$ is generated by $A\theta - \beta$ and $\text{in}_w(I_A) = \langle \partial_2^2 \rangle$, and the two exponents are $(31/2, 0, -51/2)$ and $(15, 1, -26)$. The canonical solution basis consists of the two logarithm-free hypergeometric series $\phi_{(31/2,0,-51/2)}$ and $\phi_{(15,1,-26)}$. The second series makes sense because the exponent $(15, 1, -26)$ has minimal negative support; see Theorem 3.4.14.

Now, since C is a very big cycle, we may assume that it encloses all three poles of the integrand in the complex t-plane. This implies that $\Phi_C(x_1, x_2, x_3)$ is a rational function, and hence its series expansion does not contain any monomials with fractional exponents. This implies

$$\Phi_C \;\; = \;\; c \cdot \phi_{(15,1,-26)} \qquad \text{for some } c \in \mathbb{C}^*. \tag{5.40}$$

This hypergeometric series is terminating. It has precisely 16 monomials:

$$\phi_{(15,1,-26)} \;\; = \;\; \frac{x_1^{15} x_2}{x_3^{26}} \; - \; 65 \frac{x_1^{14} x_2^3}{x_3^{27}} \; + \; \frac{2457}{2} \frac{x_1^{13} x_2^5}{x_3^{28}} \; + \; \cdots \; - \; \frac{962}{115} \frac{x_2^{31}}{x_3^{41}}.$$

The constant c can now be determined by setting $x_1 = 0$ in (5.40).

In general, our luck in finding the complex multipliers $c_1, \ldots, c_r$ in (5.39) will depend highly on whatever information is available about the (twisted) cycle C. It is not clear how such cycles should be represented on the computer in higher dimensions $m \geq 2$. No algebraic algorithms dealing with this problem are known to us. Finding the relevant canonical series solutions $s_1(x), \ldots, s_r(x)$, which would yield the asymptotic expansion of $\Phi_C(x)$ for any C, however, is a doable task of computer algebra. It involves only Gröbner bases in the Weyl algebra D. Here is an example to show this.

Example 5.6.4. (Period integral of an elliptic curve from a reflexive polytope) For an unknown homology cycle C, we examine the hypergeometric integral

$$\int_C f(x,t)^{-1} \frac{dt_1 dt_2}{t_1 t_2} \tag{5.41}$$

where $f(x,t) = x_1 t_1 + x_2 t_1 t_2 + x_3 t_2 + x_4 t_1^{-1} t_2^{-1} + x_5$. The integral satisfies the rank 4 GKZ system for A in Example 2.6.4 and $\beta = (-1, 0, 0)^T$. For

$$g(x,t) \quad = \quad x_1 t_1^2 t_2 + x_2 t_1^2 t_2^2 + x_3 t_1 t_2^2 + x_4 + x_5 t_1 t_2, \tag{5.42}$$

the integral can be written as $\int_C g^{-1} dt_1 dt_2$. Using Algorithm 5.5.12, we see that $\dim_{\mathbf{C}} H^2((\mathbf{C}^*)^2 \backslash V(f), \mathbf{C}) = 6$ and $\dim_{\mathbf{C}} H^2(\mathbf{C}^2 \backslash V(g), \mathbf{C}) = 5$ for fixed generic values of the coefficients x_1, x_2, x_3, x_4, x_5. Hence there are five linearly independent homology cycles. Our cycle C lies in their span.

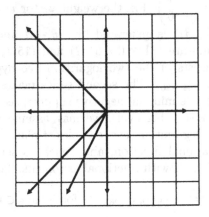

Fig. 5.1. Small Gröbner fan

Let us apply our integration algorithm. The b-function of g is $s + 1$. Compute the integral of $D/(\text{Ann}\,(g^{-1}))$ with respect to t_1 and t_2 by Algorithm 5.3.6, Theorem 5.3.13, and Algorithm 5.5.4. It follows from the output that

$$(\text{Ann}\,(g^{-1}) + \partial_{t_1} D + \partial_{t_2} D) \cap \mathbf{C}\langle x_1, \ldots, x_5, \partial_1, \ldots, \partial_5 \rangle$$

is generated by the ten operators

$$\partial_1 \partial_3 - \partial_2 \partial_5, \quad \partial_2 \partial_4 - \partial_5^2,$$
$$2x_2 \partial_2 + 3x_3 \partial_3 + x_5 \partial_5 + 1,$$
$$x_1 \partial_1 + x_2 \partial_2 + x_3 \partial_3 + x_4 \partial_4 + x_5 \partial_5 + 1,$$
$$x_3 \partial_3 + 2x_4 \partial_4 + x_5 \partial_5 + 1,$$

$$3x_1 x_3 \partial_2 - 8x_4 x_5 \partial_4 - 4x_2 x_4 \partial_5 - 3x_5^2 \partial_5 - 3x_5,$$
$$x_5 \partial_1 \partial_4 + 3x_3 \partial_2 \partial_4 + 2x_2 \partial_1 \partial_5,$$
$$x_1 \partial_2^2 + x_5 \partial_2 \partial_3 + 2x_4 \partial_3 \partial_5,$$
$$x_5 \partial_1 \partial_2 + x_3 \partial_2^2 + 2x_4 \partial_1 \partial_5,$$
$$x_5 \partial_3 \partial_4 + 2x_2 \partial_3 \partial_5 + 3x_1 \partial_5^2.$$

The holonomic rank of this D-ideal is 2. This is less than $\mathrm{rank}(H_A(\beta)) = 4$.

The small Gröbner fan of the holonomic system above agrees with the Gröbner fan of the toric ideal I_A. It consists of 7 maximal-dimensional cones in $\mathbf{R}^5$. In fact, consider the two-dimensional fan $\{\sigma\}$ generated by

$$(1,0), (0,-1), (-1,-2), (-1,-1), (-1,0), (-1,1), (0,1)$$

in Figure 5.1. The small Gröbner fan consists of cones $\sigma' + \mathrm{Im}\, A^T \subset \mathbf{R}^5$ where

$$\sigma' \quad = \quad \left\{ (p,0,0,q,0) \in \mathbf{R}^5 \mid (p,q) \in \sigma \right\}.$$

Fix the weight vector $w = (1,0,0,1,0)$. Then σ is the first quadrant of the small Gröbner fan in Figure 5.1. The indicial ideal with respect to w is

$$\langle\, 3\theta_5 + 3 + 8\theta_4,\ \theta_5 + 1 + 4\theta_3,\ \theta_5 + 1 + 8\theta_2,\ \theta_5 + 1 + 4\theta_1,\ \theta_5^2 + 2\theta_5 + 1 \,\rangle. \tag{5.43}$$

These equations have only one root of multiplicity 2. The unique exponent is $(0,0,0,0,-1)$. The solution space of the indicial equations is spanned by

$$x_5^{-1} \quad \text{and} \quad x_5^{-1} \cdot \log \frac{x_5^8}{x_1^2 x_2 x_3^2 x_4^3}.$$

The integral points in the polar dual of our Gröbner cone σ are

$$\left\{ (m, -m+n, m, n, -m-2n) \in \mathbf{Z}^5 \mid pm + qn \geq 0 \text{ for all } (p,q) \in \sigma \right\}.$$

We conclude that, for any cycle C, the asymptotic expansion of the integral (5.41) with respect to $w = (1,0,0,1,0)$ is a $\mathbf{C}$-linear combination of

$$x_5^{-1} \left(1 + O\left(\frac{x_1 x_3}{x_2 x_5}, \frac{x_2 x_4}{x_5^2} \right) \right)$$

$$\text{and} \quad x_5^{-1} \left(\log \frac{x_5^8}{x_1^2 x_2 x_3^2 x_4^3} + O\left(\frac{x_1 x_3}{x_2 x_5}, \frac{x_2 x_4}{x_5^2} \right) \right).$$

These two series can be written explicitly using our formulas in Section 3.6. Passing to other cones in Figure 5.1, we can compute all seven possible asymptotic expansions of the hypergeometric integral (5.41).

The expansion we selected is the one which is relevant in toric mirror symmetry (see Hosono, Lian, Yau [49]). Indeed, the integral (5.41) can be regarded as a period of a Calabi-Yau manifold. Here it is an elliptic curve, which explains the rank 2. In fact, the annihilating ideal of any period integral arising in toric mirror symmetry strictly contains its GKZ system. It has been determined by Stienstra [95]. Stienstra gives an explicit combinatorial description of its indicial ideal in [95, Theorem 10 (iv)]. Since the indicial ideal (5.43) is equal to the ideal $(\text{ind}_w(H_A(\beta)) : (\theta_5 + 1))$, the above list of ten operators for (5.41) is consistent with his general geometric results.

Appendix—Computer Systems for D-modules

In this appendix we describe current computer systems for D-modules and sketch their design. The emphasis is on kan/sm1 [100]. It is mostly to summarize key ideas of current systems for future mathematical programmers.

We have now (June 1999) three implementations for D-modules: kan/sm1, D-Macaulay (classic), which is based on Macaulay by Bayer and Stillman [10], and a maple package Ore_algebra, They are obtainable from the internet: http://www.math.kobe-u.ac.jp/KAN for the first two and http://www-rocq.inria.fr/algo/chyzak, ftp://ftp.inria.fr/INRIA/Projects/algo/programs/Mgfun

There are also several general purpose non-commutative Gröbner engines and they can also be used for computations in the ring of differential operators. Implementations in Singular and Macaulay 2 are in the test phase. Note that D-Macaulay (classic) is no longer supported.

Example 5.6.5. The design of kan/sm1 Version 2 is based on the homogenized Weyl algebra, which is the fundamental ring in the system. Orders are specified by weight vectors. The Gröbner basis of the Gauss hypergeometric equations in Example 1.2.9 with $a = b = c = 0$ can be computed as follows.

```
(1)    [(x1,x2,x3,x4) ring_of_differential_operators
(2)     [[(x1) -1 (Dx1) 1 (x2) -2 (Dx2) 2 (x3) -1 (Dx3) 1]]
        weight_vector
(3)     0] define_ring
(4)     [(Dx2 Dx3 - Dx1 Dx4).
(5)     (x1 Dx1 - x4 Dx4 + h^2).
(6)     (x2 Dx2 + x4 Dx4 ).
(7)     (x3 Dx3 + x4 Dx4 ).
(8)     ] /ff set
(9)     [ff] groebner  ::
```

The homogenized Weyl algebra $\mathbf{Q}\langle x_1, \ldots, x_4, \partial_1, \ldots, \partial_4, h \rangle$ is defined in lines (1), (2), (3). Here x1 stands for the variable x_1, Dx1 stands for the variable ∂_1 and so on. In line (2), a weight vector $(u, v) = (-1, -2, -1, 0, 1, 2, 1, 0)$ is given and all terms are sorted by that weight vector. Generators are given in lines (4) – (7) and a Gröbner basis in the homogenized Weyl algebra is obtained in line (9) and is printed.

Example 5.6.6. Compare the two-dimensional Weyl algebra with the ring of polynomials in two variables. Two random polynomials of two variables will typically generate a zero-dimensional ideal in the polynomial ring. However, if we consider the left ideal I in the Weyl algebra generated by two random elements, then the chance of getting the trivial ideal is almost 100 percent. Let us compute in kan/sm1 the Gröbner basis for the two "random generators":

$$x\partial_x + \partial_y^2 + y, \quad y\partial_y^2 + x$$

The output is a constant, which means that they generate the trivial ideal.

```
(1)  [(x,y) ring_of_differential_operators 0] define_ring /R set
(2)  [(Homogenize) 0] system_variable
(3)  /ff [ (x Dx + Dy^2 + y). (y Dy^2+x).] def
(4)  [ff] groebner_sugar ::
```

Lines (1) and (2) define the Weyl algebra $\mathbf{Q}\langle x, y, \partial_x, \partial_y \rangle$. The ring structure is put in the variable R. The variable ∂_x is denoted by Dx and ∂_y is denoted by Dy. The elements will be ordered by a default order (the graded reverse lexicographic order) in this example. Line (2) means that we will not use the homogenized Weyl algebra and h should be regarded as 1 in the sequel. In lines (3) and (4), the reduced Gröbner basis is computed and is printed.

Example 5.6.7. We demonstrate the implementation by Chyzak [29] in Maple V release 5. For the GKZ system of Example 1.4.23 we compute a Gröbner basis and evaluate the holonomic rank using the method in Corollary 1.4.14.

```
(0)    with(Ore_algebra): with(Groebner):
(1)    N:=4:
(2)    V:=[seq(x[i],i=1..N)],[seq(D[i],i=1..N)]:
(3)    l[1]:=D[2]*D[3]-D[1]*D[4]:
(4)    l[2]:=x[1]*D[1]-x[4]*D[4]:
(5)    l[3]:=x[2]*D[2]+x[4]*D[4]+1/2:
(6)    l[4]:=x[3]*D[3]+x[4]*D[4]+1/2:
(7)    A:=diff_algebra(seq([D[i],x[i]],i=1..N)):
(8)    T:=termorder(A,lex(op(V[2]))):
(9)    G:={seq(l[i],i=1..4)}:
(10)   GB:=gbasis(G,T);
(11)   H:=hilbertseries(GB,T,s);
(12)   subs(s=1,H);
```

In line (7) the ring of differential operators with rational function coefficients $\mathbf{Q}(x_1, \ldots, x_4)\langle \partial_1, \ldots, \partial_4 \rangle$ is defined. x[i] stands for x_i and D[i] stands for ∂_i. In lines (3) – (6), generators for the GKZ system are given. In line (8) the lexicographic order is given for our computation of Gröbner bases. In line (10) the reduced Gröbner basis is computed. In line (11) the Hilbert-Poincaré series is computed to evaluate the rank.

Let us return to kan/sm1 and take a closer look at that system. There has been no prior publications about the system itself, but it has been used for about a decade to develop new implementation ideas, to develop new algorithmic ideas in D, to verify new conjectures, and to generate examples.

The system **kan/sm1** consists of a Gröbner engine with arithmetic in D, a subset of the Postscript language with an extension for object oriented programming (**sm1**), and a module to communicate with other systems. It is designed to be a back-end engine for a distributed computing system. The language **sm1** is served as a virtual machine language for the engine.

Inside of the engine:

1. *Arithmetic in D:* It uses the Leibnitz formula (Theorem 1.1.1) for multiplications since version 2. All elements are expressed in the normally ordered form as a list of exponent vectors (see [8] for a general discussion on expressions of monomials). This part is written in C.

2. *Buchberger algorithm:* The system employs two techniques from the commutative case to make the Buchberger algorithm 1.1.9 more efficient: At (1.9), we have strategies to choose a pair; the pairs should be sorted by a "degree" and a selection should be done degree by degree. At (1.10), we can predict that the normal form is zero, see [21]. A general reference for the implementation is the book [12]. This part is written in C.

3. *D-module algorithms* are mainly written in the virtual machine language **sm1**. Some of the codes were generated by an experimental compiler, but rewritten in **sm1** by hand to get a better performance.

At present, no higher level programming language is provided for **kan/sm1**. However, the system **kan/sm1** provides one line commands for several constructions in D-modules. These are understandable without a computer science knowledge about the virtual machine. Here are some examples:

Example 5.6.8. Compute a Gröbner basis and the initial ideal of the D-ideal $D \cdot \{(x\partial_x)^2 + (y\partial_y)^2 - 1, xy\partial_x\partial_y - 1\}$ for the weight vector $(u, v) = (0, 0, 1, 1)$:

```
[ [( (x Dx)^2 + (y Dy)^2 -1) ( x y Dx Dy -1)] (x,y)
         [ [ (Dx) 1 (Dy) 1] ] ] gb pmat ;
```

Output:
```
[[x^2*Dx^2+y^2*Dy^2+x*Dx+y*Dy-1, x*y*Dx*Dy-1, y^3*Dy^3+3*y^2*Dy^2+x*Dx]
[x^2*Dx^2+y^2*Dy^2, x*y*Dx*Dy, y^3*Dy^3]]
```
The first line is the Gröbner basis and the second line is a set of generators of the initial ideal with respect to the weight vector $(0, 0, 1, 1)$.

Example 5.6.9. Find the GKZ system for $A = \begin{pmatrix} 1 & 1 & 1 & 1 \\ 0 & 1 & 3 & 4 \end{pmatrix}$ and $\beta = \begin{pmatrix} 1 \\ 2 \end{pmatrix}$.

```
[ [[1 1 1 1] [0 1 3 4]] [1 2]] gkz  ::
```

Output:
```
[ x1*Dx1+x2*Dx2+x3*Dx3+x4*Dx4-1 , x2*Dx2+3*x3*Dx3+4*x4*Dx4-2 ,
  Dx2*Dx3-Dx1*Dx4 , -Dx1*Dx3^2+Dx2^2*Dx4 , Dx2^3-Dx1^2*Dx3 ,
  -Dx3^3+Dx2*Dx4^2 ]
```

Example 5.6.10. Evaluate the holonomic rank of the GKZ systems for $A = \begin{pmatrix} 1 & 1 & 1 & 1 \\ 0 & 1 & 3 & 4 \end{pmatrix}$ with $\beta = \begin{pmatrix} 1 \\ 2 \end{pmatrix}$ and $\beta = \begin{pmatrix} 0 \\ 0 \end{pmatrix}$. Show also the execution times:

```
{ [ [[1 1 1 1] [0 1 3 4]] [1 2]] gkz  rrank ::} timer
{ [ [[1 1 1 1] [0 1 3 4]] [0 0]] gkz  rrank ::} timer
```

Output:

```
    5
 User time:1.000000 seconds, System time:0.010000 s, Real time:1 s
    4
 User time:1.320000 seconds, System time:0.000000 s, Real time:1 s
```

Example 5.6.11. Compute the b-function of $f = x^3 - y^2 z^2$ and the annihilating ideal of f^{r_0} where r_0 is the minimal integral root of the b-function:

```
(oxasir.sm1) run
[(x^3 - y^2 z^2) (x,y,z)] annfs /ff set
ff message
ff 1 get 1 get fctr ::
```

Output:

```
[ [ -y*Dy+z*Dz , 2*x*Dx+3*y*Dy+6 , -2*y*z^2*Dx-3*x^2*Dy ,
   -2*y^2*z*Dx-3*x^2*Dz , -2*z^3*Dx*Dz-3*x^2*Dy^2-2*z^2*Dx ]  ,
  [-1,-139968*s^7-1119744*s^6-3802464*s^5-7107264*s^4
   -7898796*s^3-5220720*s^2-1900500*s-294000]]
  [[ -12 , 1 ] , [ s+1 , 1 ], [3*s+5 , 1], [ 3*s+4, 1],
   [6*s+7, 2], [6*s+5, 2]]
```

The first two rows of the output give generators of the annihilating ideal of $(x^3 - y^2 z^2)^{-1}$. The b-function is $(s+1)(3s+5)(3s+4)(6s+7)^2(6s+5)^2$.

Example 5.6.12. Compute the de Rham cohomology of $X = \mathbf{C}^2 \backslash V(x^3 - y^2)$:

```
(cohom.sm1) run
[(x^3-y^2) (x,y)] deRham ;
```

Output:

```
  0-th cohomology:  [    0 , [    ] ]
 -1-th cohomology:  [    1 , [    ] ]
 -2-th cohomology:  [    1 , [    ] ]
 [1 , 1 , 0 ]
```

This means that $H^2(X, \mathbf{C}) = 0$, $H^1(X, \mathbf{C}) = \mathbf{C}^1$, $H^0(X, \mathbf{C}) = \mathbf{C}^1$.

Example 5.6.13. Compute the integral of $I = D \cdot \{\partial_t - (3t^2 - x), \partial_x + t\}$, which annihilates the function $e^{t^3 - xt}$, with respect to the variable t:

```
(cohom.sm1) run
[ [(Dt - (3 t^2-x)) (Dx + t)] [(t)]
  [ [(t) (x)] [ ]] 0] integration
```

Output:

```
 [    [    1 , [    3*Dx^2-x ]  ]  ]
```

References

1. Adams, W.W., Loustaunau, P. (1994): *An Introduction to Gröbner Bases.* Graduate Studies in Mathematics, Vol.III, American Mathematical Society, Providence.
2. Adolphson, A. (1994): Hypergeometric functions and rings generated monomials. Duke Mathematical Journal **73**, 269–290.
3. Aomoto, K. (1977): On the structure of integrals of power products of linear functions. Scientific Papers of the College of General Education, University of Tokyo **27**, 49–61.
4. Aomoto, K., Kita, M. (1994): *Theory of Hypergeometric Functions.* Springer, Tokyo (in Japanese).
5. Appell, P., Kampé de Fériet, J. (1926): *Fonctions Hypergéometrique et Hypersphériques - Polynomes d'Hermite.* Gauthier-Villars, Paris.
6. Assi, A., Castro-Jiménez, F.J., Granger, M. (1996): How to calculate the slopes of a *D*-module. Compositio Mathematica **104**, 107–123.
7. Assi, A., Castro-Jiménez, F.J., Granger, M. (1998): The standard fan of a D-module. To appear.
8. Bachmann, O., Schönemann, H. (1998): Monomial representations for Gröbner basis computations. Proceedings of the International Symposium in Symbolic and Algebraic Computations '98, ACM Press, 309–318.
9. Barvinok, A. (1993): Computing the volume, counting integral points and exponential sums. Discrete and Computational Geometry **10**, 123–141.
10. Bayer, D., Stillman, M.: Macaulay: a computer algebra system for algebraic geometry. Available by anonymous ftp from zariski.harvard.edu
11. Bayer, D., Stillman, M. (1987): A criterion for detecting *m*-regularity. Inventiones Mathematicae **87**, 1–11.
12. Becker, T., Weispfenning, V. (1993): *Gröbner Bases.* Springer, New York.
13. Bernstein, I.N. (1972): The analytic continuation of generalized functions with respect to a parameter. Functional Analysis and its Applications **6**, 273–285.
14. Bieberbach, L. (1965): *Theorie der Gewöhnlichen Differentialgleichungen.* 2nd edition, Springer, Heidelberg.
15. Björk, J.E. (1979): *Rings of Differential Operators.* North-Holland, New York.
16. Billera, L.J., Filliman, P., Sturmfels, B. (1990): Constructions and complexity of secondary polytopes. Advances in Mathematics **83**, 155–179.
17. Borel, A., Grivel, P.P., Kaup, B., Haefliger, A., Malgrange, B., Ehlers, F. (1987): *Algebraic D-modules.* Academic Press, Boston.
18. Bott, R., Tu, L. W. (1982): *Differential Forms in Algebraic Topology.* Springer, New York.
19. Brylinski, J.-L. (1986): *Transformations Canoniques, Dualité Projective, Théorie de Lefschetz, Transformations de Fourier et Sommes Trigonométriques.* Astérisque **140-141**, 3–134.

20. Bruns, W., Herzog, J. (1993): *Cohen-Macaulay Rings*. Cambridge Studies in Advanced Mathematics 39, Cambridge University Press, Cambridge.
21. Capani, A., De Dominicis, G., Niesi, G., Robbiano, L. (1997): Computing minimal finite free resolutions. Journal of Pure and Applied Algebra **117**, **118**, 105–117.
22. Capani, A., Niesi, G., Robbiano, L. (1999): CoCoA Version 3.6.: computing in commutative algebra, http://cocoa.dima.unige.it
23. Castro, F. (1984): Théorème de division pour les opérateurs différentiels et calcul des multiplicités. Thèse de 3ème cycle, Université de Paris 7. See also, Castro, F. (1987): Calculs effectifs pour les idéaux d'opérateurs différentiels. Travaux en Cours, **24**, 1–19, Hermann, Paris.
24. Cattani, E., D'Andrea, C., Dickenstein, A. (1998): Rational solutions of the A-hypergeometric system associated with a monomial curve. math.AG/9805005, to appear in Duke Mathematical Journal.
25. Collart, S., Kalkbrener, M., Mall, D. (1997): Converting bases with the Gröbner walk. Journal of Symbolic Computation **24**, 465–469.
26. Coutinho, S.C. (1995): *A Primer of Algebraic D-modules*. London Mathematical Society Student Texts 33, Cambridge University Press, Cambridge.
27. Cox, D., Little, J., O'Shea, D. (1991): *Ideals, Varieties and Algorithms*. Springer, New York.
28. Cox, D., Little, J., O'Shea, D. (1998): *Using Algebraic Geometry*. Springer, New York.
29. Chyzak, F. (1998): Fonctions holonomes en calcul formel. Thèse de 3ème cycle. http://www-rocq.inria.fr/algo/chyzak.
30. Deligne, P. (1970): Équations Différentielles à Points Singuliers Réguliers. Springer Lecture Notes in Mathematics **163**.
31. Dimca, A. (1992): *Singularities and Topology of Hypersurfaces*. Universitext. Springer, New York.
32. Eisenbud, D. (1995): *Commutative Algebra with a View Toward Algebraic Geometry*. Springer, New York.
33. Erdélyi, A. (Editor) (1953): *Higher Transcendental Functions, vol. I*. MacGraw Hill, New York.
34. Gabber, O. (1981): The integrability of the characteristic variety. American Journal of Mathematics **103**, 445–468.
35. Galligo, A. (1985): Some algorithmic questions on ideals of differential operators. EUROCAL '85, Springer Lecture Notes in Computer Science **204**, 413–421.
36. Gel'fand, I.M. (1986): General theory of hypergeometric functions. Soviet Mathematics Doklady **33**, 573–577.
37. Gel'fand, I.M., Graev, M.I., Postnikov, A. (1997): Combinatorics of hypergeometric functions associated with positive roots. The Arnold-Gelfand Mathematical Seminars, Birkhäuser, Boston, 205–221.
38. Gel'fand, I.M., Zelevinsky, A.V., Kapranov, M.M. (1989): Hypergeometric functions and toral manifolds. Functional Analysis and its Applications **23**, 94–106.
39. Gel'fand, I.M., Kapranov M.M., Zelevinsky, A.V. (1990): Generalized Euler Integrals and A-Hypergeometric Functions. Advances in Mathematics **84**, 255–271.
40. Gel'fand, I.M., Zelevinsky, A.V., Kapranov, M.M. (1993): Correction to the paper: "Hypergeometric functions and toric varieties". Functional Analysis and its Applications **27**, 295. See also Math. Reviews 95a:22010.
41. Gel'fand, I.M., Kapranov M.M., Zelevinsky, A.V. (1994): *Discriminants, Resultants and Multidimensional Determinants*. Birkhäuser, Boston.

42. Grayson, D., Stillman, M. (1999): Macaulay 2: a computer algebra system for algebraic geometry, Version 0.8.52, http://www.math.uiuc.edu/Macaulay2

43. Greuel, G.-M., Pfister, G., Schönemann, H. (1998): Singular 1.2.2, http://www.mathematik.uni-kl.de/~zca

44. Gritzmann, P., Sturmfels, B. (1993): Minkowski addition of polytopes. Computational complexity and applications to Gröbner bases. SIAM Journal of Discrete Mathematics **6**, 246–269.

45. Gunning, R.C., Rossi, H. (1965): *Analytic Functions of Several Complex Variables*, Prentice-Hall, Englewood Cliffs.

46. Hartshorne, R. (1966): Connectedness of the Hilbert scheme. Publications I.H.E.S. **29**, 261–304.

47. Hartshorne, R. (1977): *Algebraic Geometry*. Springer, New York.

48. Hibi, T. (1995): *Commutative Algebra and Combinatorics*. Springer, Tokyo (in Japanese).

49. Hosono, S., Lian, B.H., Yau, S.-T. (1996): GKZ-generalized hypergeometric systems in mirror symmetry of Calabi-Yau hypersurfaces. Communications in Mathematical Physics **182**, 535–577.

50. Hosono, S., Lian, B.H., Yau, S.-T. (1997): Maximal degeneracy points of GKZ systems. Journal of the American Mathematical Society **10**, 427–443.

51. Hoşten, S., Thomas, R.R. (1998): Standard pairs and group relaxations in integer programming, to appear in Journal of Pure and Applied Algebra.

52. Hoşten, S., Thomas, R.R. (1999): Associated primes of initial ideals of lattice ideals, Mathematical Research Letters **6**, 83–97.

53. Hotta, R. (1991): Equivariant D-modules. Proceedings of ICPAM Spring School in Wuhan, 1991, edited by P. Torasso, Travaux en Cours, Hermann, Paris, to appear, math.RT/9805021.

54. Hotta, R., Tanisaki, T. (1994): *D-modules and Algebraic Groups*. Springer, Tokyo (in Japanese).

55. Ince, E.L. (1956): *Ordinary Differential Equations*. (Reprinted paperback edition), Dover Publications, New York.

56. John, F. (1938): The ultrahyperbolic equation with 4 independent variables. Duke Mathematical Journal **4**, 300–322.

57. Kaneko, J. (1998): The Gauss-Manin connection of the integral of the deformed difference product. Duke Mathematical Journal **92**, 355–379.

58. Kashiwara, M. (1976): B-functions and holonomic systems — rationality of roots of B-functions. Inventiones Mathematicae **38**, 33–53.

59. Kashiwara, M. (1978): On the holonomic systems of linear differential equations, II. Inventiones Mathematicae **49**, 121–135.

60. Kashiwara, M., Kawai, T. (1981): On holonomic systems of microdifferential equations. III — Systems with regular singularities. Publications of the Research Institute of Mathematical Sciences, Kyoto University **17**, 813–979.

61. Kashiwara, M. (1982): Vanishing cycle sheaves and holonomic systems of differential equations. Algebraic Geometry, Proceedings, Tokyo/Kyoto, 1982. Edited by M.Raynaud and T.Shioda. Springer Lecture Notes in Mathematics **1016**, 134–142.

62. Kashiwara, M. (1983): *Systems of Microdifferential Equations*. Birkhäuser, Boston.

63. Lazard, L. (1983): Gröbner bases, Gaussian elimination, and resolution of systems of algebraic equations. Proceedings EUROCAL '83, Springer Lecture Notes in Computer Science, **162**, 146–156.

64. Laurent, Y. (1987): Polygône de Newton et b-fonctions pour les modules microdifférentiels. Annals Scientifique Ecole Normale Supérieure, **20**, 391–441.

65. Laurent, Y., Monteiro Fernandes, T., (1988): Systémes différentiels fuchsiens le long d'une sous-variété. Publications of the Research Institute of Mathematical Sciences, Kyoto University **24**, 397–431.

66. Majima, H. (1984): *Asymptotic Analysis for Integrable Connections with Irregular Singular Points.* Springer Lecture Notes in Mathematics, **1075**.

67. Malgrange, B. (1975): Polynômes de Bernstein d'une singularité isolée. Springer Lecture Notes in Mathematics, **459**, 98–119.

68. Matsumura, H. (1986): *Commutative Ring Theory.* Cambridge University Press, Cambridge.

69. Mebkhout, Z. (1989): *Le Formalisme des Six Opérations de Grothendieck pour les D_X-modules Cohérents.* Travaux en Cours **35**, Hermann, Paris.

70. Mora, T. (1987): Seven variations on standard bases. Preprint.

71. Mora, T. (1997): Gröbner duality and multiple points in linearly general position. Proceedings of the American Mathematical Society **125**, 1273–1282.

72. Mora, T., Robbiano, L. (1988): The Gröbner fan of an ideal. Journal of Symbolic Computation **6**, 183–208.

73. Nilsson, N. (1965): Some growth and ramification properties of certain integrals on algebraic manifolds. Arkiv för Matematik **5**, 463–476.

74. Noro, T. et al. (1993, 1995): Risa/Asir, ftp://endeavor.fujitsu.co.jp/pub/isis/asir

75. Oaku, T. (1994): *Gröbner Bases and Systems of Linear Differential Equations — An Introduction to Computational Algebraic Analysis.* Sophia University Lecture Note Series, Tokyo (in Japanese).

76. Oaku, T. (1994): Computation of the characteristic variety and the singular locus of a system of differential equations with polynomial coefficients. Japan Journal of Industrial and Applied Mathematics **11**, 485–497.

77. Oaku, T. (1997): Algorithms for the b-function and D-modules associated to a polynomial. Journal of Pure and Applied Algebra **117** & **118**, 495–518.

78. Oaku, T. (1997): Algorithms for b-functions, restrictions, and algebraic local cohomology groups of D-modules. Advances in Applied Mathematics **19**, 61–105.

79. Oaku, T., Takayama, N. (1997): An algorithm for de Rham cohomology groups of the complement of an affine variety via D-module computation. To appear in Journal of Pure and Applied Algebra.

80. Oaku, T., Takayama, N. (1998): Algorithms for D-modules — Restrictions, tensor product, localization and algebraic local cohomology groups. math.AG/9805006.

81. Oaku, T., Takayama, N. (1999): Five operations on holonomic functions. In preparation.

82. Oaku, T., Takayama, N., Walther, U. (1998): A localization algorithm for D-modules. math.AG/9811039.

83. Oberst, U. (1996): Finite-dimensional systems of partial differential or difference equations. Advances in Applied Mathematics **17**, 337–356.

84. Oda, T. (1988): *Convex Bodies and Algebraic Geometry: An Introduction to the Theory of Toric Varieties.* Ergebnisse der Mathematik und ihrer Grenzgebiete, 3. Folge, Band 15, Springer, Heidelberg.

85. Ohara, K. (1997): Computation of the monodromy of the generalized hypergeometric function $_pF_{p-1}(a_1, \ldots, a_p; b_2, \ldots, b_p; z)$. Kyushu Journal of Mathematics **51**, 101–124.

86. Petkovsek, A.K., Wilf, P., Zeilberger, D. (1996): $A = B$, A.K.Peters, Ltd., Wellesley.

87. Reid, G. (1991): Algorithms for reducing a system of PDEs to standard form, determining the dimension of its solutions space and calculating Taylor series solutions. European Journal of Applied Mathematics **2**, 293–318.

88. Saito, M. (1992): Parameter shift in normal generalized hypergeometric systems. Tohoku Mathematical Journal **44**, 523–534.

89. Saito, M., Sturmfels, B., Takayama, N. (1999): Hypergeometric polynomials and integer programming. Compositio Mathematica, **155**, 185–204.

90. Saito, M., Takayama, N. (1994): Restrictions of $\mathcal{A}$-hypergeometric systems and connection formulas of the $\triangle_1 \times \triangle_{n-1}$- hypergeometric function. International Journal of Mathematics **5**, 537–560.

91. Sato, M., Kawai, T., Kashiwara, M. (1973): *Microfunctions and Pseudodifferential Equations*. Springer Lecture Notes in Mathematics, **287**.

92. Schrijver, A. (1986): *Theory of Linear and Integer Programming*. Wiley Interscience, Chichester.

93. Stafford, J.T. (1977): Weyl algebras are stably free. Journal of Algebra **48**, 297–304.

94. Stanley, R. (1996): *Combinatorics and Commutative Algebra*, second edition. Progress in Mathematics **41**, Birkhäuser, Boston.

95. Stienstra, J. (1998): Resonant hypergeometric systems and mirror symmetry. Integrable Systems and Algebraic Geometry (Proceedings of the Taniguchi Symposium 1997), 412–452. Edited by Saito, M.-H., Shimizu, Y., Ueno, K., World Scientific, Singapore.

96. Sturmfels, B. (1995): *Gröbner Bases and Convex Polytopes*. University Lecture Notes, Vol. 8. American Mathematical Society, Providence.

97. Sturmfels, B. (1996): Solving algebraic equations in terms of $\mathcal{A}$-hypergeometric series, to appear in Discrete Mathematics.

98. Sturmfels, B., Trung, N.V., Vogel, W. (1995): Bounds for degrees of projective schemes. Mathematische Annalen **302**, 417–432.

99. Sturmfels, B., Weismantel, R., Ziegler, G. (1995): Gröbner bases of lattices, corner polyhedra, and integer programming. Beiträge zur Algebra und Geometrie **36**, 281-298.

100. Takayama, N. (1991, 1994, 1999): Kan: A system for computation in algebraic analysis. Source code available at http://www.math.kobe-u.ac.jp/KAN/. Version 1 (1991), Version 2 (1994), The latest version is 2.990612 (1999).

101. Walther, U. (1997): Algorithmic computation of local cohomology modules and the cohomological dimension of algebraic varieties. To appear in Journal of Pure and Applied Algebra.

102. Walther, U. (1998): Algorithmic computation of de Rham cohomology of complements of complex affine varieties, math.AG/9807176.

103. Yoshida, M. (1997): *Hypergeometric Functions, My Love: Modular Interpretations of Configuration Spaces*. Vieweg Verlag, Wiesbaden. Japanese translation by Kyoritsu Publishing Company, Tokyo.

104. Yoshida, M., Takano, K. (1976): On a linear system of pfaffian equations with regular singular points. Funkcialaj Ekvacioj **19**, 175–189.

105. Ziegler, G. (1995): *Lectures on Polytopes*. Springer, New York.

Preprints in "math.AG/" and "math.RT" are obtainable from http://xxx.lanl.gov

Index